AF849494

Bioresources for Managing and Treating Dementia

Globally, dementia is a health challenge with remarkable socio-economic implications. In spite of advances in conventional treatment, there is an urgent need for more accessible, effective, and safer therapeutic options. Bioresources—including microbial metabolites, medicinal plants, and marine-derived compounds—offer a promising avenue for dementia treatment and management. Phytochemicals like terpenoids, flavonoids, and alkaloids have shown neuroprotective properties, like anti-inflammatory, antioxidative, and cholinesterase-inhibitory effects. An overview of the potential of various bioresources in dementia treatment and management is discussed in this book. However, the translation of bioresource-based therapies into clinical practice faces several obstacles. Limitations are variability in low bioavailability, concentrations of bioactive compounds, problems with regulations, and the need for rigorous clinical validation. Surmounting these obstacles entails advanced biotechnological approaches, interdisciplinary collaboration, and supportive policy frameworks. This book explores the potential of bioresources in dementia management, focusing on their mechanisms of action and key prospects and addressing the critical obstacles that must be navigated to harness their full therapeutic potentials.

Key Features:

- Discusses pathophysiology and prevalence of dementia as well as the significance of the applications of diverse bioresources for managing and treating dementia
- Highlights the application of medicinal plants, phytochemicals, microbial products, and marine-derived bioproducts for the management and treatment of dementia, with emphasis on their mechanisms of action
- Discusses challenges, considerations, and future directions on the use of bioresources for managing and treating dementia

This book is ideal for clinicians, biotechnologists, biomedical scientists, dementia caregivers, researchers, and academicians.

Neuroscience of Neurodegenerative Disorders

Series Editor Wael Mohamed

Wael Mohamed Yousef Mohamed
BBCH (Egypt), MMSc (Egypt), MD/PhD (USA)

Dr. Mohamed earned his PhD from Penn State University (USA) and currently is an Associate Professor at IIUM[International Islamic University Malaysia] Medical School in Malaysia. He serves as an editor for *Frontiers in Neurology* and PLOS-ONE Journals. His research is well supported by extramural funding. He is the founder and chair of the AfrAbia-PD-Genomic Consortium.

About This Series

This book series aims to publish volumes focusing on both basic and clinical research in the field of Neurodegenerative Disorders with a focus on delineating the recent advances in various disorders including Parkinson's Disease, Alzheimer's Disease, ALS, Huntington's Disease, among other common Neurodegenerative disorders. The books will examine different disorders aspects including pathophysiology, diagnostic, treatment, and precision medicine. Volumes in the series will also highlight the recent advances in the management of Neurodegenerative disorders especially in OMICS and Genomic era and precision medicine. Volumes will be useful for neuroscientists, neurologists, psychiatrists, psychologists, neurosurgeons, and those interested in preventive medicine.

Integrating Neuroimaging, Computational Neuroscience, and Artificial Intelligence
Edited by Indranath Chatterjee and Nasrallah Moradikor

Non-Pharmacological Interventions for Dementia: A Comprehensive Approach
Edited by Abraham Olufemi Asuku, Ebele Joan Ajagun, and Gbonjubola Oyinlola Ogungbangbe

Bioresources for Managing and Treating Dementia: Prospects and Challenges
Edited by Abraham Olufemi Asuku, Ayodeji Johnson Ajibare, and Olugbemi Tope Olaniyan

Bioresources for Managing and Treating Dementia

Prospects and Challenges

Edited by

Abraham Olufemi Asuku, Ayodeji Johnson Ajibare and Olugbemi Tope Olaniyan

CRC Press
Taylor & Francis Group
Boca Raton London New York

CRC Press is an imprint of the
Taylor & Francis Group, an **informa** business

Designed cover image: Author, includes AI generated content.

First edition published 2026
by CRC Press
2385 NW Executive Center Drive, Suite 320, Boca Raton FL 33431

and by CRC Press
4 Park Square, Milton Park, Abingdon, Oxon, OX14 4RN

CRC Press is an imprint of Taylor & Francis Group, LLC

© 2026 selection and editorial matter, Abraham Olufemi Asuku, Ayodeji Johnson Ajibare, and Olugbemi Tope Olaniyan; individual chapters, the contributors

Reasonable efforts have been made to publish reliable data and information, but the author and publisher cannot assume responsibility for the validity of all materials or the consequences of their use. The authors and publishers have attempted to trace the copyright holders of all material reproduced in this publication and apologize to copyright holders if permission to publish in this form has not been obtained. If any copyright material has not been acknowledged please write and let us know so we may rectify in any future reprint.

Except as permitted under U.S. Copyright Law, no part of this book may be reprinted, reproduced, transmitted, or utilized in any form by any electronic, mechanical, or other means, now known or hereafter invented, including photocopying, microfilming, and recording, or in any information storage or retrieval system, without written permission from the publishers.

For permission to photocopy or use material electronically from this work, access www.copyright.com or contact the Copyright Clearance Center, Inc. (CCC), 222 Rosewood Drive, Danvers, MA 01923, 978-750-8400. For works that are not available on CCC please contact mpkbookspermissions@tandf.co.uk

For Product Safety Concerns and Information please contact our EU representative GPSR@taylorandfrancis.com. Taylor & Francis Verlag GmbH, Kaufingerstraße 24, 80331 München, Germany.

Trademark notice: Product or corporate names may be trademarks or registered trademarks and are used only for identification and explanation without intent to infringe.

ISBN: 978-1-041-10914-3 (hbk)
ISBN: 978-1-041-10915-0 (pbk)
ISBN: 978-1-003-65738-5 (ebk)

DOI: 10.1201/9781003657385

Typeset in Times
by KnowledgeWorks Global Ltd.

Contents

Foreword

It is a privilege to introduce this book, *Bioresources for Managing and Treating Dementia: Prospects and Challenges.* Dementia continues to rise globally, with major medical, social, and economic consequences. Despite some progress in understanding its biology and improving clinical care, effective therapies remain limited. This reality requires sustained exploration of alternative and complementary approaches. This book focuses on one such approach: the use of natural bioresources. Plants, phytochemicals, microbial products, and marine-derived bioproducts are discussed here in detail. Many of these bioproducts have long been known in traditional systems of medicine. Recent studies have confirmed that they may act through antioxidant, anti-inflammatory, and neuroprotective pathways, all of which are relevant to the mechanisms that underlie dementia.

The contributors have organized the material in a way that brings both the promise and the limits of these resources into view. While natural products such as flavonoids, alkaloids, and terpenoids have shown encouraging effects in laboratory and animal studies, moving these findings into clinical application is not straightforward. Low bioavailability, poor absorption, and variability in compound content are well-recognized difficulties. Similarly, microbial and marine-derived compounds such as fucoidan, docosahexaenoic acid, and chitosan have been studied with interesting outcomes but still face barriers in production, regulation, and clinical testing. The book is structured to cover both the scientific and practical aspects. Early chapters address the pathophysiology and global burden of dementia, together with current pharmacotherapies and their limits. Also, the book draws attention to regulatory frameworks and the responsibilities of agencies that oversee approval and safe use. Later chapters review specific plant-based agents—*Ginkgo biloba*, *Bacopa monnieri*, curcumin, *Withania somnifera*, *Ocimum gratissimum*, and others—highlighting the evidence supporting their potential role. Additional chapters examine compounds derived from fungi and marine sources, expanding the discussion beyond plant material.

The editors have also emphasized an interdisciplinary approach. Biotechnology, pharmacology, neuroscience, and clinical sciences are all represented. This allows readers to see how molecular pathways, therapeutic targets, and health policy connect. Such integration is essential for advancing new strategies in dementia care. An important feature of this work is its recognition of obstacles. The book outlines not only the biological mechanisms but also the regulatory, clinical, and translational challenges that remain. This balance between opportunities and difficulties ensures that readers understand both the potential and the work that lies ahead. The relevance of this subject goes beyond academic research. Dementia affects individuals, families, and health systems. Exploring bioresources is not only about adding more treatment options but also about identifying solutions that may be more affordable, accessible, and sustainable. This makes the discussion in the book significant for low- and middle-income regions as well as high-income countries.

The audience for this volume is wide. Researchers will find a comprehensive survey of bioresources and their mechanisms. Clinicians will gain insight into possible complementary approaches. Policymakers will be able to see the importance of regulatory clarity. Students and educators will benefit from the structured review of existing knowledge and the identification of gaps for future study. I acknowledge the effort of the editors and contributors in bringing this book in at the right time. It offers an organized account of what is known, what remains uncertain, and what directions are most promising. The material presented here can serve as a reference point for the next stage of research and development.

It is my expectation that this book will encourage further investigation, foster collaboration across scientific and medical disciplines, and guide policies that support responsible innovation. In the long term, the insights gathered here may help shape more effective strategies for those living with dementia and for the societies that support them.

Professor Baskar Gurunathan
Professor, Department of Biotechnology
Head, Global Networking
St. Joseph's College of Engineering, Chennai, India

Preface

Dementia is a significant global health challenge in the present time, as it affects millions of individuals physically, along with their loved ones. It cannot be described as a single disease but as a collective of various symptoms affecting areas of the brain involved in memory, cognition, and behavior, leaving the patient an empty shell of what they once were. With the aging population growing larger worldwide, there is a fear that we might lose our older ones far too soon. Studies have focused on pharmacological treatments that can be provided for those who suffer from it, and despite decades of research, these treatments remain limited to helping curb or reverse the progression of dementia. But an emerging spotlight has been placed on the use of bioresources to, at the very least, manage symptoms and prevent the neurodegeneration that accompanies it. They do not only reduce the pain and distress of dementia but also help build up the lost autonomy, dignity, emotional well-being, and overall quality of life of the patient.

This book, *Bioresources for Managing and Treating Dementia: Prospects and Challenges*, delves into the current understanding of alternative strategies regarding dementia care and was conceived as a resource for clinicians, biotechnologists, biomedical scientists, dementia caregivers, researchers, and academicians, and the families of dementia patients who all seek to find ways to alternatively understand strategies involved in dementia care.

Across 18 chapters, this volume introduces its readers to the present knowledge of dementia and its global prevalence, as well as the pharmacotherapies of dementia and their limitations. It brings an understanding of the bioresources intervention into significant view and then presents evidence-based interventions, including the use of *Ginkgo biloba*, *Bacopa monnieri*, curcumin, Huperzine A, *Withania somnifera*, *Ocimum gratissimum*, green tea, 6-gingerol, quercetin, lion's mane, *Ganoderma lucidum*, fucoidan, docosahexaenoic acid from fish oil and algae, and chitosan. It concludes with discussing the challenges involved in the applications of these bioresources as interventions for dementia and what the future could hold for the application of bioresources for managing and treating dementia.

The chapters were contributed by 67 academicians and researchers from nine different countries (Nigeria, Rwanda, South Africa, the United Kingdom, France, Brazil, South Korea, Canada, and the United States of America) across the world. We extend our heartfelt thanks to all the contributors that helped bring this volume to life and also the many patients and family members whose documented journeys with dementia continue to shape and inspire more like this. The aim of this book is not only to inform but also to bring a different perspective to how we perceive and respond to dementia. It is hoped that the knowledge shared will empower readers to have a more nuanced response to the management and treatments that go beyond our conventional mindset.

Abraham Olufemi Asuku, Ph.D.
Bioresources Development Centre, Department of Medical Biotechnology, National Biotechnology Research and Development Agency, Ogbomoso, Nigeria

Ayodeji Johnson Ajibare, Ph.D.
Department of Physiology, Faculty of Basic Medical Sciences, Lead City University, Ibadan, Nigeria

Olugbemi Tope Olaniyan, Ph.D.
Department of Physiology, Faculty of Basic Medical Sciences, Kwara State University, Malete, Nigeria/Department of Physiology, College of Medicine, Korea University, Seoul, South Korea

About the Editors

Abraham Olufemi Asuku is a Ph.D. holder in neurophysiology in the Department of Physiology, Faculty of Basic Medical Sciences, University of Ilorin, Ilorin, Nigeria. He obtained his first degree (B.Sc.) and master's degree (M.Sc.) from the same department. Dr. Asuku is a passionate neuroscientist and medical biotechnologist. He currently works with the National Biotechnology Research and Development Agency, Nigeria, as a chief scientific officer. His area of interest is on mitigation of neurotoxicity, neurodegenerative diseases, and neurodevelopmental disorders by phytochemicals, and also on the applications of nanotechnology for neurodegenerative disease drug delivery. He is a member of the Neuroscience Society of Nigeria and the Physiological Society, United Kingdom. He is an award-winning researcher and has over 60 publications, including book chapters and research and review articles of good international repute with a high impact factor. He has served as a reviewer for many reputable journals.

He has also presented seminars at many conferences, both locally and internationally. He was one of the invited guest speakers at the VII International Conference of Universidad Tecnica De Manabi, Ecuador, in 2023. Dr. Asuku was an editor of the book titled *Non-pharmacological Intervention for Dementia: A Comprehensive Approach*, and he is currently editing two other books, *Apitherapy for Neurodegenerative Diseases* and *Evolution and Innovation in Parkinson's Disease Therapies in Under-Resourced Settings*, to be published by CRC Taylor and Francis Publisher and IGI Global Scientific Publishing, respectively. His long-term goal is to discover therapeutic targets and help create focused treatments for neurodegenerative diseases. By bridging the gap between basic and translational neuroscience, he seeks to further knowledge of neuronal health and provide insights that may result in novel treatments.

Ayodeji Johnson Ajibare holds an MBBS, an M.Sc., and a Ph.D. in Physiology from the University of Ilorin. He is a senior lecturer in the Department of Physiology at Lead City University, Ibadan, where he also serves in several leadership and clinical roles, including head of the Department of Physiology, director of clinical services at Lead City University Hospital, and acting dean of the Faculty of Basic Medical Sciences. Dr. Ajibare's research program sits at the intersection of neuroscience, endocrinology, and reproductive health. His work explores mechanisms of neurotoxicity, neurodegeneration, oxidative stress signaling, and endocrine-linked brain dysfunction, with particular emphasis on nutritional and phytochemical neuroprotection. He has published more than 50 scholarly works, with 28 indexed in Scopus, and has coauthored multiple book chapters. In addition to his research, Dr. Ajibare is actively involved in medical education, mentoring, and clinical service. He has coordinated MBBS training, contributed to university ethics and academic committees, and presented his findings at both national and international scientific meetings. He is also a member of professional bodies, including the Nigeria Medical Association and the Association of Specialist Medical Doctors in Academics. As a coeditor of *Bioresources for Managing and Treating Dementia: Prospects and Challenges*, Dr. Ajibare brings together clinical insight and translational research expertise. His long-term goal is to bridge basic neuroscience and clinical practice to advance innovative therapeutic strategies for dementia and related disorders.

Olugbemi Tope Olaniyan is Associate Professor of Applied Physiology and Biotechnology, Faculty of Basic Medical Sciences, College of Health Sciences, Kwara State University, Malete, Nigeria, and Visiting Professor in Department of Physiology, College of Medicine, Korea University, Seoul, South Korea. He holds a B.Sc. (University of Ilorin, Nigeria), M.Sc., Ph.D. (University of Ibadan, Nigeria) in Physiology and Postdoctoral Fellowship in Biotechnology from Institute de Biotecnologico, Centro de Investigacion Biomedica, University of Granada, Spain. Consistently over the years, he has been ranked among the top 500 scientists in Nigeria with scholarly output from Scopus SciVal/Elsevier between 2020 and 2025, holding seventh position in Africa in Medical Sciences and Stanford University World's Top 2% Scientists.

Dr Olaniyan has experience in cell culture, medical imaging, histology, immunohistochemistry, molecular and cellular biology techniques, ELISA, fluorescence microscopy, western blot, mitochondria physiology (sea horse), nanotechnology, biomedical statistics (STATA, GraphPad Prism, and SPSS), animal handling techniques such as mice, rats, *C. elegans*, zebra fish, and drosophilae. He also has vast experience in the utilization of modern technology for teaching and research such as Anatomage table, PowerLab, High and Low fidelity 3G SimMan Mannequins, and learning management system (LMS). He has over 15 years of cognate experience in teaching and research and has trained over 60 graduate students and co-supervised five master's and two Ph.D. postgraduates. In addition, he has published over 90 journals, including 250 book chapters in reproductive biology, applied physiology, biotechnology, bioinformatics, and biophysics in international reputable journals. He is a member of several professional bodies, including the Physiological Society of Nigeria, Nigerian Bioinformatics and Genomics Networks, Drosophila Africa, Anti-Vec Network, Genetics Society of America, West Africa Genetics Society. His future goals will be to focus on applied physiology, molecular biology, and biotechnology techniques through the adoption and utilization of high-resolution microarray, next-generation sequencing, and current techniques in molecular biology and imaging that could help in the analysis of different factors that affect human health and the environment in order to achieve sustainable development goals.

Contributors

Patrick Oluwole Abolarin
Department of Physiology
School of Basic Medical Sciences
Babcock University
Ilishan-Remo, Nigeria

Uthman Okikiola Adebayo
Department of Medical Laboratory Science
Neuropsychiatric Hospital
Abeokuta, Nigeria

Oluwafemi Abidemi Adedotun
Department of Human Anatomy
Elizade University
Ilara-mokin, Nigeria

Oyaname Miracle Adejumo
Neuro-Reproductive and Metabolism Unit
Department of Physiology
Faculty of Basic Medical and Health Sciences
College of Medicine
Lead City University
Ibadan, Nigeria

Kehinde Oluwaseyi Adeniji
Department of Anatomy
Benjamin Carson (Snr)
College of Health and Medical Sciences
Babcock University
Ilishan-Remo, Nigeria

Temitope Oluwatobi Adeniji
Faculty of Pharmacy
Olabisi Onabanjo University
Ago-Iwoye, Nigeria

Philip Adeyemi Adeniyi
Neuroscience Section
Department of Paediatrics
Oregon Health & Science University
Portland, Oregon

Eniola Oluwatomisin Adesanya
Department of Human Anatomy
College of Medicine
Lead City University
Ibadan, Nigeria

Abdulmajeed Opeyemi Agboola
Department of Public Health
Kwara State University
Malete, Nigeria

Faith Ewaoluwa Ajayi
FdSc Assistant Practitioner Associate Member
Faculty of Public Health
Colchester, United Kingdom

Duyilemi Chris Ajonijebu
Department of Human Anatomy and Physiology
University of Johannesburg
Doornfontein, South Africa

Favour Ayomide Akinpelu
Department of Anatomy
Faculty of Basic Medical Sciences
University of Lagos
Lagos, Nigeria

Tolulope Oluwatimilehin Alabi
Department of Anatomy
Faculty of Basic Medical Sciences
University of Ilorin
Ilorin, Nigeria

Odugbesan Olawunmi Anthonia
Department of Physiology
Faculty of Basic Medical Sciences
Afe Babalola University
Ado-Ekiti, Nigeria

Oluwapelumi Lydia Apanisile
Department of Anatomy
Faculty of Basic Medical Sciences
University of Ilorin
Ilorin, Nigeria

Kafilat Temidayo Aransi-ola
Neuro-Reproductive and Metabolism Unit
Department of Physiology
Faculty of Basic Medical and Health Sciences
College of Medicine
Lead City University
Ibadan, Nigeria

Victor Bassey Archibong
Department of Human Anatomy
College of Medicine and Health Sciences
University of Rwanda, Butare campus
Butare, Rwanda

Tolulope Timothy Arogundade
Department of Human Anatomy
Neurobiology Division
Faculty of Basic Medical Sciences
Redeemers University
Ede, Nigeria

Gideon Opeyemi Ayilara
Department of Physiology
Faculty of Basic Medical Sciences
Adeleke University
Ede, Nigeria

Maryam Tayo Ayinla
Department of Physiology
Faculty of Basic Medical Sciences
College of Health Sciences
University of Ilorin
Ilorin, Nigeria

Abosede Mary Ayoola
Department of Anatomy
Lead City University
Ibadan, Nigeria

Olabisi Elizabeth Ayoola
Department of Physiology
Faculty of Basic Medical Sciences
College of Health Sciences
Kwara State University
Malete, Nigeria

Ahmed Olalekan Bakare
Anaesthesiology and Critical Care Medicine
Johns Hopkins University School of Medicine
Baltimore, Maryland

Abdulwarith Ayomide Bamidele
Department of Public Health
Faculty of Allied Health Sciences
Kwara State University
Malete, Nigeria

Victor Jesulayomi Bamisaye
Department of Plant Molecular Biology
Faculty of Pharmacy
Université de Tours
Tours, France

Adedamola Aminat Bayo-Olugbami
Department of Physiology
Faculty of Basic Medical Sciences
Osun State University
Osogbo, Nigeria

Abosede Oreoluwa Bolaji
Department of Physiology
Faculty of Basic Medical Sciences
Adeleke University
Ede, Nigeria

Moyosoluwa Mary Dada
Department of Anatomy
Benjamin Carson (Snr)
College of Health and Medical Sciences
Babcock University
Ilishan-Remo, Nigeria

Ambali Olanrewaju Danwahab
Department of Anatomy
Faculty of Basic Medical Sciences
College of Health Sciences
Kwara State University
Malete, Nigeria

Grace Ayobami Fajemidagba
Department of Physiology
Faculty of Basic Medical Sciences
Redeemer's University
Ede, Nigeria

Emmanuel Taiwo George
Department of Physiology
Faculty of Basic Medical Sciences
Osun State University
Osogbo, Nigeria

Babatunde Oluwaseun Ibitoye
Department of Human Anatomy
Federal University of Technology
Akure, Nigeria

Ruqayyah Yetunde Ibiyeye
Department of Anatomy
Faculty of Basic Medical Sciences
College of Health Sciences
Kwara State University
Malete, Nigeria

Hidaayah Oluwamayowa Jimoh-Abdulghaffaar
Department of Physiology
Faculty of Basic Medical Sciences
College of Health of Sciences
University of Ilorin
Ilorin, Nigeria

Ryu Ki-Jin
Reproductive Endocrinology
Obstetrics and Gynecology, Medicine
Korea University Seoul
Seoul, South Korea

Rhoda Mama Kolo
Developmental & Neuro-Biology Research Unit
Department of Anatomy
Faculty of Basic Medical Sciences
College of Health Sciences
University of Ilorin
Ilorin, Nigeria

Abdullahi Mohammed
Department of Human Anatomy
School of Medicine and Pharmacy
College of Medicine and Health Sciences
University of Rwanda
Kigali, Rwanda

Olawale Olaleye Obembe
Department of Physiology
Faculty of Basic Medical Sciences
Osun State University
Osogbo, Nigeria

Gbonjubola Oyinlola Ogungbangbe
Memorial University of Newfoundland
Newfoundland and Labrador, Canada

Usman Okeme
Department of Chemical Sciences
College of Natural and Applied Sciences, Summit University
Offa, Nigeria

Olutayo Folajimi Olaseinde
Department of Neuroscience
University of Minnesota
Minneapolis, Minnesota

Obafemi Samuel Olayinka
Department of Public Health Science
Faculty of Allied Health Sciences
Kwara State University
Malete, Nigeria

Akintayo Christopher Oloruntoba
Department of Medicine and Surgery
Obafemi Awolowo University Teaching Hospital Complex
Ile-Ife, Nigeria

Femi Oloye
Department of Chemistry
Division of Physics and Computational Sciences
University of Pittsburgh at Bradford
England, United Kingdom

Olatoun Motunrayo Olutunde
Department of Anatomy
Lead City University
Ibadan, Nigeria

Olabode Oluwadare Akintoye
Neuro-Reproductive and Metabolism Unit
Department of Physiology
Faculty of Basic Medical and Health Sciences
College of Medicine
Lead City University
Ibadan, Nigeria

Olowomeye Mary Oluwakemi
Department of Physiology
Faculty of Basic Medical Sciences
College of Medicine and Health Sciences
Achievers University
Owo, Nigeria

Niyi Ebenezer Omojadegbe
Department of Chemical Sciences
Bingham University
Karu, Nigeria

Oluwafemi Abiola Onaolapo
Department of Physiology
Faculty of Health Sciences
Atiba University
Oyo, Nigeria

Kelechi Roselyn Onyema
Department of Anatomy
Benjamin Carson (Snr)
College of Health and Medical Sciences
Babcock University
Ilishan-Remo, Nigeria

Olaleye Tobi Opeyemi
Department of Physiology
Faculty of Basic Medical Sciences,
College of Medicine and Health Sciences
Achievers University
Owo, Nigeria

Omoruyi Gloria Osayamen
Department of Internal Medicine
Obafemi Awolowo University Teaching Hospital Complex
Ile-Ife, Nigeria

David Temidayo Ovioke
Department of Anatomy
Faculty of Basic Medical Sciences
Lead City University
Ibadan, Nigeria

Bamidele Victor Owoyele
Neuroscience & Inflammation Unit
Department of Physiology
Faculty of Basic Medical Science College of Health Sciences
University of Ilorin
Ilorin, Nigeria

Precious Adeoye Oyedokun
Department of Physiology
Faculty of Basic Medical Sciences
Ladoke Akintola University of Technology
Ogbomoso, Nigeria

Mosunmola Busayo Oyeleke
Department of Physiology
Faculty of Basic Medical Sciences
Afe Babalola University
Ado-Ekiti, Nigeria

Fabunmi A. Oyesanmi
Section Sports Medicine
Faculty of Health Science
University of Pretoria
Pretoria, South Africa

Rukayat Omowumi Salahudeen
Department of Public Health
Faculty of Allied Health Sciences
Kwara State University
Malete, Nigeria

Olaniyi Kehinde Samuel
Department of Physiology
College of Medicine and Health Sciences
Afe Babalola University
Ado-Ekiti, Nigeria

Onyekweli Samuel
Department of Radiation Oncology
Lagos University Teaching Hospital
Idi-Araba, Nigeria

Anne Fiyinfoluwa Sanwo
Development Neurobiology & Forensic Unit
Department of Anatomy
Faculty of Basic Medical Sciences
University of Ibadan
Ibadan, Nigeria

Akinsomisoye Olumide Stephen
Department of Morphology, Genetics
Orthodontics and Pediatric Dentistry
School of Dentistry, Sao Paulo State University (UNESP)
Araraquara, Brazil

Fatimo Ajoke Sulaimon
Department of Anatomy
Faculty of Basic Medical Sciences
College of Health Sciences
University of Ilorin
Ilorin, Nigeria

Azeezat Dagbo Yakub
Department of Anatomy
Faculty of Basic Medical Sciences
University of Ilorin
Ilorin, Nigeria

Acknowledgments

We are deeply grateful for the support of all the contributors to this book, whose prompt response and valuable input have been instrumental in the success of this book. We sincerely appreciate each contributor's enthusiasm and great support in ensuring that this project was completed on time. Their partnership and collaboration are commendable. We also show our immense gratitude to the anonymous reviewers whose constructive feedback helped improve the quality of this book. We deeply appreciate the support of Professor Wael Mohamed, who is the series editor for the volume of this book. We would like to acknowledge the great support of the CRC Press and production team, whose guidance was vital in ensuring the timely completion of this project. In conclusion, we would like to express our immense gratitude to the staff, management, and colleagues at the National Biotechnology Research and Development Agency (Nigeria); the Department of Physiology, Faculty of Basic Medical Sciences, Lead City University, Ibadan (Nigeria); the Department of Physiology, Faculty of Basic Medical Sciences, Kwara State University, Malete (Nigeria); and the Department of Physiology, College of Medicine, Korea University, Seoul (South Korea) for their support throughout this book project.

Abraham Olufemi Asuku, Ph.D.
Bioresources Development Centre, Department of Medical Biotechnology, National Biotechnology Research and Development Agency, Ogbomoso, Nigeria

Ayodeji Johnson Ajibare, Ph.D.
Department of Physiology, Faculty of Basic Medical Sciences, Lead City University, Ibadan, Nigeria

Olugbemi Tope Olaniyan, Ph.D.
Department of Physiology, Faculty of Basic Medical Sciences, Kwara State University, Malete, Nigeria/Department of Physiology, College of Medicine, Korea University, Seoul, South Korea

1 Dementia

Pathophysiology, Types, and Global Prevalence

Akintayo Christopher Oloruntoba,
Olowomeye Mary Oluwakemi, Olaleye Tobi Opeyemi,
Omoruyi Gloria Osayamen, Onyekweli Samuel,
Akinsomisoye Olumide Stephen, and Olaniyi Kehinde Samuel

1.1 INTRODUCTION

Dementia, a neurological condition that is deteriorating good health, functional dependence, progressive cognitive impairment, and adverse health outcomes, is a common condition among the elderly (Chin, 2023). In the field of public health, it is a common problem. Dementia affects about 47 million individuals across the globe, and there is a tendency for this case to double (Chin, 2023). In this disease condition, the areas in the brain mostly affected are the temporal lobe and frontal lobe; other parts might also be involved. People who have dementia frequently have abnormalities in their brains that come from a variety of sources. Dementia is a progressive impairment of memory, thinking, and behavior, all of which together constitute human consciousness. Usually, emotional problems, communication difficulties, and a decrease in motivation accompany these symptoms (Błaszczyk, 2022). It is commonly known that computing in cognitive brain functions, including learning, motor control, behavior regulation, and sensory perception, is carried out via axonal-dendritic chemical synapses in cortical networks (Olugbemi et al., 2023a, b; Adetunji et al., 2023). Dementia is not a sickness. When a group of symptoms are together, they are called a syndrome. The negative impact impairs affected individuals in interacting with others and prevents them from having a perception of and comprehending their surroundings. In dementia, a condition that is frequently progressive in nature, cognitive ability declines more than may be expected from normal aging. It affects memory, thinking, orientation, comprehension, calculation, learning capacity, language, and judgment (Deleon & Miller, 2018; Ayinla et al., 2020; Asuku et al., 2024a). Cognitive function impairment is often preceded or coexists with deterioration in motivation, social behavior, or emotional control (World Health Organization, 2017). Although different forms of dementia have been identified by scientists, the ones that commonly affect the aged have been reported (Dementia UK, 2019).

1.2 PATHOPHYSIOLOGY OF DEMENTIA

It is well acknowledged that misfolded proteins, referred to as proteinopathies, and/or cerebrovascular disease (CVD) caused the pathophysiology of dementia (Chin, 2023). Frontotemporal dementia (FTD), vascular dementia, dementia with Lewy bodies (DLB), and Alzheimer's disease (AD) are the most common causes of late-onset dementia (Chin, 2023; Asuku et al., 2025). The neocortex, limbic system, thalamic systems, and brain stem regulatory system are the primary neurological substrates of cognition and awareness, according to Tucker et al. (2022). Consciousness depends on the positive connections between the brain stem and midbrain. The cytoarchitectonic features

DOI: 10.1201/9781003657385-1

of limbic regions and their cortico-cortical networks render them more susceptible to neurodegeneration and aging because of their high activity and flexibility (García-Cabezas et al., 2019). It has been demonstrated that as people age, their prefrontal limbic regions become less stable in activity due to a decrease in the number of inhibitory parvalbumin neurons, which block nearby pyramidal neurons (García-Cabezas et al., 2019). There are two recognized pathological causes for the aging brain. First, as people age, their energy metabolism breaks down, leading to neurodegeneration and eventual neuronal death. The second is that as individuals age, their cognitive and motor capacities deteriorate, causing the brain to gradually involute (Błaszczyk, 2022). Myelinated connections, which link disparate brain areas and are crucial for sustaining higher-level cognitive processes, are particularly significant for cognitive, motor, and behavioral functions (O'Brien & Thomas, 2015). Myelination is crucial for maintaining, modifying, and enhancing overall brain function (Cividini et al., 2021). Most importantly, myelin reduces the energy cost of neuronal transmission by spatially distributing mitochondria around the nodes of Ranvier. The part of the brain matter (white) is involved in a crucial role in nearly every aspect of cognition, brain evolution, and functioning (Bang et al., 2015). The myelination process is critical to the efficacy of higher function and how they are modified and maintained. However, axonal myelination reduces the energy cost of neuronal transmission by allowing mitochondria to be distributed locally close to the Ranvier nodes. Oxidative stress increases the amount of proteins that are damaged and irreparable because it decreases the production of nicotinamide adenine dinucleotide (NAD)-dependent enzymes. Several factors, including reaction stoichiometry, metabolite utilization rate, and the speed of molecule translocation across lipid bilayers, influence pathways of metabolism in cells. NAD deficiency causes energy and metabolic crises to accumulate in aging neurons (Brazill et al., 2017).

1.3 DEMENTIA TYPES

1.3.1 Alzheimer's Disease

This is the most common dementia, and its cases are increasing on a yearly basis. It can occur alone or in conjunction with other illnesses, in which case it's called "mixed dementia"(see Section 1.3.5). AD, which already affects individuals, is predicted to have affected 13.8 million people by 2050. AD is the sixth most prevalent death cause and fifth most common cause in persons more than 65, according to the Centers for Disease Control and Prevention (CDC) (2015). The most common early symptoms include trouble identifying words and forgetfulness of recent events. Because it is progressive, the symptoms worsen and the brain changes with time. There are individual variations in the talents affected and the pace of advancement from one year to the next. Affected individuals may be able to complete everyday chores more easily and have much greater cognitive ability. On bad days, people may find it harder to perform everyday chores, and their cognitive abilities may decline. As the disease progresses, more verbal difficulties and memory loss become apparent in the absence of acetylcholine, which is essential for memory and learning (Cividini et al., 2021).

Masters et al. (2015) stated that what causes Alzheimer's is categorically tangles of neurofibrils as well as plaques of Aβ. One of the characteristics is caused by phosphorylated tau proteins (Lane et al., 2018). It is commonly accepted that tau hyperphosphorylation, which results in microtubule destabilization, is the pathogenic mechanism that produces downstream neurodegenerative damage, leading to activation of microglia, synapse loss, and inactivity of the neurons (Masters et al., 2015; Holper et al., 2022). According to the Braak method, tau—rather than Aβ—is the cause of reduction in brain size, impairment in cognitive activity, and deterioration among people with AD. Tau pathology often begins in the temporal cortices (Aschenbrenner et al., 2018). AD is multicausal, despite the amyloid cascade theory being the most common pathogenic explanation for the disease for many years. The proteins tau and Aβ are commonly seen in AD patients. It is now more widely accepted that neuroinflammation, coupled with microglial activation, contributes to AD, which is associated with Aβ buildup, cell death, and neuronal damage (Calsolaro & Edison, 2016). A comprehensive investigation found prevalence increased with age and APOE ε4 carrier status, with

51% (95% CI: 33%, 69%) of those with DLB having Aβ–PET–positive results (Ossenkoppele et al., 2015). Investigation using a statewide platform showed that larger individuals with either diffuse or transitional disease of Lewy body had a considerable tau/Aβ burden at autopsy (Graff-Radford et al., 2017; Chin et al., 2020).

1.3.2 Vascular Dementia

In this type there is an obstruction in the flow of blood to the higher centers, thus limiting activity of the neurons, resulting in dormancy of cells in the brain (De Reuck et al., 2018). Diabetes, smoking, diet, obesity, hypertension, and hyperlipidemia are just a few of the many vascular risk factors that may be at play (O'Brien & Thomas, 2015). Diabetes raises the risk of dementia in addition to vascular disease because it causes chemicals generated from the hormone amylin to build up in the brain. Vascular dementia can occur following a stroke; however, it often advances gradually. Vascular dementia can manifest with symptoms due to the illness's site, including its nature. In addition, memory and language issues, depression, anxiety, apathy, and a slowdown of mental processes are common (O'Brien & Thomas, 2015).

1.3.3 Lewy Body-Associated Dementia

This is the third most common dementia, and it is closely related to Parkinsonism and/or AD due to the numerous features they have in common. For instance, cognitive impairment and, eventually, dementia can result from Parkinson's disease. Cells in the cerebral cortex contain tiny clusters of a protein called alpha-synuclein in Lewy bodies associated with dementia (Chin et al., 2020). These clusters serve as diagnostic markers for this class of diseases. Impairment of memory is a clinical feature of this disease. There are problems with planning, losing attention, and getting lost in space. Parkinson's disease-like symptoms, including shaking in the limbs, stumbling when walking, and a limited capacity to articulate oneself, are caused by Lewy body dementia, which ranks third in terms of prevalence. Visual hallucinations, frequent falls, substantial changes in conscious awareness, and nightmares are also symptoms of Lewy body dementia. In addition to one or more of the four main clinical characteristics of DLB—recurrent well-formed visual hallucinations, spontaneous motor parkinsonism, rapid eye movement (REM) sleep behavior disorder, and cognitive fluctuations—people with DLB typically show progressive cognitive impairment (Masters et al., 2015; Asuku et al., 2024b).

1.3.4 Frontotemporal Dementia

An extremely uncommon kind of dementia known as FTD includes various states that nightmares impact the central area of the brain involved in motivation, language, and planning. FTD can be classified into several types based on whether the frontotemporal lobe is primarily involved (Warren et al., 2013).

Significant behavioral changes and difficulties with making the day span out can have an impact on the personality. Changes in eating patterns include excessive eating/desiring meals high in sugar. Individuals with FTD also present with speech problems such as difficulty in communicating, they lack conceptualization and cannot interpret speech. A wide range of clinical neurodegenerative symptoms is included in FTD, which is characterized by gradual abnormalities in functionality at higher centers. The broad spectrum of neuropathologies that underpin the clinical symptoms of FTD are collectively said to be degenerational frontotemporal lobe. The utility of several medication classes is examined. Medication for AD cannot help people with FTD and may even make neuropsychiatric symptoms worse. Safety concerns, peer and caregiver support, and lifestyle changes are examples of non-pharmacological treatment approaches (Nisar et al., 2021). With so many active clinical trials focusing on various pathogenetic pathways, there are tremendous opportunities for transformative breakthroughs in the cure for FTD/disorder.

1.3.5 Mixed Dementia

Most of the different types of dementia are coexisting in this case. The most common kind combines the clinical characteristics and brain changes in disease patterns in Alzheimer's with associated vasculature. A combination of AD and vascular pathology is commonly seen during post-mortem examination, and this is increasingly common with increased age, above 75–80 years (De Reuck et al., 2018). In this type of dementia, there is associated progressive loss of skills and comorbidity, diabetes, smoking, and increased cholesterol/ischemic heart disease (Pennington et al., 2020). Kawasi et al. (2015) stated that mixed dementia is more common in later age. For instance, those who are 85 years old have evidence of two or more causes of dementia (De Reuck et al., 2018). A person's lifetime risk of getting dementia symptoms is increased if they have Alzheimer's brain changes in addition to brain changes from another kind of dementia, as opposed to Alzheimer's brain changes alone (Kawasi et al., 2015). Figure 1.1 shows the different types of dementia.

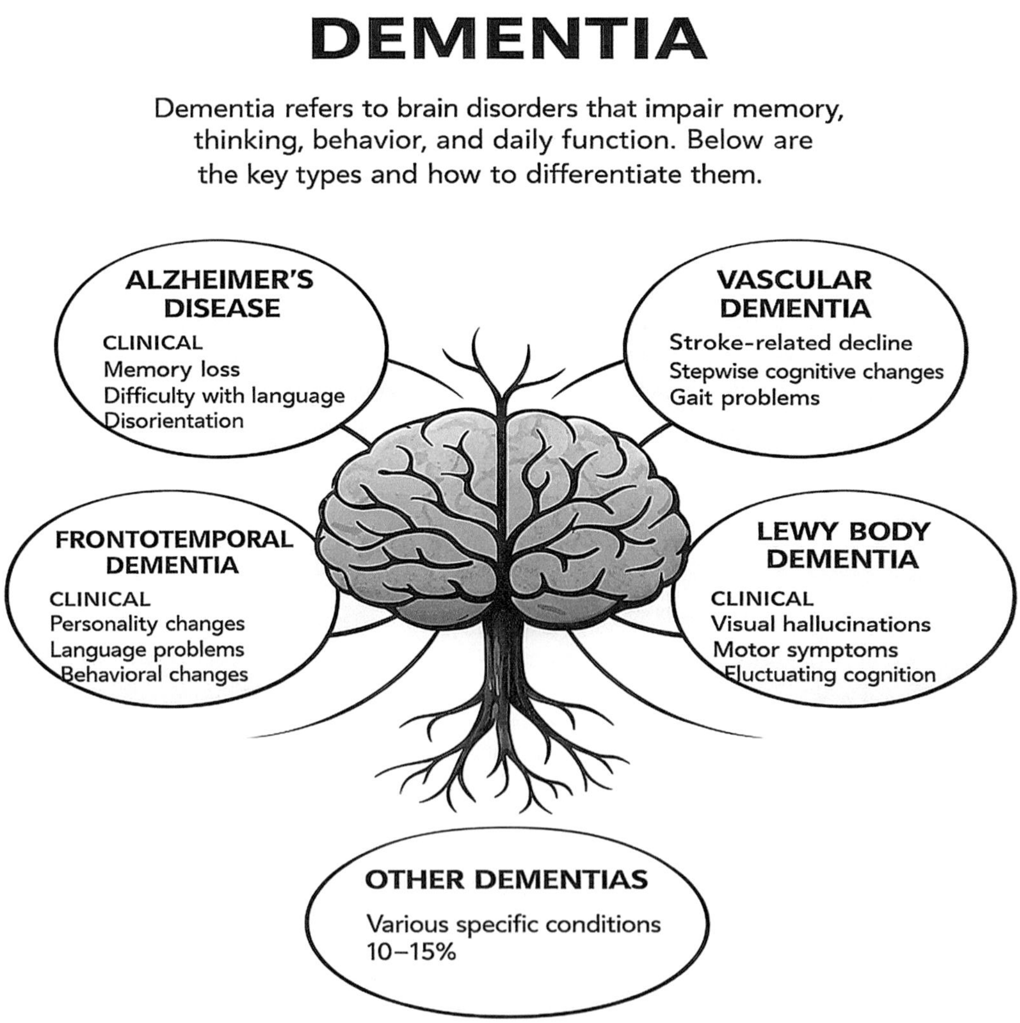

FIGURE 1.1 Dementia tree: showing the different types of dementia.

1.4 PROGNOSIS AND TREATMENT OF DEMENTIA

Numerous management options are said to improve quality of life and lessen symptoms. Galantamine (Reminyl), rivastigmine (Exelon), and donepezil (Aricept) are examples of cholinesterase inhibitors that can improve nerve cell communication and reduce cognition in mild to severe AD (Livingston et al., 2020). Memantine and having contact with people are very beneficial in the treatment of dementia. Supportive strategies, including routine establishment, caregiver education, and maintaining a secure environment, are necessary for comprehensive dementia care. The symptoms of dementia usually worsen with time, making it a progressive disorder. The rate of progress might vary significantly from person to person. Early detection and treatment in some circumstances can improve life quality. Hence, it progressively impairs cognitive and physical functioning, necessitating more intensive care. Establishing a routine might make people feel less confused and more stable. Using clear signs and simplifying the living space are two environmental adjustments that promote safety and independence. Help groups and counseling programs are beneficial for people with dementia because they offer both emotional and practical help.

1.5 GLOBAL PREVALENCE OF DEMENTIA

It is well known that the increasing data of dementia has surged; in the next 20 years, this figure is predicted to triple (Błaszczyk, 2022). In keeping with past estimates, women with dementia will surpass that of males. By 2050, this pattern was expected to continue (Jorm et al., 2017; Nichols & Szoeke, 2019). These tendencies persist even though vascular risks are more prevalent in men than in women, suggesting that there may be strong opposing mechanisms at play. Considering that women make up 65% of dementia patients and men make up 35% (Matthews et al., 2016). According to Public Health England (2018), it was the cause of 15.8% of deaths in women and 8.3% in men. Erol and colleagues (2015) revealed that women are more prone to developing severe dementia than men do.

1.6 CONCLUSION

Dementia affects the majority of the people who are growing older, and the effects on the economy could be worrisome. Knowing its many manifestations and potential treatment choices is important for those who provide health support and manage the affected ones. Early detection with pharmaceutical treatments will greatly enhance individuals impacted, even if there is presently no cure. There is promise for improved dementia treatments and, potentially, preventive measures due to ongoing scientific and medical studies.

REFERENCES

Adetunji, C.O., Olugbemi, O.T., Olorunsola, A., Ayobami, D., Mayowa, J., & Alex, E. (2023). Classification of Neurodegenerative Disorders Using Machine Learning Techniques. Artificial Intelligence for Neurological Disorders, 261–73. Academic Press.

Aschenbrenner, A.J., Gordon, B.A., Benzinger, T.L., Morris, J.C., & Hassenstab, J.J. (2018). Influence of tau PET, amyloid PET, and hippocampal volume on cognition in Alzheimer's disease. *Neurology*. 91(9):e859–66. doi: 10.1212/WNL.0000000000006075.

Asuku, A.O., Ayinla, M.T., Gbonjubola, O.O., Babatunde, S.S., Olajide, T.S., & Oyerinde, T.O. (2024a). Neurocognitive Aspects of Dementia. In Nutrition in Brain Aging and Dementia. Nutritional Neurosciences. Springer.

Asuku, A.O., Ayinla, M.T., Olajide, T.S., Oyerinde, T.O., Yusuf, J.A., Bayo-Olugbami, A.A., & Fajemidagba, G.A. (2024b). Coffee and Parkinson's Disease. In Progress in Brain Research, 289: Neuroscience of Coffee. Elsevier. doi: 10.1016/bs.pbr.2024.06.001.

Asuku, A.O., Ogungbangbe, G.O., & Olajide, T.S. (2025). Effect of Happiness and Positive Emotions on Dementia. In Neuroscience of Feelings. Progress in Brain Research. Elsevier. doi: 10.1016/bs.pbr.2025.03.001.

Ayinla, M.T., Asuku, A.O., Bayo-Olugbami, A., Ayeni, A.O., Abiola, A.A., & Owoyele, B.V. (2020). Melatonin and vitamin C modulate cholinergic neurotransmission and oxidative stress in scopolamine-induced rat model of memory impairment. *J Afr Assoc Physiol Sci.* 8(1):20.

Bang, J., Spina, S., & Miller, B.L. (2015). Frontotemporal dementia. *Lancet.* 386:1672–82.

Błaszczyk, J.W. (2022). Pathogenesis of dementia. *Int J Mol Sci.* 24(1):543. doi: 10.3390/ijms24010543.

Brazill, J.M., Li, C., Zhu, Y., & Zhai, R.G. (2017). NMNAT: It's an NAD+ synthase. It's a chaperone. It's a neuroprotector. *Curr Opin Genet Dev.* 44:156–62.

Calsolaro, V., & Edison, P. (2016). Neuroinflammation in Alzheimer's disease: Current evidence and future directions. *Alzheimers Dement.* 12(6):719–32. doi: 10.1016/j.jalz.2016.02.010.

Centers for Disease Control and Prevention (CDC). Health, United States, 2015: With Special Feature on Racial and Ethnic Health Disparities. CDC website. https://www.cdc.gov/nchs/data/hus/hus15.pdf.

Chin, K.S. (2023). Pathophysiology of dementia. *Aust J Gen Pract.* 52(8). doi: 10.31128/AJGP-02-23-6736.

Chin, K.S., Yassi, N., Churilov, L., Masters, C.L., & Watson, R. (2020). Prevalence and clinical associations of tau in Lewy body dementias: A systematic review and meta-analysis. *Parkinsonism Relat Disord.* 80:184–93. doi: 10.1016/j.parkreldis.2020.09.030.

Cividini, C., Basaia, S., & Spinelli, E.G. (2021). Amyotrophic lateral sclerosis-frontotemporal dementia: Shared and divergent neural correlates across the clinical spectrum. *Neurology.* 98:e402–15.

Deleon, J., & Miller, B.L. (2018). Frontotemporal dementia. *Handb Clin Neurol.* 148:409–30.

Dementia UK (2019). What is dementia? Dementia UK, London. https://www.dementiauk.org/book.

De Reuck, J., Maurage, C.A., & Deramecourt, V. (2018). Aging and cerebrovascular lesions in pure and in mixed neurodegenerative and vascular dementia brains: A neuropathological study. *Folia Neuropathol.* 56(2):81–7.

Erol, R., Brooker, D., & Peel, E. (2015). Women and dementia. A global research review. *Alzheimers Dis Int.* https://eprints.worc.ac.uk/id/eprint/3744.

García-Cabezas, M., Zikopoulos, B., & Barbas, H. (2019). The structural model: A theory linking connections, plasticity, pathology, development and evolution of the cerebral cortex. *Anat Embryol.* 224:985–1008. doi: 10.1007/s00429-019-01841-9.

Graff-Radford, J., Aakre, J., & Savica, R. (2017). Duration and pathologic correlates of Lewy body disease. *JAMA Neurol.* 74(3):310–5. doi: 10.1001/jamaneurol.2016.4926.

Holper, S., Watson, R., & Yassi, N. (2022). Tau as a biomarker of neurodegeneration. *Int J Mol Sci.* 23(13):23. doi: 10.3390/ijms23137307.

Jorm, A.F., Dear, K.B.G., & Burgess, N.M. (2017). Projections of future numbers of dementia cases in Australia with and without prevention. *Aust N Z J Psychiatry.* 39:959–63.

Kawasi, C.H., Kim, R.C., Sonnen, J.A., Bullain, S.S., Trieu, T., & Corrada, M.M. (2015). Multiple pathologies are common and related to dementia in the oldest-old: The 90+ study. *Neurology.* 85(6):535–42.

Lane, C.A., Hardy, J., & Schott, J.M. (2018). Alzheimer's disease. *Eur J Neurol.* 25(1):59–70. doi: 10.1111/ene.13439.

Livingston, G., Huntley, J., & Sommerlad, A. (2020). Dementia prevention, intervention, and care: 2020 report of the Lancet commission. *Lancet*, 396(10248):413–46. doi: 10.1016/S0140-6736(20)30367-6.

Masters, C.L., Bateman, R., Blennow, K., Rowe, C.C., Sperling, R.A., & Cummings, J.L. (2015). Alzheimer's disease. *Nat Rev Dis Primers.* 1(1):15056. doi: 10.1038/nrdp.2015.56.

Matthews, F.E., Stephan, B., Robinson, C.M., Jagger, L., Barnes, C., Arthur, L.E., Brayne, A.C., & Cognitive Function and Ageing Studies (CFAS) Collaboration (2016). A two decade dementia incidence comparison from the Cognitive Function and Ageing Studies I and II. *Nat Commun.* 7:11398. https://www.nature.com/articles/ncomms11398.

Nichols, E., & Szoeke, C.E. (2019). Global, regional, and national Burden of Alzheimer's Disease and other dementias, 1990–2016: A systematic analysis for the Global Burden of Disease Study 2016. *Lancet Neurol.* 18:88–106.

Nisar, M., Abubaker, Z.J., & Nizam, M.A. (2021). Behavioral and cognitive response to selective serotonin reuptake inhibitors in frontotemporal lobar degeneration: Asystematic review and meta-analysis. *Clin Neuropharmacol.* 44:175–83.

O'Brien, J.T., & Thomas, A. (2015). Vascular dementia. *Lancet.* 386(10004):1698–706. doi: 10.1016/S0140-6736(15)00463-8.

Olugbemi, O.T., Adetunji, C.O., Ayobami, D., Olorunsola, A., Mayowa, J.A., & Alex, E. (2023a). Clinical Applications of Deep Learning in Neurology and Its Enhancements with Future Predictions. In Artificial Intelligence for Neurological Disorders, 9–224. Academic Press.

Olugbemi, O.T., Adetunji, C.O., Ayobami, D., Olorunsola, A., Mayowa, J.A., & Alex, E. (2023b). New Trends in Deep Learning for Neuroimaging Analysis and Disease Prediction. In Artificial Intelligence for Neurological Disorders, 275–87. Academic Press.

Ossenkoppele, R., Jansen, W.J., & Rabinovici, G.D. (2015). Prevalence of amyloid PET positivity in dementia syndromes: A meta-analysis. *JAMA*. 313(19):1939–49. doi: 10.1001/jama.2015.4669.

Pennington, C., Marini, L., Coulthard, E., & Love, S. (2020). Mixed neuropathology in frontotemporal lobar degeneration. *Amyotroph Lateral Scler Frontotemporal Degener*. 21:301–8.

Public Health England (2018). Research and analysis. Chapter 2: Trends in mortality, section 5. https://www.gov.uk/government/publications/health-profile-for-england-2018/chapter-2-trends-in-mortality.

Tucker, D.M., Luu, P., & Johnson, M.M. (2022). Neurophysiological mechanisms of implicit and explicit memory in the process of consciousness. *J Neurophysiol*. 128:872–91. doi: 10.1152/jn.00328.2022.

Warren, J.D., Rohrer, J.D., & Rossor, M.N. (2013). Clinical review. Frontotemporal dementia. *BMJ*. 6(347):F4827. doi: 10.1136/bmj.f4827.

World Health Organization (2017). Dementia: Key Facts. WHO. https://www.who.int/news-room/fact-sheets/detail/dementia.

2 Pharmacotherapies of Dementia and Their Limitations

Hidaayah Oluwamayowa Jimoh-Abdulghaffaar

2.1 INTRODUCTION

The term dementia describes the loss of social and cognitive skills that impact an individual's day-to-day functions and life (thinking, remembering, and reasoning). It could also have an impact on an individual's personality and feelings. It can be minor when it first begins to interfere with a person's functioning, or severe, when a person must rely only on others to do basic daily functions, including eating. Dementia affects millions of people worldwide.

However, it is not a normal part of ageing, as some individuals grow old beyond 85 years of age without any symptoms of dementia. As of 2018, the worldwide prevalence of dementia was put at 50 million. By 2030, this figure was expected to rise to 82 million, and by 2050, it was expected to reach 152 million (Patterson et al., 2004).

Dementia can be progressive and irreversible, such as Alzheimer's disease (AD) (by far the commonest type of dementia), vascular dementia, Lewy body dementia, frontotemporal dementia, and mixed dementia, which is two or more forms of dementia combined; reversible, as seen in infections and immune disorders (e.g. multiple sclerosis), metabolic or endocrine disorders (e.g. thyroid disorders, hypoglycaemia, hypo- or hyper-natriaemia/calcaemia, low vitamin-1, B-6, B-12, E or copper, normal pressure hydrocephalus, sub-dural bleeding, brain tumours, or side-effect of medications); or associated with other disease conditions like Huntington's disease, Creutzfeldt-Jakob disease, traumatic brain injury, and Parkinson's disease. Researchers have recently characterised a recently discovered type of dementia called limbic-predominant age-related TDP-43 encephalopathy (LATE).

An essential component of dementia care is pharmacological treatment. To control the underlying disease, alleviate cognitive symptoms, and lessen related behavioural issues, it is crucial to appropriately assess dementia patients for the acceptability of pharmaceutical treatment. Following a dementia diagnosis, the aetiology and severity of the dementia are the primary determinants of the symptomatic treatment. The effective use of pharmacotherapy in the symptomatic management of dementia requires a solid understanding of the dosage and pharmacokinetic characteristics, adverse effects, and anticipated advantages of the drugs to be prescribed. Because the medications are typically costly, the patient and carers typically decide when to start symptomatic treatment. Three general categories can be used to group the pharmacological treatment of dementia. These include medications that (i) improve cognitive symptomatology, (ii) reverse or stabilise the underlying disease, and (iii) treat mood, behavioural, or psychiatric symptoms related to dementia.

2.2 PATHOPHYSIOLOGICAL BASIS FOR PHARMACOTHERAPY

The hallmarks of AD include neuronal and synaptic loss, neurofibrillary tangles, and Aβ deposition, which are linked to neurotransmitter dysfunction (Asuku et al., 2024, 2025). Additionally, preclinical and clinical research indicate that this neurotransmitter disruption is a major contributing element to the illness's cognitive and neuropsychiatric symptoms (Ayinla et al., 2020). Cholinesterase

DOI: 10.1201/9781003657385-2

inhibitors were first used to treat the symptoms of cognitive loss in AD due to a cholinergic deficit; however, most patients only see a slight improvement from this therapy approach. Compounds that may help AD patients regain their cognitive function and memory are subjects of current research (Solas et al., 2015).

2.2.1 Neurochemical Deficits in Dementia

Part of the explanation for the behavioural and cognitive signs of AD is an imbalance in different neurotransmitter levels. Enhanced comprehension of neurotransmitter system imbalances in various behavioural and psychiatric symptoms of dementia (BPSD) may serve as the foundation for logical therapeutic approaches. Aggressive behaviour is associated with decreased serotonergic and increased noradrenergic activity, while cholinergic–monoaminergic imbalance is likely a contributing factor to the pathophysiology of affective disorders (Cummings & Kaufer, 1996). Serotonergic blockage causes significant behavioural abnormalities, whereas selective cholinergic dysfunction frequently only causes modest impairments in spatial memory and other behavioural tasks. Accordingly, 5-HT has been suggested to be involved in preserving behavioural abilities when cholinergic transmission is diminished (Dringenberg & Zalan, 1999). While substantial serotonergic denervation has been documented in AD in the past (Chen et al., 1996, 2000), its clinical importance has not yet been fully established. Additionally, serotonin has been linked to many BPSD-related functions, including aggressiveness, sadness, anxiety, and psychosis (Chen et al., 1996; Lanctôt et al., 2001).

2.2.1.1 Cholinergic

According to research, acetylcholine (ACh) contributes significantly to the pathophysiology of cognitive symptomatology in AD. The cholinergic system is the brain's neuromodulatory neurotransmitter system of greatest importance (Lanari et al., 2006). Selective attention is mostly regulated by cholinergic projections to the cortex and thalamus from the basal forebrain. According to research on the neurobiology of neuropsychiatric symptoms in AD, cholinergic deficit is implicated in a number of the most common neuropsychiatric symptoms of the disease; cholinergic agents have psychotropic qualities in this context and lessen the behavioural abnormalities of AD patients by influencing the temporal and frontal structures that mediate emotional behaviours (Cummings & Kaufer, 1996). Cholinergic deficiencies are strongly implicated in the development of AD psychosis, according to pharmacologic, biochemical, and structural data. The temporal lobes exhibit the most pronounced cholinergic abnormalities in AD; metabolic, as well as perfusion imaging studies using PET and SPECT show that delusions are associated with dysfunction of the temporal and frontal lobes. The pathophysiology of psychosis in numerous illnesses is linked to the amygdala and frontotemporal areas. Psychosis is associated with a cholinergic deficit, perfusion, and metabolic abnormalities in these areas, and less with histopathologic alterations, which are more prominent in the parietal cortex. The nucleus basalis's position between limbic system afferents and neocortical efferents may account for AD patients' diminished ability to elaborate emotional cues and develop apathy. Lesions in the nucleus basalis essentially cut the neocortex off from a significant input from the limbic system, which impairs the brain's capacity to appropriately interpret the emotional meaning of inputs (Cummings & Kaufer, 1996). Apathy in AD and other illnesses has been associated with reduced perfusion of the medial frontal cortex. All these findings provide credence to the theory that behavioural abnormalities in AD could be influenced by modified cholinergic pathways. Although about 40% of AD patients experience depressive symptoms, episodes of major depression, and suicide are extremely uncommon. AD's cholinergic deficit may change the brain processes that underlie symptoms of depression. This helps reduce the incidence of depressive symptoms and explains why serious depression and suicide are uncommon in AD patients.

2.2.1.2 Glutamatergic

While neuronal survival and synaptic plasticity depend on glutamatergic neurotransmission, which is excitatory, through the N-methyl-D-aspartate receptor (NMDAR), overactivity of NMDAR leads to excitotoxicity and encourages cell death (Wang & Reddy, 2017). This has been suggested as one of the possible mechanisms for the neurodegeneration that happens in AD. Both downstream signalling pathways and the receptor's location (synaptic or extrasynaptic) affect NMDAR-mediated responses. While extrasynaptic NMDAR stimulation causes cell death and hence contributes to the aetiology of AD, synaptic NMDAR activity begins plasticity and stimulates cell survival (Wang & Reddy, 2017). Extrasynaptic NMDAR function is specifically blocked by memantine, an NMDAR antagonist used to treat AD.

2.2.1.3 Noradrenergic

The primary origin of the nucleus for noradrenergic fibres in the brain of mammals is the locus coeruleus (LC) (Lanari et al., 2006). Many studies have shown that depressive individuals with AD have a larger loss of LC neurones (Forstl et al., 1994), and there is a change in AD (Hoogendijk et al., 1995). It's unclear exactly how the NA system contributes to the behavioural symptoms of AD, though.

2.2.1.4 Dopaminergic

Dopamine (DA) is produced by the cell bodies of nerves in the midbrain. Additionally, it is found throughout the forebrain, specifically in the basal ganglia. It is in charge of controlling movement and regulating limbic centre activity. In certain regions of AD patients' brains, lower DA levels are seen. D2 receptor inactivation causes behavioural alterations in mice that are linked to limbic system activity. Psychosis and aggressive behaviour are probably linked to dopaminergic system malfunction (Lanari et al., 2006). The psychostimulant action of dopamine agonists is contrasted with the selective blocking of DA receptors by antipsychotics, which regulate schizophrenia symptoms. This prevents DA from contributing to more holistic elements of behaviour, mostly via the mesolimbic dopaminergic system (Lanari et al., 2006).

2.2.1.5 GABAergic

GABA is the central nervous system's principal inhibitory neurotransmitter. Cholinergic and glutamatergic neurones are extensively innervated by GABAergic neurones. It has been shown that the GABAergic system malfunction could be involved in human cognitive impairment. According to Solas et al. (2015), severe cases of AD have been shown to exhibit a notable decrease in GABA levels, which may be the cause of the behavioural and psychiatric symptoms of the disorder.

The excitatory/inhibitory (E/I) network's inhibitory component, namely, malfunction in the gamma-aminobutyric acid (GABA) signalling system, has received comparatively little attention. GABAergic remodelling in the AD brain may have started early in the disease's aetiology, according to mounting data, making it a viable molecular target for pharmaceutical treatments and drug development. The effectiveness of some GABAergic medications in reducing or reversing different aspects and symptoms of AD has been investigated; this may offer a new avenue for addressing the increasing demand for safer and more effective treatments (Calvo-Flores Guzmán et al., 2018).

2.2.1.6 Serotonergic

Patients with AD show a marked reduction in serotonin (5-HT) levels in several brain regions. Certain BPSD, including depression, anxiety, agitation, restlessness, and aggression, have been particularly linked to abnormal serotonergic systems. Patients with AD who exhibit psychotic symptoms are known to have lower levels of 5-HT in the presubiculum (Lanari et al., 2006).

2.2.2 Role of Amyloid Plaques and Tau Tangles

2.2.2.1 Amyloid Plaques

Genetic mutations in the presenilin (PS1, PS2) and amyloid precursor protein (APP) genes cause early-onset familial AD. These mutations alter the pathogenic pathway for APP synthesis and proteolysis, resulting in increased amyloid β (Aβ) production (Wang & Reddy, 2017).

2.2.2.2 Tau Tangles

Pathological alterations in AD frequently include neurofibrillary tangles generated from phosphorylated tau. Neurofibrillary tangles exacerbate mitochondrial dysfunction and oxidative stress, which damage synapses and cause neurodegeneration. AD symptoms include severe synapse loss and neuronal death.

2.3 CURRENT APPROVED PHARMACOTHERAPIES

There is currently no known cure for dementia (Ritchie et al., 2015); thus, dementia treatment remains a huge issue because current medications merely delay but do not stop the course of dementia. Management focuses on treating neuropsychiatric and neurologic symptoms to improve the quality of life for patients and their carers. Environmental adaptations, cognitive rehabilitation, pharmacological and nonpharmacological treatment of behavioural symptoms, symptomatic pharmacological therapy for cognitive deficiencies, and complication prevention and treatment are all part of the suggested treatment plan (Trotta et al., 2023). Although both pharmaceutical and nonpharmacological therapies have been proposed for dementia treatment, the data for their efficacy and safety are confusing and conflicting. Pharmacotherapy is frequently the primary intervention used to alleviate symptoms or slow the course of dementia disorders. The existing agents have variable therapeutic effects and efficacy (Takramah & Asem, 2022). Pharmacological options for reversing or stabilising dementia include treating identifiable reversible causes, lowering proven risk factors, and using disease-modifying interventions to reduce disease development (Table 2.1).

TABLE 2.1
Pharmacological Approaches Targeting Underlying Disease

Underlying Disease	Drugs
A. Identifiable Reversible Causes	
i. Depression (pseudodementia)	Antidepressants, e.g. MAO inhibitors
ii. Deficiency state, e.g. Vit. B_{12} deficiency, hypothyroidism	Treat the specific condition
iii. Infection, e.g. HIV-associated dementia, neurosyphilis	Treat the specific condition
iv. Metabolic disorders, e.g. hypoglycaemia, hypercalcaemia	Treat the specific condition
B. Vascular Factors	
i. Hyperlipidaemia, hypertension, diabetes mellitus, smoking, and obesity	Treat the specific condition
ii. Homocysteine-lowering agents	Folate, pyridoxine, vitamin B_{12}
iii. Secondary stroke (prevention)	Antiplatelets
iv. Atrial fibrillation and cardioembolic strokes	Anticoagulants
C. Ancillary Treatment	
i. Anti-inflammatory agents	NSAIDs, COX-2 inhibitors
ii. Antioxidants	Alpha tocopherol
iii. Multi-modal mechanisms	Statins, omega-3 fatty acids, oestrogen (post-menopausal women)

Source: Shiong (2008).

Currently, the only FDA-approved drugs for AD patients are those that target either the cholinergic system (e.g. donepezil, tacrine, rivastigmine, and galantamine) (Calvo-Flores Guzmán et al., 2018) or the glutamatergic system (e.g. memantine) (Geldenhuys & Darvesh, 2015; Greig, 2015; Alzheimer's Association Report, 2017). None of these medications is curative or disease-modifying; they only provide short, symptomatic relief to a portion of Alzheimer's sufferers. Researchers are always searching for novel treatment targets because existing drugs are ineffective, unable to stop or reverse cognitive loss, and have a host of unfavourable side effects (Casey et al., 2010; Greig, 2015). It is believed that abnormal neuronal hyperexcitability plays a significant role in AD cell death. GABAergic neurotransmission upregulation with $GABA_AR$ agonists may thus be a potential technique for reversing AD's hyperexcitability and preventing cell death (Calvo-Flores Guzmán et al., 2018).

The only symptomatic drugs available for cognitive, behavioural, and global functioning in dementia patients are cholinesterase inhibitors and memantine, an NMDA receptor antagonist (NICE, 2018). Galantamine, rivastigmine, and donepezil are three internationally acknowledged and accepted cholinesterase inhibitors used in the treatment of dementia (Table 2.2). Together with memantine, they are classified as symptomatic therapy and are not believed to have neuroprotective properties or to change the course of AD or other forms of dementia (NICE, 2018).

They have been licensed and approved by the NICE HTA for use in mild-to-moderate AD (donepezil, rivastigmine, galantamine) and moderate to severe AD (memantine) due to numerous trials showing their effectiveness in the disease. Cholinesterase inhibitors have also been shown to be useful in certain disorders, such as Parkinson's disease dementia, and dementia with Lewy bodies (Stinton et al., 2015); however, they are not effective in frontotemporal and vascular dementia (O'Brien et al., 2017). Memantine has been tested for Parkinson's disease and dementia with Lewy bodies; however, there is no solid proof of its effectiveness at this time (Stinton et al., 2015).

Antidepressants, antipsychotics, and mood stabilisers are used to manage behavioural and psychological symptoms of dementia (BPSD). Various GABAergic drugs have been shown to improve cognitive impairment by preventing some of Aβ's harmful effects (Mark et al., 1995; Louzada et al., 2004). Some medications targeting the GABAergic system, such as agonists, antagonists, inverse agonists, and other modulators, have demonstrated cognitive improvements. According to DeBoer and Westerink (1994), these drugs may influence cognition directly via ionotropic or metabotropic GABARs or indirectly via GABAergic modulation of glutamatergic and cholinergic activity. Although many GABAergic medications have already been tested in human clinical trials and animal models, none have become effective pharmaceutical treatments for dementia or cognitive dysfunction. Meanwhile, a significant number of potential treatments are either undergoing testing or require further evaluation. To fully explore the potential of this approach, additional research is necessary (Calvo-Flores Guzmán et al., 2018).

TABLE 2.2
Dementia Drugs in Clinical Use

Class	Examples
1. Cholinesterase inhibitors	Donepezil (Aricept®)
	Rivastigmine (Exelon®)
	Galantamine (Reminyl®), (Razadyne®)
2. NMDA antagonists	Memantine (Exiba®), (Namenda®)
3. Anti-amyloid antibody	Aducanumab (Aduhelm®)
	Lecanemab

2.3.1 Emerging and Investigational Therapies

2.3.1.1 Disease-Modifying Agents (e.g. Anti-Amyloid Monoclonal Antibodies)

Anti-amyloid monoclonal antibodies (mAbs), e.g. **aducanumab** and **lecanemab**, have been approved to start treating individuals with early-stage AD who have demonstrated β-amyloid pathology (Aβ) in the U.S. They produce a substantial decrease in total brain Aβ revealed by amyloid positron emission tomography (PET) (Cummings et al., 2023). **Remternetug** is currently in a phase III trial, while **gantenerumab** was discontinued from a phase III clinical trial following negative outcomes.

2.3.1.2 Anti-Tau Therapies

Table 2.3 summarises anti-tau therapies.

2.3.1.3 Neuroprotective and Anti-inflammatory Agents

$GABA_AR$ modulators that improve cognition by providing neuroprotection and preventing amyloid formation.

Non-amyloidogenic processing involves APP proteolysis by α- and γ-secretase enzymes, which play a physiological role in healthy neurones. Amyloidogenic processing occurs by successive cleavage activity of β- and γ-secretases (Chow et al., 2010). Taurine, etazolate, and valproic acid have all been shown to increase non-amyloidogenic processing.

Through the non-amyloidogenic pathway for Aβ synthesis, **etazolate**, a $GABA_AR$ positive allosteric modulator (Barnes et al., 1983), enhances cognition in aged rats (Drott et al., 2010) and AD patients when used in conjunction with other medications (Vellas et al., 2011). Etazolate therapy was associated with decreased inflammation and improved behavioural task performance in rats following brain injury, according to Siopi et al. (2013). Further investigation is necessary to ascertain whether etazolate on its own is safe and beneficial in enhancing human intellect.

Taurine has been found to produce tau polymerisation, one of the primary characteristics of AD, despite its high tolerability profile, neuroprotective efficacy in preventing the generation of Aβ, and capacity to enhance cognition (Santa-Maria et al., 2007). Taurine's appropriateness as an anti-amyloidogenic cognitive enhancer for the treatment of AD may therefore need to be confirmed by more studies in patients with AD and other animal models of AD.

TABLE 2.3
Tau-Targeting Therapies

S/No.	Drug type	Example
1	Antisense oligonucleotides	Lecanemab
2	Phosphatase activators	Phosphoprotein phosphatase 2A
3	Kinase inhibitors	Tyrosine kinase inhibitor (bosutinib), JNK3 inhibitors, ERK1/2 inhibitors, CDK5 inhibitors, p38 MAPK inhibitors, and GSK-3β inhibitors (Tideglusib) (Mathuram et al., 2018; Li et al., 2023)
4	Acetylation inhibitors	Donepezil (Aricept), rivastigmine (Exelon), and galantamine (Razadyne)
5	Deglycosylation inhibitors	Empagliflozin and dapagliflozin
6	Caspase inhibitors	
7	Tau aggregation inhibitors	Tramiprosate, clioquinol (Vahid et al., 2024)
8	Microtubule stabilisers	Taxanes, epothilones (Ballatore et al., 2012)
9	Immunotherapy	
	a. Active	AN1792 (beta-amyloid [Abeta]1-42) vaccine (Alshamrani, 2023)
	b. Passive immunotherapy	Aβ42 (Alshamrani, 2023)

Source: Congdon et al. (2023).

2.3.1.4 Nutraceuticals and Repurposed Drugs

2.3.1.4.1 $GABA_A$ Receptor Agonists

4,5,6,7-Tetrahydroisoxazolo[5,4-c]pyridin-3-ol (THIP) reduced NMDA-induced neuronal mortality by roughly 30% in hippocampus slice cultures, demonstrating neuroprotective benefits against N-methyl-D-aspartate receptor (NMDAR)-induced excitotoxicity (Kristensen et al., 2003). When compared to individuals receiving a placebo, it did not significantly alter tests of verbal and visuospatial cognition in AD patients with mild to moderate dementia (Mohr et al., 1986).

When administered at least 24 hours before Aβ treatment, **muscimol** demonstrated a suppression of Aβ-induced neuronal apoptosis that is concentration-dependent, as well as increases in caspase 3 activity, reactive oxygen species production, glutamate release, and cytosolic Ca^2+, in pre-treated cultured rat cortical neurons (Lee et al., 2005). This suggests that chronic activation of $GABA_A$Rs may be necessary to mitigate the toxicity caused by Aβ (Lee et al., 2005).

2.3.1.4.2 $GABA_A$ Receptor Antagonists

In old wild-type and APP/PS1 mice, **picrotoxin (PTX)** restored memory deficits by reversing the enhanced $GABA_A$R-mediated inhibition, which may be exacerbated by ageing and mutant APP and PS1 expression (Yoshiike et al., 2008). Age, APP, and PS1 gene mutations are the two main risk factors for AD, and they reduce cognitive decline linked to these factors (Calvo-Flores Guzmán et al., 2018).

2.3.1.4.3 Allosteric Modulators of $GABA_A$ Receptors

Neuroactive steroids, β-carbolines, and benzodiazepines (BZDs) are examples of allosteric modulators of $GABA_A$R activity (Sigel & Steinmann, 2012).

2.3.1.4.4 Benzodiazepines

According to some research, BZDs may be able to treat dementia and AD by (1) preventing excitotoxicity by increasing GABAergic activity through an anti-glutamatergic mechanism. (2) Preventing brain Aβ buildup, since γ-secretase contributes to the amyloidogenic pathway's manufacture of Aβ (Churcher et al., 2003). But according to some observational studies, long-term BZD usage may raise the incidence of dementia, including AD (Ator et al., 2010; Gallacher et al., 2012; Koh et al., 2013). Moreover, a case–control study by Hamann et al. (2008) on the connection between chronic BZD use and vascular dementia risk failed to establish any correlation after adjusting for prodromal BZD use.

2.3.1.4.5 β-Carbolines

β-Carboline alkaloids, a class of β-carbolines, protect against AD, not through direct interaction with $GABA_A$Rs but suppression of several forms of tau phosphorylation by inhibiting a specific tau kinase – DYRK1A (Frost et al., 2011). **Harmine**, a specific β-carboline alkaloid, enhances short-term memory in rats but causes motor impairment (Mennenga et al., 2015). While it is proven that some β-carbolines enhance cognition through their actions on $GABA_A$R in rodents, evidence supporting the same claim in humans is scarce.

2.3.1.4.6 Other $GABA_A$R Modulators

2.3.1.4.6.1 Lutein Lutein, the primary carotenoid present in the human brain tissue, is a $GABA_A$R positive allosteric modulator. It has antioxidant and anti-inflammatory properties in rhesus monkeys (Mohn et al., 2016).

2.3.1.4.6.2 Resveratrol A phytoalexin derived from plants, resveratrol (3,5,4′-trihydroxy-trans-stilbene) has anti-ageing, anti-inflammatory, anti-diabetic, and antioxidant qualities (Kizmazoglu et al., 2015). By modifying $GABA_AR$-induced currents, this chemical and its dimer trans-ε-viniferin may enhance cognition (Moussa et al., 2017). In a rat model of vascular dementia, it has also been demonstrated to enhance memory and learning (Ma et al., 2013). It has also been discovered that certain resveratrol compounds offer protection against Aβ-induced neurotoxicity (Kim et al., 2007).

2.3.1.4.6.3 Sevoflurane Sevoflurane, a common anaesthetic, has been linked to tau hyperphosphorylation (Tao et al., 2014) through hippocampal protein kinase B (Akt) activation and extracellular signalling-regulated kinase (Erk) in mice. Its chronic use causes irreversible deficits in spatial memory on the MWM test (Le Freche et al., 2012).

2.3.1.4.7 Drugs Targeting $GABA_B$ Receptors

2.3.1.4.7.1 $GABA_B$ Receptor Agonists Combining the effects of the $GABA_BR$ agonist – baclofen, with the glutamatergic neurotransmission inhibitor – acamprosate would address the excitatory-inhibitory imbalance in AD (Chumakov et al., 2015). At modest doses, the combination medication – PXT-864 shielded neurons and endothelial structures against inflammation, oxidative stress, tau hyperphosphorylation, and apoptosis caused by Aβ oligomers in vitro (Chumakov et al., 2015). In a variety of hippocampus-dependent learning and memory paradigms, it dramatically reverses Aβ-induced deficits (Chumakov et al., 2015).

2.3.1.4.7.2 $GABA_B$ Receptor Antagonists CGP36742 (SGS742), the first available orally active $GABA_BR$ antagonist, demonstrates cognition-enhancing effects in mice, rats, and rhesus monkeys across various cognitive and learning tasks (Mondadori et al., 1996) – possibly by reducing hippocampal CREB2 (ATF4) activity (Helm et al., 2005). It also exhibited anti-depressant effects in rats, mediated by increased nerve growth factor and brain-derived neurotrophic factor levels in the hippocampus and cortex (Froestl et al., 2004). In early phase II trials, it showed notable improvements in attention, working memory, and visual information processing when administered to patients with mild cognitive impairment (Froestl et al., 2004; Tomlinson et al., 2004). However, in a phase IIb monotherapy trial, it did not meet its efficacy endpoints, preventing further development (Tomlinson et al., 2004). Another $GABA_BR$ antagonist, CGP55845, demonstrates potential cognition-enhancing effects by fully reversing olfactory discrimination learning deficits in aged rats with learning impairments (Lasarge et al., 2009). However, it has not been tested in humans.

2.3.1.4.8 Drugs Targeting GABA Transporters

According to a number of studies, GABA transporters show varied expression profiles in AD and may be involved in or make up for the abnormal neuronal excitability seen in the disease (Wu et al., 2014; Fuhrer et al., 2017). Although there is ongoing research to determine the ramifications of these AD-related alterations, data point to GABA transporters as a possible target for improving GABAergic impairments in AD. Memory deficits were restored, and LTP impairment was reversed by SNAP-5114 – a GAT-4 inhibitor (Wu et al., 2014), while neuroprotective and cognition-enhancing benefits were shown with NNC-711, a drug with a similar structure that selectively targets GAT-1 (O'Connell et al., 2001).

2.3.1.4.8.1 GABA-Synthesising Enzymes **GAD65** and **GAD67** are crucial for controlling extracellular and synaptic GABA levels, which in turn affect GABAergic tone (Martin and Rimvall, 1993). Aβ42 plaque load is significantly reduced in different brain areas in 5xFAD mice with artificially induced GAD67 haploinsufficiency (Wang et al., 2017). It would be possible to further explore the pharmaceutical down-regulation of GAD67 activity as a means of treating amyloidogenesis.

2.3.1.4.8.2 Nav1.1 Voltage-Gated Sodium Channel Nav1.1 expression is significantly reduced in the inferior parietal cortices of AD patients (Verret et al., 2012). As a result, modulating these channels in GABAergic neurons may be a unique way to treat AD.

2.3.1.4.8.3 Optogenetics Aβ40 and Aβ42 species levels can be decreased by optogenetic activation of gamma oscillatory activity at 40 Hz via interneurons expressing parvalbumin (Iaccarino et al., 2016). Genes linked to microglial anti-amyloidogenic activity seem to be induced in this process (Iaccarino et al., 2016). The discovery that these 40-Hz oscillations could be produced in the primary visual cortex using a non-invasive, light flickering stimulus was highly intriguing from a therapeutic standpoint. This resulted in decreased Aβ plaque load in the visual cortex and modifications in the morphology of microglia that were comparable to those seen in the hippocampus with optogenetic stimulation (Iaccarino et al., 2016). This might serve as the foundation for an optogenetic treatment that targets the aggregation of amyloid.

2.4 LIMITATIONS OF CURRENT PHARMACOTHERAPIES

Extracellular Aβ deposition and the formation of intracellular neurofibrillary tangles made of hyperphosphorylated tau are the two main pathological hallmarks of AD that have been the focus of much research in recent years. The amyloid cascade hypothesis has dominated AD research in recent decades, and many attempts to develop AD therapies have either focused on pharmacological or immunological compounds that target Aβ formation and deposition or targeted cytotoxic neurofibrillary tangles, which are believed to disrupt axonal transport and contribute to synaptic dysfunction. These approaches are faced with various limitations, including:

2.4.1 Disease Progression

Unfortunately, none of these compounds has improved cognition in AD patients (Mohandas et al., 2009; Pedersen & Sigurdsson, 2015). The drugs do not halt underlying neurodegeneration.

2.4.2 Efficacy Limitations:

The drugs provide modest symptomatic relief, no cure, and have produced little or no significant clinical benefit so far (Doody et al., 2014; Salloway et al., 2014).

2.4.3 Side Effects and Tolerability

Furthermore, the drugs are fraught with numerous side effects and a tolerability profile ranging from GI upset, dizziness, cardiovascular risks, and neuropsychiatric adverse effects.

2.4.4 Access and Cost Issues

In addition, dementia drugs are not readily available, expensive, and subject to high levels of healthcare disparities.

2.4.5 Ethical Considerations

Because dementia mostly affects the elderly, engaging in polypharmacy with minimal benefit to the quality of life remains a huge challenge.

2.4.6 Research Challenges

Given the rigorous clinical trial design, patient heterogeneity, and limited translation from animal models, dementia studies are difficult and expensive. This has caused many large corporations with substantial resources to scale back their research into treatments for neurodegeneration regularly, posing a significant research challenge.

2.5 CONCLUSION

In conclusion, current pharmacotherapies for dementia offer symptomatic benefits but neither neuroprotection nor disease modification. Further basic research is essential to evaluate potential therapies and investigate the pathological mechanisms of AD and other tauopathies. This will aid in optimising current candidates and uncover novel targets for treatment. Future research should focus on combination therapies, personalised medicine, and biomarker-guided treatment.

REFERENCES

Alshamrani, M. (2023). Recent trends in active and passive immunotherapies of Alzheimer's disease. *Antibodies*, *12*(2), 41. https://doi.org/10.3390/antib12020041

Alzheimer's Association Report. (2017). 2017 Alzheimer's disease facts and figures. *Alzheimer's & Dementia*, 13, 325–373.

Asuku, A.O., Ayinla, M.T., Gbonjubola, O.O., Babatunde, S.S., Olajide, T.S., Oyerinde, T.O. (2024). Neurocognitive Aspects of Dementia. In: Nutrition in Brain Aging and Dementia. *Nutritional Neurosciences*, Springer, Singapore.

Asuku, A.O., Ogungbangbe, G.O., Olajide, T.S. (2025). Effect of Happiness and Positive Emotions on Dementia. In Neuroscience of Feelings. Progress in Brain Research, Elsevier, https://doi.org/10.1016/bs.pbr.2025.03.001

Ator, N.A., Atack, J.R., Hargreaves, R.J., Burns, H.D., Dawson, G.R. (2010). Reducing abuse liability of GABAA/benzodiazepine ligands via selective partial agonist efficacy at alpha1 and alpha2/3 subtypes. *Journal of Pharmacology and Experimental Therapeutics*, 332, 4–16.

Ayinla, M.T., Asuku, A.O., Bayo-Olugbami, A., Ayeni, A.O., Abiola, A.A., Owoyele, B.V. (2020). Melatonin and vitamin C modulate cholinergic neurotransmission and oxidative stress in Scopolamine–induced rat model of memory impairment. *Journal of African Association of Physiological Sciences*, *8*(1), 20.

Ballatore, C., Brunden, K.R., Huryn, D.M., Trojanowski, J.Q., Lee, Y. (2012). Microtubule stabilizing agents as potential treatment for Alzheimer's disease and related neurodegenerative tauopathies. *Journal of Medicinal Chemistry*, *55*(21), 8979. https://doi.org/10.1021/jm301079z

Barnes, D.M., White, W.F., Dichter, M.A. (1983). Etazolate (SQ20009): Electrophysiology and effects on [3H] flunitrazepam binding in cultured cortical neurons. *Journal of. Neuroscience*, *3*, 762–772.

Calvo-Flores Guzmán, B., Vinnakota, C., Govindpani, K., Waldvogel, H.J., Faull, L.M., Kwakowsky, A. (2018). The GABAergic system as a therapeutic target for Alzheimer's disease. *Journal of Neurochemistry*, *146*(6), 649–669. https://doi.org/10.1111/jnc.14345

Casey, D.A., Antimisiaris, D., O'Brien, J. (2010). Drugs for Alzheimer's disease: Are they effective? *P T.* 35, 208–211.

Chen, C.P.L.H., Alder, J.T., Bowen, D. M. (1996). Presynaptic serotonergic markers in community-acquired cases of Alzheimer's disease: Correlations with depression and neuroleptic medication. *Journal of Neurochemistry*, 66, 1592–1598.

Chen, C.P.L.H., Eastwood, S.L., Hope, T., McDonald, B., Francis, P.T., Esiri, M.M. (2000). Immnunocytochemical study of the dorsal and median raphe nuclei in patients with Alzheimer's disease prospectively assessed for behavioural changes. *Neuropathology and Applied Neurobiology*, 26, 1–10.

Chow, V.W., Mattson, M.P., Wong, P.C., Gleichmann, M. (2010). An overview of APP processing enzymes and products. *Neuromolecular Medicine*, *12*(1), 1–12.

Chumakov, I., Nabirotchkin, S., Cholet, N. *et al.* (2015). Combining two repurposed drugs as a promising approach for Alzheimer's disease therapy. *Scientific Reports*, 5, 7608.

Churcher, I., Ashton, K., Butcher, J.W. *et al.* (2003). A new series of potent benzodiazepine gamma-secretase inhibitors. *Bioorganic & Medicinal Chemistry Letters*, 13, 179–183.

Congdon, E.E., Ji, C., Tetlow, A.M., Jiang, Y., Sigurdsson, E.M. (2023). Tau-targeting therapies for Alzheimer's disease: Current status and future directions. *Nature Reviews. Neurology*, *19*(12), 715. https://doi.org/10.1038/s41582-023-00883-2

Cummings, J.L., Kaufer, D. (1996). Neuropsychiatric aspects of Alzheimer's disease: The cholinergic hypothesis revisited. *Neurology*, 47(4), 876–883.

Cummings, J., Leisgang Osse, A.M., Cammann, D., Powell, J., Chen, J. (2023). Anti-amyloid monoclonal antibodies for the treatment of Alzheimer's disease. *Biodrugs*, *38*(1), 5. https://doi.org/10.1007/s40259-023-00633-2

DeBoer, P., Westerink, B.H.C. (1994). GABAergic modulation of striatal cholinergic interneurons: An in vivo microdialysis study. *Journal of Neurochemistry*, *62*, 70–75.

Dementia: Assessment, management and support for people living with dementia and their carers. London: National Institute for Health and Care Excellence (NICE); 2018 Jun. (NICE Guideline, No. 97.) 11, Cholinesterase inhibitors and memantine for dementia. Available from: https://www.ncbi.nlm.nih.gov/books/NBK536484/

Doody, R.S., Thomas, R.G., Farlow, M., Iwatsubo, T., Vellas, B., Joffe, S., …, Mohs, R. (2014). Phase 3 trials of solanezumab for mild-to-moderate Alzheimer's disease. *New England Journal of Medicine*, *370*(4), 311–321.

Dringenberg, H.C., Zalan, R.M. (1999). Serotonin-dependent maintenance of spatial performance and electroencephalography activation after cholinergic blockade: Effects of serotonergic receptor antagonists. *Brain Research*, 837, 242–253.

Drott, J., Desire, L., Drouin, D., Pando, M., Haun, F. (2010). Etazolate improves performance in a foraging and homing task in aged rats. *European Journal of Pharmacology*, *634*, 95–100.

Forstl, H., Burns, A., Levy, R., Cairns, N. (1994). Neuropathological correlates of psychotic phenomena in confirmed Alzheimer's disease. *British Journal of Psychiatry*, 165, 53–59.

Froestl, W., Gallagher, M., Jenkins, H., Madrid, A., Melcher, T., Teichman, S., Mondadori, C.G., Pearlman, R. (2004). SGS742: The first GABA(B) receptor antagonist in clinical trials. *Biochemical Pharmacology*, 68, 1479–1487.

Frost, D., Meechoovet, B., Wang, T., Gately, S., Giorgetti, M., Shcherbakova, I., Dunckley, T. (2011). beta-carboline compounds, including harmine, inhibit DYRK1A and tau phosphorylation at multiple Alzheimer's disease-related sites. *PLoS ONE* 6, e19264.

Fuhrer, T.E., Palpagama, T.H., Waldvogel, H.J., Synek, B.J.L., Turner, C., Faull, R.L., Kwakowsky, A. (2017). Impaired expression of GABA transporters in the human Alzheimer's disease hippocampus, subiculum, entorhinal cortex and superior temporal gyrus. *Neuroscience*, *351*, 108–118.

Gallacher, J., Elwood, P., Pickering, J., Bayer, A., Fish, M., Ben-Shlomo Y. (2012). Benzodiazepine use and risk of dementia: Evidence from the Caerphilly Prospective Study (CaPS). *Journal of Epidemiology and Community Health*, 66, 869–873.

Geldenhuys, W.J., Darvesh, A.S. (2015) Pharmacotherapy of Alzheimer's disease: Current and future trends. *Taylor & Francis*, 15, 3–5.

Greig S.L. (2015). Memantine ER/donepezil: A review in Alzheimer's disease. *CNS Drugs*, 29, 963–970.

Hamann, J., Wernicke, C., Lehmann, J., Reichmann, H., Rommelspacher, H., Gille, G. (2008). 9-Methyl-beta-carboline up-regulates the appearance of differentiated dopaminergic neurones in primary mesencephalic culture. *Neurochemistry International*, 52, 688–700.

Helm, K.A., Haberman, R.P., Dean, S.L., Hoyt, E.C., Melcher, T., Lund, P.K., Gallagher, M. (2005). GABA B receptor antagonist SGS742 improves spatial memory and reduces protein binding to the cAMP response element (CRE) in the hippocampus. *Neuropharmacology*, 48, 956–964.

Hoogendijk, W.J., Pool, C.W., Troost, D., *et al.* (1995). Image analyser-assisted morphometry of the locus coeruleus in Alzheimer's disease, Parkinson's disease and amyotrophic lateral sclerosis. *Brain*, 118, 131–143. http://dx.doi.org/10.1093/brain/118.1.131.

Iaccarino, H.F., Singer, A.C., Martorell, A.J. *et al* (2016). Gamma frequency entrainment attenuates amyloid load and modifies microglia. *Nature*, *540*, 230.

Kim, H.J., Lee, K.W., Lee, H.J. (2007). Protective effects of piceatannol against beta-amyloid–induced neuronal cell death. *Annals of the New York Academy of Sciences*, 1095, 473–482.

Kizmazoglu, C., Aydin, H.E., Sevin, I.E., Kalemci, O., Yuceer, N., Atasoy, M.A. (2015). Neuroprotective effect of resveratrol on acute brain ischemia reperfusion injury by measuring annexin V, p53, Bcl-2 levels in rats. *Journal of Korean Neurosurgical Society*, 58, 508–512.

Koh, M.T., Rosenzweig-Lipson, S., Gallagher, M. (2013). Selective GABA(A) $\alpha5$ positive allosteric modulators improve cognitive function in aged rats with memory impairment. *Neuropharmacology*, 64, 145–152.

Kristensen, B.W., Noraberg, J., Zimmer, J. (2003). The GABA A receptor agonist THIP is neuroprotective in organotypic hippocampal slice cultures. *Brain Research*, *973*, 303–306.

Lanari, A., Amenta, F., Silvestrelli, G., Tomassoni, D., Parnetti, L. (2006). Neurotransmitter deficits in behavioural and psychological symptoms of Alzheimer's disease. *Mechanisms of Ageing and Development*, *127*(2), 158–165. https://doi.org/10.1016/j.mad.2005.09.016

Lanctôt, K.L., Herrmann, N., Mazzotta, P. (2001). Role of serotonin in the behavioral and psychological symptoms of dementia. *The Journal of Neuropsychiatry and Clinical Neurosciences*, 13(1), 5–21.

Lasarge, C.L., Banuelos, C., Mayse, J.D., Bizon, J.L. (2009). Blockade of GABA(B) receptors completely reverses age-related learning impairment. *NeuroSci*, 164, 941–947.

Lee, B.Y., Ban, J.Y., Seong, Y.H. (2005). Chronic stimulation of the $GABA_A$ receptor with muscimol reduces amyloid beta protein (25–35)-induced neurotoxicity in cultured rat cortical cells. *Neuroscience. Research*, *52*, 347–356.

Le Freche, H., Brouillette, J., Fernandez-Gomez, F.J. *et al.* (2012). Tau phosphorylation and sevoflurane anesthesia: An association to postoperative cognitive impairment. *Anesthesiology*, 116, 779–787.

Li, Z., Yin, B., Zhang, S., Lan, Z., Zhang, L. (2023). Targeting protein kinases for the treatment of Alzheimer's disease: Recent progress and future perspectives. *European Journal of Medicinal Chemistry*, *261*, 115817. https://doi.org/10.1016/j.ejmech.2023.115817

Louzada, P.R., Paula Lima, A.C., Mendonca-Silva, D.L., Noel, F., de Mello, F.G., Ferreira, S. (2004). Taurine prevents the neurotoxicity of beta-amyloid and glutamate receptor agonists: Activation of GABA receptors and possible implications for Alzheimer's disease and other neurological disorders. *FASEB J*, *18*, 511–518.

Ma, X., Sun, Z., Liu, Y., Jia, Y., Zhang, B., Zhang, J. (2013). Resveratrol improves cognition and reduces oxidative stress in rats with vascular dementia. *Neural Regeneration Research*, 8, 2050–2059.

Mark, R.J., Ashford, J.W., Goodman, Y., Mattson, M.P. (1995). Anticonvulsants attenuate amyloid beta-peptide neurotoxicity, Ca^{2}+ deregulation, and cytoskeletal pathology. *Neurobiology of Ageing*, *16*, 187–198.

Martin, D.L., Rimvall, K. (1993). Regulation of γ-aminobutyric acid synthesis in the brain. *Journal of Neurochemistry*, 60, 395–407.

Mathuram, T.L., Reece, L.M., Cherian, K.M. (2018). GSK-3 inhibitors: A double-edged sword? – An update on Tideglusib. *Drug Research*, *68*(8), 436–443.

Mennenga, S.E., Gerson, J. E., Dunckley, T., Bimonte-Nelson, H.A. (2015). Harmine treatment enhances short-term memory in old rats: Dissociation of cognition and the ability to perform the procedural requirements of maze testing. *Physiology & Behavior*, 138, 260–265.

Mohandas, E., Rajmohan, V., Raghunath, B. (2009). Neurobiology of Alzheimer's disease. *Indian Journal of Psychiatry*, *51*(1), 55–61.

Mohn, E.S., John, W., Erdman, J., Neuringer, M., Kuchan, M.J., Johnson, E.J. (2016), Differential expression of genes involved in inflammatory immune response and protein ubiquitination in the prefrontal cortex of rhesus macaque with high and low lutein content. *FASEB Journal*, 30, 913–918–913.918.

Mohr, E., Bruno, G., Foster, N., Gillespie, M., Cox, C., Hare, T.A., Tamminga, C., Fedio, P., Chase, T.N. (1986). GABA-agonist therapy for Alzheimer's disease. *Clinical Neuropharmacology*, *9*, 257–263.

Mondadori, C., Möbius, H.J., Borkowski, J. (1996). The GABA(B) receptor antagonist CGP 36,742 and the nootropic oxiracetam facilitate the formation of long-term memory. *Behavioural Brain Research*, 77, 223–225.

Moussa, C., Hebron, M., Huang, X., Ahn, J., Rissman, R.A., Aisen, P.S., Turner R.S. (2017). Resveratrol regulates neuro-inflammation and induces adaptive immunity in Alzheimer's disease. *Journal of Neuroinflammation*, 14, 1.

O'Brien, J.T., Holmes, C., Jones, M., Jones, R., Livingston, G., McKeith, I., … Burns, A. (2017). Clinical practice with anti-dementia drugs: A revised (third) consensus statement from the British Association for Psychopharmacology. *Journal of Psychopharmacology*, 31(2), 147–168.

O'Connell, A.W., Fox, G.B., Kjoller, C., Gallagher, H.C., Murphy, K.J., Kelly, J., Regan, C.M. (2001). Anti-ischemic and cognition-enhancing properties of NNC-711, a gamma-aminobutyric acid reuptake inhibitor. *European Journal of Pharmacology*, *424*, 37–44.

Patterson, C.E., Passmore, A.P., Crawford, V.L.S. (2004). A 6-month open-label study of the effectiveness and tolerability of galantamine in patients with Alzheimer's disease. *International Journal of Clinical Practice*, *58*(2), 144–148.

Pedersen, J.T., Sigurdsson, E.M. (2015). Tau immunotherapy for Alzheimer's disease. *Trends in Molecular Medicine*, *21*(6), 394–402.

Ritchie, C.W., Terrera, G.M., Quinn, T.J. (2015). Dementia trials and dementia tribulations: Methodological and analytical challenges in dementia research. *Alzheimer's Research & Therapy*, *7*(1), 31.

Salloway, S., Sperling, R., Fox, N.C., Blennow, K., Klunk, W., Raskind, M., ..., Brashear, H.R. (2014). Two phase 3 trials of bapineuzumab in mild-to-moderate Alzheimer's disease. *New England Journal of Medicine*, *370*(4), 322–333.

Santa-Maria, I., Hernandez, F., Moreno, F.J., Avila, J. (2007). Taurine, an inducer for tau polymerisation and a weak inhibitor for amyloid-beta-peptide aggregation. *Neuroscience Letters*, *429*, 91–94.

Shiong, L.W. (2008). Pharmacological treatments of dementia, in: Dementia update. *SFP 2008*, *35*(1), 78–83.

Sigel, E., Steinmann, M.E. (2012). Structure, function, and modulation of GABAA receptors. *Journal of Biological Chemistry*, *287*(48), 40224–40231.

Siopi, E., Llufriu-Daben, G., Cho, A.H., Vidal-Lletjos, S., Plotkine, M., Marchand-Leroux, C., Jafarian-Tehrani, M. (2013). Etazolate, an alpha-secretase activator, reduces neuroinflammation and offers persistent neuroprotection following traumatic brain injury in mice. *Neuropharmacology*, *67*, 183–192.

Solas, M., Puerta, E., Ramírez, M.J. (2015). Treatment options in Alzheimer´s disease: The GABA story. *Current Pharmaceutical Design*, *21*(34), 4960–4971. https://doi.org/10.2174/1381612821666150914121149

Stinton, C., McKeith, I., Taylor, J.P., Lafortune, L., Mioshi, E., Mak, E., ..., O'Brien, J.T. (2015). Pharmacological management of Lewy body dementia: A systematic review and meta-analysis. *American Journal of Psychiatry*, 172(8), 731–742.

Takramah, W.K., Asem, L. (2022). The efficacy of pharmacological interventions to improve cognitive and behavioural symptoms in people with dementia: A systematic review and meta-analysis. *Health Science Reports*, *5*(6), e913. https://doi.org/10.1002/hsr2.913

Tao, G., Zhang, J., Zhang, L., Dong, Y., Yu, B., Crosby, G., Culley, D.J., Zhang, Y., Xie Z. (2014). Sevoflurane induces tau phosphorylation and glycogen synthase kinase 3beta activation in young mice. *Anesthesiology*, 121, 510–527.

Tomlinson, J., Cummins, H., Wendt, J., Margolin, D., Pahl, J., Jenkins, H., Pearlman, R., Teichman, S. (2004). Sgs742, a novel Gaba receptor antagonist, improves cognition in patients with mild cognitive impairment. *Neurology*, 62, A128.

Trotta, F., Biscetti, L., Cherubini, A. (2023). Pharmacological Treatment of Cognitive and Behavioural Disorders in Dementia. In: Cherubini, A., Mangoni, A.A., O'Mahony, D., Petrovic, M. (eds) Optimizing Pharmacotherapy in Older Patients. Practical Issues in Geriatrics. Springer, Cham. https://doi.org/10.1007/978-3-031-28061-0_20

Vahid, Z.F., Eskandani, M., Dadashi, H., Vandghanooni, S., Rashidi, M. (2024). Recent advances in potential enzymes and their therapeutic inhibitors for the treatment of Alzheimer's disease. *Heliyon*, *10*(23), e40756. https://doi.org/10.1016/j.heliyon.2024.e40756

Vellas, B., Sol, O., Snyder, P.J. *et al* (2011). EHT0202 in Alzheimer's disease: A 3-month, randomised, placebo-controlled, double-blind study. *Current Alzheimer Research*, *8*, 203–212.

Verret, L., Mann Edward, O., Hang Giao, B. *et al* (2012). Inhibitory interneuron deficit links altered network activity and cognitive dysfunction in Alzheimer's model. *Cell*, *149*, 708–721.

Wang, R., Reddy, P.H. (2017). Role of glutamate and NMDA receptors in Alzheimer's disease. *Journal of Alzheimer's Disease: JAD*, *57*(4), 1041. https://doi.org/10.3233/JAD-160763

Wang, Y., Wu, Z., Bai, Y.-T., Wu, G.-Y., Chen, G. (2017). Gad67 haploinsufficiency reduces amyloid pathology and rescues olfactory memory deficits in a mouse model of Alzheimer's disease. *Molecular Neurodegeneration*, *12*, 73.

Wu, Z., Guo, Z.Y., Gearing, M., Chen, G. (2014). Tonic inhibition in dentate gyrus impairs long-term potentiation and memory in an Alzheimer's disease model. *Nature Communications*, *5*, 4159.

Yoshiike, Y., Kimura, T., Yamashita, S., Furudate, H., Mizoroki, T., Murayama, M., Takashima, A. (2008). GABAA Receptor-mediated acceleration of aging-associated memory decline in APP/PS1 mice and its pharmacological treatment by picrotoxin. *PLoS ONE*, 3, e3029.

3 Bioresources

Global Regulatory Bodies and Agencies

Babatunde Oluwaseun Ibitoye, Oluwafemi Abidemi Adedotun, and Femi Oloye

3.1 INTRODUCTION

Dementia is a condition in which an individual may suffer a loss in cognitive ability more than is expected in normal aging, and it is a healthcare issue of global concern (Ayinla et al., 2020; Asuku et al., 2025). The prevalence of dementia in the world is around 55 million people, out of whom nearly 10 million are diagnosed annually according to the sources released by the World Health Organization (WHO) (2021). The quest for effective therapies has given rise to more interest in natural bioresources like plants, microorganisms, and ocean organisms as medicines. This integration of the bioresource information to the main practice of dementia care, however, must be well regulated in terms of safety, efficacy, and ethical therapeutic use. Numerous international and national regulations organizations, such as the World Health Organization (WHO), Convention on Biological Diversity (CBD), National Agency for Food and Drug Administration and Control (NAFDAC), are actively involved in the ethics of access to such resources, which should be maintained and regenerated where feasible. The big question remains whether equity has been observed in the distribution of the limited resources to benefit mankind, biosafety, and international law adherence. The dementia burden globally is on the rise, and interventions with the potential to be effective must be put in place through a multidisciplinary and cross-sectoral approach. Bioresources offer a crucial alternative in managing neurodegenerative disorders due to their bioactive components and therapeutic properties (WHO, 2023). However, to harness their potential responsibly, a well-built regulatory framework is essential. In the development of drugs and their traditional medicines, regulatory bodies provide regulation in the conservation of bioresources, research, commercialization, and safe use of bioresources.

3.2 CONCEPTUALIZING BIORESOURCES AND THE NEED FOR REGULATION

Bioresources include biological materials with potential health or commercial value, including naturally occurring biological resources in the form of medicinal plants, microbial strains, marine organisms, and genetically derived materials that hold great therapeutic values. These resources have historically been utilized in ethnomedicine and are currently under investigation in the scientific domain of neurodegenerative disease treatment, such as dementia. Regulatory control over the use of these resources in medical practice (to treat dementia) is required to prevent bio-piracy, advance the safety of products, standardize their formulation, and protect biodiversity.

Common problems associated with the unregulated use of bioresources include contamination, misidentification, unsafe processing procedures, and unproven therapeutic claims. Lack of proper regulation has resulted in many instances of patient harm due to adulterated herbal products or misidentified species. Regulatory bodies mitigate these risks by implementing Good Agricultural and Collection Practices (GACP), pharmacopoeia standards, and rigorous quality control (Lee et al., 2020).

DOI: 10.1201/9781003657385-3

In high-income countries, regulatory bodies such as the U.S. Food and Drug Administration (FDA) and the European Medicines Agency (EMA) enforce strict quality, safety, and efficacy guidelines. Conversely, countries like Nigeria face gaps in regulatory implementation due to limited resources and unregulated traditional practices.

3.3 DEFINING BIORESOURCES IN DEMENTIA RESEARCH

Bioresources and their products are used as therapeutic agents or otherwise to benefit mankind. Howes and Perry (2011) identified that flavonoids and alkaloids found in many ethnomedicinal plants exhibit antioxidant activity and may modulate neurotransmitter systems relevant to dementia. Similarly, Kennedy et al. (2010) demonstrated that *Ginkgo biloba* extract improved working memory and cognitive processing speed in older adults with mild cognitive impairment. Other bioresources under active investigation include *Bacopa monnieri*, known for its memory-enhancing effects; *Panax ginseng*, which exhibits neuroprotective and anti-fatigue properties; 6-gingerol that improves cognitive and neuroprotective functions (Ayinla & Asuku, 2025a, 2025b); and omega-3 fatty acids, from marine sources, shown to reduce neuroinflammation and improve synaptic function in Alzheimer's models (Aguiar & Borowski, 2013; Dyall, 2015). Despite promising preclinical data, many plant-derived compounds remain untested in large-scale clinical trials due to regulatory challenges, inconsistent phytochemical profiles, lack of standardized dosing, and limited funding for traditional medicine research (Adetunji et al., 2022).

The limited effectiveness of current pharmacological agents like donepezil and memantine in halting disease progression highlights the need for alternative therapies. Bioresources offer a complementary pathway by targeting oxidative stress, inflammation, and neurogenesis, key mechanisms implicated in dementia pathogenesis. Given the vulnerability of dementia patients to drug interactions and neurotoxicity, unregulated bioresource-based products can pose significant health risks. Without proper standardization, there is a danger of variability in active compounds that could affect treatment outcomes.

3.4 GENERAL ROLE OF BIORESOURCES REGULATION

Despite the global abundance of diverse bioresources, ranging from region-specific plants, animals, and microorganisms to genetic materials with immense potential for disease prevention and treatment, including dementia, there is a compelling need for robust regulation of these resources at both local and international levels. Such regulation serves multiple purposes, from ensuring biosafety to conserving the biome.

Bioprospecting, the exploration of bioresources for the development of nutraceuticals, bioactive compounds, and medicines, aims to promote sustainable use of biological materials through biotechnology while ensuring their conservation and scientific and economic utilization (Olaniyan & Adetunji, 2022). At the core of bioresource regulation lies the principle of equitable benefit sharing among nations. Uncontrolled exploitation of bioresources, particularly the removal of keystone species, can destabilize ecosystems, reduce biodiversity, and trigger secondary extinctions. This ecological imbalance threatens global ecosystem stability and, ultimately, human survival. Regulations are therefore essential to prevent such destructive outcomes.

International institutions increasingly play a legitimizing role in transactions involving bioresources (Bößner et al., 2020). However, concerns remain that economically dominant nations may influence multilateral agreements to serve their own strategic and economic interests. This underscores the necessity for stringent, transparent, and fair regulations at both national and international levels.

Due to increasing bioresource utilization, bio-based product trade, and transboundary environmental effects, bioresource governance challenges are also emerging. Although most bioresource regulation strategies remain national in scope, these challenges underscore the importance of international cooperation and collaboration. Regional cooperation offers a promising pathway for addressing common challenges and seizing shared opportunities. However, the wide variation in the objectives and priorities of national strategies complicates the formation of universal agreements. International environmental agreements are more likely to succeed when mutual incentives exist, especially in cases where potential environmental damage outweighs the costs of participation, as with climate change. Transboundary impacts such as land-use changes resulting from increased biomass and bioenergy trade may necessitate new international agreements or governance structures. A cross-cutting, cross-sectoral approach supported by adequate regulations and policies will not only minimize the risks of unregulated bioresource use but also promote sustainable development, safeguard biodiversity, and foster international cooperation for the benefit of humanity.

3.5 MAJOR BROAD REGULATORY ORGANIZATIONS AND THEIR ROLES

3.5.1 World Health Organization (WHO)

WHO is critical in the establishment of international standards of traditional and complementary medicine (Figure 3.1). It had developed its traditional medicine strategy 2014–2023 to assist member states in formulating proactive policies and initiating action plans to reinforce the position of traditional medicine in health systems (WHO, 2013). The next plan (2025–2034) will focus on research, innovation, regulation, equity, cooperation, and integration of traditional medicine into health services of all countries (WHO, 2025).

3.5.2 U.S. Food and Drug Administration (FDA)

FDA controls botanical drug activities under its Center of Drug Evaluation and Research (CDER). The Botanical Drug Development Guidance describes the regulatory process of botanical products; it focuses on careful clinical studies and quality assurance (Food and

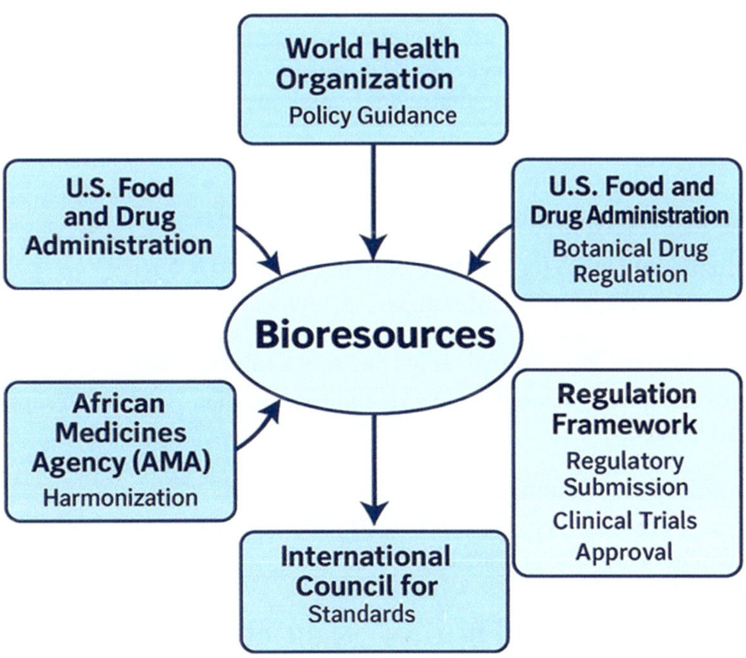

FIGURE 3.1 Graphical Representation of International Regulatory Organizations governing bioresources.

Drug Administration (FDA), 2016). The Botanical Review Team of the FDA offers scientific advice regarding the matters of raw material control and pharmacognosy of natural products (Figure 3.1) (FDA, 2023).

3.5.3 European Medicines Agency (EMA)

EMA's Committee on Herbal Medicinal Products (HMPC) is the body that evaluates and gives a scientific opinion about herbal products (Figure 3.1). The HMPC prepares the monograph, based on which herbal medicinal goods can be approved in the European Union (European Medicines Agency (EMA), 2021).

3.5.4 African Medicines Agency (AMA)

The African Medicines Agency (AMA) was established in 2021 to synchronize medical regulation in the African Union, including traditional medicines regulation. It aims to enhance the availability of quality-assured medicines, as well as to fight the spread of substandard and counterfeited products (Figure 3.1) (African Union, 2021).

3.5.5 International Council for Harmonization (ICH)

The International Council for Harmonization (ICH) combines the representatives of regulatory agencies and the pharmaceutical industry to negotiate over medical-scientific and technical questions of drug registration. It is essential in the harmonization of procedures involved in the development and registration of pharmaceuticals, including pharmaceuticals obtained through bioresources (Figure 3.1; International Council for Harmonisation (ICH), 2021).

3.5.6 Convention on Biological Diversity (CBD) and Nagoya Protocol

The CBD and its Nagoya Protocol are international treaties that govern the access to genetic resources and equitable sharing of benefits. They ensure countries retain sovereignty over their bioresources and that research and commercial gains are fairly distributed (United Nations, 2021).

3.5.7 Other International Bodies

3.5.7.1 International Union for Conservation of Nature (IUCN)

International Union for Conservation of Nature (IUCN) focuses on biodiversity protection.

3.5.7.2 United Nations Educational, Scientific and Cultural Organization (UNESCO)

United Nations Educational, Scientific and Cultural Organization (UNESCO) supports bioethical frameworks and indigenous knowledge systems.

3.5.7.3 Pharmacopoeia Commissions

Commissions, such as the U.S. Pharmacopeia (USP) and European Pharmacopoeia (Ph. Eur.), provide monographs for bioactive compounds.

3.6 NIGERIAN REGULATORY BODIES ON BIORESOURCES

Nigeria, rich in medicinal biodiversity, has developed its regulatory framework for managing bioresources.

3.6.1 National Agency for Food and Drug Administration and Control (NAFDAC)

NAFDAC oversees the regulation and registration of herbal medicines, nutraceuticals, and other natural products. The agency ensures safety, efficacy, and quality control through laboratory testing, GMP inspections, and product labeling regulations (National Agency for Food and Drug Administration and Control (NAFDAC), 2023).

3.6.2 National Institute for Pharmaceutical Research and Development (NIPRD)

National Institute for Pharmaceutical Research and Development (NIPRD) supports the research, standardization, and commercialization of medicinal plants. It plays a key role in validating traditional medicine for neurodegenerative conditions like dementia (National Institute for Pharmaceutical Research and Development (NIPRD), 2023).

3.6.3 Nigerian Natural Medicine Development Agency (NNMDA)

This agency focuses on the integration and modernization of indigenous medical knowledge systems. It supports documentation, clinical studies, and capacity-building for local herbal practitioners.

3.6.4 National Biotechnology Development Agency (NABDA)

National Biotechnology Development Agency (NABDA) promotes biotechnology research and bioresource conservation in Nigeria. It facilitates biodiversity mapping and ethical gene access initiatives.

3.6.5 Forestry Research Institute of Nigeria (FRIN)

Forestry Research Institute of Nigeria (FRIN) is responsible for sustainable forest resource use, including medicinal plant conservation and propagation.

3.7 ACCESS AND ETHICAL CONSIDERATIONS

3.7.1 Nagoya Protocol and Biodiversity Access

The Nagoya Protocol on access and benefit sharing, to the CBD, is a set of decisions that provide a legal framework for the sharing of benefits arising out of the access and utilization of genetic resources in a way that is considered fair and equitable. This generally applies to the case of bioresources utilized in dementia research with proper compensation paid to source countries and indigenous communities (Convention on Biological Diversity (CBD), 2011).

3.7.2 Intellectual Property and Patents

Intellectual property rights form an important consideration in the commercialization of products of bioresource products. Many types of intellectual property regulation are governed by the minimum requirements established by the Trade-Related Aspects of Intellectual Property Rights (TRIPS) Agreement of the World Trade Organization, which has implications on the patenting of bioresources (WTO, 1994).

TABLE 3.1
Selected Global Regulatory Agencies and Their Focus Areas in Bioresource Oversight

Agency	Region	Mandate/Focus
WHO	Global	Traditional medicine strategy, policy frameworks, global collaboration
FDA (USA)	United States	Botanical drug regulation, clinical trial oversight, drug safety, and efficacy
EMA (HMPC)	European Union	Herbal monographs, standardization, authorization of herbal medicinal products
AMA	Africa	Harmonization of drug regulation, traditional medicine oversight
ICH	Global	Harmonization of technical drug development and approval guidelines

3.8 REGULATORY CHALLENGES AND PROSPECTS IN BIORESOURCE DEVELOPMENT

Despite significant advances, regulatory gaps persist in areas such as enforcement, harmonization of standards, limited clinical data, and inadequate policy implementation. For instance, many herbal remedies for dementia are marketed without clinical backing, leading to safety concerns.

Strengthening collaboration among global and national bodies, investing in research infrastructure, and empowering local communities remain essential. The AMA, recently ratified by the African Union, promises to unify continental bioresource regulation and enhance cross-border standardization (African Union (AU), 2023) as shown in Table 3.1.

3.9 ROLE OF REGULATORY BODIES IN ACHIEVING SDGs THROUGH BIORESOURCE UTILIZATION

Regulatory bodies directly contribute to the SDGs through the aspects discussed in following sections.

3.9.1 SDG 3 (Good Health and Well-being)

Ensuring safe bioresource-based therapies for diseases like dementia.

3.9.2 SDG 9 (Industry, Innovation, and Infrastructure)

Promoting bioeconomy development via standardization and innovation.

3.9.3 SDG 15 (Life on Land)

Protecting biodiversity through sustainable harvesting regulations.

3.9.4 SDG 17 (Partnerships for the Goals)

Encouraging international cooperation in bioresource governance.

3.10 RESEARCH AND DEVELOPMENT FRAMEWORKS

3.10.1 WHO Collaborating Centers

WHO collaborating centers help to assess and integrate traditional medicine into the national health system; they offer technical assistance and capacity building (WHO, 2021).

3.10.2 Global Alliance for Chronic Diseases (GACD)

The Global Alliance for Chronic Diseases (GACD) invests in research on non-communicable disorders, such as dementia, where international investigators are encouraged to collaborate to resolve gnawing international health issues (Global Alliance for Chronic Diseases (GACD), 2021).

3.10.3 Public–Private Partnerships

Public–private partnerships such as the Innovative Medicines Initiative (IMI) in Europe encourage partnerships between government and private sector organizations to expedite the development of safer and more effective medicines, including those based on bioresources (Innovative Medicines Initiative (IMI), 2021).

3.11 CHALLENGES AND OPPORTUNITIES

3.11.1 Standardization and Quality Control

One of the significant issues in using bioresources in the context of dementia is the absence of standardization. The difference in plant materials, extraction procedures, and formulation influences both efficacy and safety (Ekor, 2014). Global standards of consistency need to be set up and enforced by the regulatory bodies.

3.11.2 Clinical Trials and Evidence Production

Most treatment solutions on bioresources lack clinical evidence. Herbal and traditional remedies are also becoming subject to randomized controlled trials (RCT) prohibited before authorization, with an example being FDA and EMA regulatory frameworks (FDA, 2016; EMA, 2021).

3.11.3 Ethical Access and Prevention of Biopiracy

Moral issues are maintained with regard to the lack of appropriate benefit-sharing in the utilization of the indigenous knowledge and biodiversity, commonly referred as biopiracy. Ethical innovation continues to depend on following the Nagoya Protocol and country biodiversity legislation.

3.11.4 How to Integrate Traditional Systems and Modern Systems

Good integration of traditional medical system into the national systems is slow because of the cases of uncertainty, infrastructures, and training. A research program within WHO is filling this gap with capacity-building programs (WHO, 2021).

3.12 CONCLUSION

Bioresources are very promising in the way dementia can be addressed using sustainable and culturally relevant therapies. Regulatory environment, consisting of institutions such as WHO, FDA, EMA, and AMA, is an essential part of navigating ethical access, scientific justification, and equal consumption of such resources. The global and Nigerian agencies ensure the maintenance of bioresource integrity, research, and protection of indigenous knowledge, as well as therapeutic development in line with the safety standards. Going forward, harmonizing global standards, empowering national regulatory bodies like NAFDAC and NBMA, and safeguarding indigenous rights will be crucial to unlocking the full potential of bioresources in treating neurodegenerative disorders such as dementia.

REFERENCES

Adetunji, C.O., Ukhurebor, K.E., Olaniyan, O.T., Michael, O.S., Oloke, J.K., & Ubi, B.E. (2022). Ethical and Social Aspects of Modern Biotechnology. In *Biosafety and Bioethics in Biotechnology*. CRC Press. Pages 51–67.

African Union (AU). (2021). *Treaty for the Establishment of the African Medicines Agency.*

Aguiar, S., & Borowski, T. (2013). Neuropharmacological review of the nootropic herb *Bacopa monnieri*. *Rejuvenation Research*, 16(4), 313–326.

Asuku, A.O., Ogungbangbe, G.O., & Olajide, T.S. (2025). Effect of Happiness and Positive Emotions on Dementia. In *Neuroscience of Feelings. Progress in Brain Research*. Elsevier. https://doi.org/10.1016/bs.pbr.2025.03.001

Ayinla, M.T., & Asuku, A.O. (2025a). The anti-apoptotic effects of 6-gingerol-rich fraction of ginger on mercury chloride-induced cerebral and cerebellar dysfunction. *Arabian Journal of Medicinal and Aromatic Plants*, 11, 2. https://doi.org/10.48347/IMIST.PRSM/ajmap-v11i2.50900

Ayinla, M.T., & Asuku, A.O. (2025b). The modulatory role of 6-gingerol-rich portion of ginger on brain oxidative stress, inflammation and neurotransmission in chronic unpredicted mild stress-induced depressed male Wistar rats. *Avicenna Journal of Phytomedicine*, Epub ahead of print.

Ayinla, M.T., Asuku, A.O., Bayo-Olugbami, A., Ayeni, A.O., Abiola, A.A., & Owoyele, B.V. (2020). Melatonin and vitamin C modulate cholinergic neurotransmission and oxidative stress in Scopolamine–induced rat model of memory impairment. *Journal of African Association of Physiological Sciences*, 8(1), 20.

Bößner, S., Johnson, F.X., & Shawoo, Z. (2020). Governing the bioeconomy: What role for international institutions? *Sustainability*, 13(1), 286.

Convention on Biological Diversity (CBD). (2011). *Nagoya Protocol on access and benefit-sharing*. Convention on Biological Diversity.

Dyall, S.C. (2015). Long-chain omega-3 fatty acids and the brain: A review of the independent and shared effects of EPA, DPA and DHA. *Frontiers in Aging Neuroscience*, 7, 52.

Ekor, M. (2014). The growing use of herbal medicines: Issues relating to adverse reactions and challenges in monitoring safety. *Frontiers in Pharmacology*, 4, 177.

European Medicines Agency (EMA). (2021). *Committee on Herbal Medicinal Products (HMPC): Overview.* European Medicines Agency.

Food and Drug Administration (FDA). (2016). *Botanical Drug Development: Guidance for Industry.* U.S. Food and Drug Administration.

Food and Drug Administration (FDA). (2023). *Botanical Review Team Activities.* U.S. Food and Drug Administration.

Global Alliance for Chronic Diseases (GACD). (2021). *Global Alliance for Chronic Diseases: Strategic Plan.*

Howes, M.J., & Perry, E. (2011). The role of phytochemicals in the treatment and prevention of dementia. *Drugs & Aging*, 28(6), 439–468.

Innovative Medicines Initiative (IMI). (2021). *Innovative Medicines Initiative: Mission and Projects.*

International Council for Harmonisation (ICH). (2021). *International Council for Harmonisation of Technical Requirements for Pharmaceuticals for Human Use.*

Kennedy, D.O., Scholey, A.B., & Wesnes, K.A. (2010). Dose dependent changes in cognitive performance and mood following acute administration of *Ginkgo biloba*. *Nutritional Neuroscience*, 5(3), 213–220.

Lee, S.H., Park, Y., & Kim, D.H. (2020). Herbal therapeutics in neurodegeneration. *Phytotherapy Research*, 34(3), 546–558.

National Agency for Food and Drug Administration and Control (NAFDAC). (2023). Mandate and functions of NAFDAC. Retrieved from https://www.nafdac.gov.ng

National Institute for Pharmaceutical Research and Development (NIPRD). (2023). Validation of Traditional Medicine in the Treatment of Neurodegenerative Diseases. In *Annual Report*. NIPRD Publications, Abuja.

Olaniyan, O.T., & Adetunji, C.O. (2022). Relevance of Natural Bioresources and Their Application in Pharmaceutical, Food and Environment. In *Agricultural Biotechnology*. CRC Press. Pages 379–384.

United Nations (2021). Convention on Biological Diversity. In *Nagoya Protocol on Access to Genetic Resources and the Fair and Equitable Sharing of Benefits Arising from Their Utilization*. United Nations Environment Programme. Retrieved from https://www.cbd.int/abs/

World Health Organization (WHO). (2013). *WHO Traditional Medicine Strategy 2014–2023.* World Health Organization.

World Health Organization (WHO). (2021). *Dementia Fact Sheet.* World Health Organization.

World Health Organization (WHO). (2023). *New Global Action Plan on Epilepsy and Other Neurological Disorders Published.* Retrieved from https://www.who.int/news/item/20-07-2023-new-global-action-plan-on-epilepsy-and-other-neurological-disorders-published

World Health Organization (WHO). (2025). *Draft WHO Traditional Medicine Strategy 2025–2034.* World Health Organization.

WTO. (1994). *Trade-Related Aspects of Intellectual Property Rights (TRIPS) Agreement.* World Trade Organization.

4 *Ginkgo biloba* for Managing and Treating Dementia

Odugbesan Olawunmi Anthonia
and Gbonjubola Oyinlola Ogungbangbe

4.1 INTRODUCTION

Dementia is characterized by cognitive impairment often associated with neurological disorders like Alzheimer's disease (AD), vascular dementia, and Lewy body due to the gradual deterioration of neuronal activity (Alzheimer's Disease International, 2020; Asuku et al., 2022; Carroll & Benca, 2024). It said about dementia that by 2050, 55 million people and above will have lived with the disease, forecasting an increase of 78 million by 2030 and 139 million by 2050. In 2019, 57.4 million people were affected worldwide, with the expected numbers reaching 152.8 million by 2050, mainly due to population aging (World Health Organization, 2023). In OECD countries, over 21 million people had dementia in 2021, and this number could increase by nearly 50% by 2040 (Institute for Health Metrics and Evaluation, 2022). In sub-Saharan Africa, about 2.10 million people live with dementia, with the highest prevalence (19.7%) among aged females from 80 years and above, which varies between regions minimally (George-Carey et al., 2015).

One of the long-lived orchards in China, 1000–4700 years old, whose botanical name is *Ginkgo biloba*, is a historical traditional medicine, also known as the maidenhair tree or living fossil. As a glyph of rich historical and cultural heritage, several nomenclatures have been attributed to *Ginkgo* leaves or the seeds' pattern over the years due to its heroic status among Asian continent trees (Ginkgo, 2025). Chinese ginkgo tree terminology includes Yinxing, Baiguoshu, Yazhangshu, Gongsunshu, Fozhijia, and Lingyan (Ginkgo, 2025). Its tree and leaves are uniquely of ancient, extant species, based on their potential medicinal benefits, durability, longevity, and exceptional tolerance to pollution. In contrast, its longevity has gained a compensatory approach, yet it maintains a balance between the processes of growth and aging (Chen et al., 2021).

Far and wide, the great influence on medicinal and ornamental value and phylogenetic traits as an ancient functional species within the division Ginkgopsida, *G. biloba*'s niche in plant evolution represents a focal point for several plants in extinction, and its genome gives factual depths that facilitate conservation of natural resources for tourist schemes (Chen et al., 2021). *Ginkgoales* is the only existing species, a relatively ancient gymnosperm that dates back to the pre-existing Permian period and is distinctly useful for investigating the evolution of gymnosperms due to its sperm motility. The *G. biloba* tree is a dioecious species whose roasted seeds are edible, whose propagation of seed is easy, and whose motile sperm swim to fertilize the egg (Pendarvis et al., 2018). An example of this reproductive trait is found in plants like cycads, ferns, mosses, and algae. The enclosed arillus, fleshy, fruitlike yellow seeds, is attached to a long stalk known as sarcotesta (Hori et al., 2012). The arillus contains butyric acid and emits an odor reminiscent of rancid butter. Also, they possess a seasonal, leathery, fan-shaped leaf, which is shed in winter (Belwal et al., 2019).

4.1.1 *Ginkgo biloba* and Its Phytochemicals

Phytochemicals are naturally occurring plant metabolites essential for the physiologic function of the body (Frank et al., 2020; Asuku et al., 2024; Phytochemicals, 2025). Phytochemicals are grouped into carotenoids and polyphenols. Furthermore, classification of flavonoids is based on their chemical

DOI: 10.1201/9781003657385-4

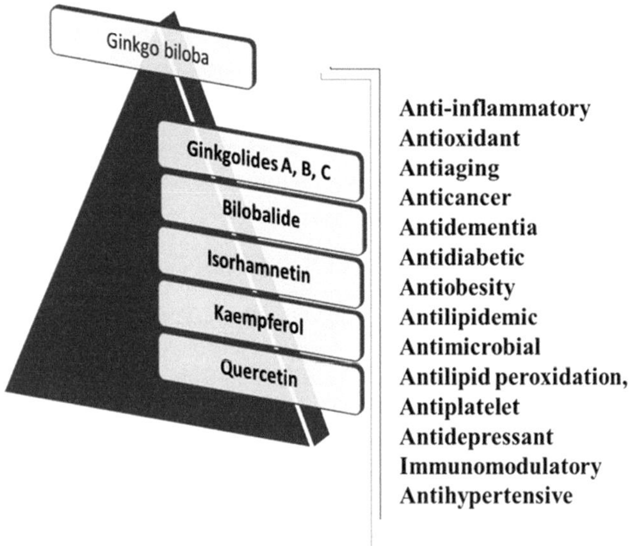

FIGURE 4.1 Component of *Ginkgo biloba.*

structure similarities, like flavanols: catechins, epicatechins, and proanthocyanidins; anthocyanins; flavones; flavanones; and isoflavones (Hassan Mohammed et al., 2020). The tree of *G. biloba* contains essential compounds like terpene trilactones, flavonoids, and polyphenols. Furthermore, terpene trilactones consist of bilobalide and ginkgolides A, B, C, J, P, and Q, which reduce inflammation, allergies, and clotting factors, while bilobalide protects brain cells (Yin et al., 2013). Flavonoids and polyphenols are solid antioxidant properties essential for brain homeostasis. It comprises biflavones and flavonol glycosides such as quercetin, kaempferol, and isorhamnetin (Figure 4.1), which act as essential antioxidants as a result of their phenolic structures (Rojas et al., 2016).

4.2 CHARACTERISTICS OF *GINKGO BILOBA*

- Environmental adaptability
- Highly resistant to insects and pathogens
- Gene duplication abilities for immunity and stress responses
- Highly resistant to pollution, e.g. air.

4.3 TAXONOMIC CLASSIFICATIONS OF *GINKGO BILOBA*

KINGDOM: Plantae
DIVISION OR PHYLUM: Gingkophyta
CLASS: Ginkgopsida or Ginkgoales
ORDER: Ginkgoales
FAMILY: Ginkgoaceae
GENUS: *Ginkgo*
SPECIES: *Ginkgo biloba*

4.4 USES OF *GINKGO BILOBA*

G. biloba has medicinal, spiritual, and horticultural importance in Chinese traditions. Savaskan & Hoerr (2020) state that *Ginkgo* fruits and fresh leaves are used in homeopathy. *G. biloba* wood is a slow-decaying, fire-resistant wood used to make furniture, chessboards, carvings, and casks. *G. biloba* leaf serves as a cognitive enhancer and treatment for and management of hypertension, menopausal-related cognitive decline, tinnitus, and post-stroke recovery (Savaskan & Hoerr, 2020). It also serves as a growing, teeny landscape and a captivating shade-providing tree in a lot of cities and gardens (Wang et al., 2023), while its extract enhances the circulatory system, acts as reactive oxygen species (ROS) scavengers, and antagonizes platelet factor.

4.5 *GINKGO BILOBA* MECHANISMS OF ACTIONS IN DEMENTIA

Neurotoxins, physical factors, and biological agents interfere with nerve cell physiology essential for transmitting signals within the nervous system. Neurotoxins cross the blood–brain barrier, destroy the glia cells, and release cytokines, resulting in inflamed brain cells, oxidative damage, mitochondrial injury, and mental disabilities (Patow et al., 2024). Extracts from medicinal plants such as dried *G. biloba* leaves are incorporated as therapeutic approaches in managing and treating memory and cognitive impairments like dementia due to the bioavailability of bioactive compounds (Rauf et al., 2022).

Ginkgolides act as platelet-activating factor antagonists, and sesquiterpene lactone is neuroprotective against mitochondrial dysfunction, strengthens synaptic plasticity by enhancing neurotransmitter processes, and contains anti-inflammatory properties in animal models (Dower et al., 2015). Essentially, ginkgolides contain a specific percentage of regulated ginkgo extract. They enhance endothelial function and scavenge neuronal free radicals. In addition, they serve as a potent platelet-activating factor that abates neuroinflammation and ischemic damage (Shi et al., 2009). Ginkgolide B is considered cathartic. The relationship between flavonoids and terpenoids is symbiotic, with flavonoids targeting oxidative stress and vascular physiology, while terpenoids offer neuroprotective properties (Dower et al., 2015).

4.5.1 The Role of *Ginkgo biloba* in Neuroplasticity and Neurogenesis

Extract of *Ginkgo* shields the body's antioxidant system against foreign bodies by promoting brain cells and axonal growth through signaling pathways like Akt/mTOR and Nrf2, which are involved in neuronal development, survival, and neuroplasticity (Li et al., 2022). It contributes to functional recovery in patients with dementia, as ginkgo stimulates the hippocampus, an essential area for learning and memory, to form neurons and stem cells (Ge et al., 2021). During cell culture, it enhances the growth of dendrites and axons necessary for neuroflexibility (Nguyen et al., 2025). In addition, it was revealed that EGb761 significantly enhances the formation and growth of neurospheres (Wang et al., 2018). *Ginkgo* upregulates growth-associated protein (GAP-43), involved in the learning process, and brain-derived neurotrophic factor (BDNF) in memory recovery, and downregulates axonal growth inhibitors that improve synaptic connection and repair worn-out axons. BDNF is responsible for neuronal differentiation, growth, and development (Ge et al., 2021).

White–gray matter remodeling is a functional activity of axonal regeneration (Venkat et al., 2017). Previous research suggests that EGb761 enhances the growth of neurites in the axons and dendrites via collapsin response mediator protein 2 (CRMP2) (Venkat et al., 2017). It promotes axonal regeneration and neuronal plasticity in severe brain damage via the protein kinase B (AKT) pathway, activating CRMP2 and mediating its downstream target glycogen synthase kinase-3β (GSK-3β) at Ser9 (Wang et al., 2017). CRMP2 is found within growing axons and regulates axonal outgrowth and guidance (Lejri et al., 2019).

Tan et al. (2015) revealed the neuroprotective effect of EGb761 in senile dementia. Similarly, it reduces anxiety-like behavior and locomotive function (Montes et al., 2015; Singh et al., 2017). Additionally, in treating dementia, it modulates inflammatory mediators and the cholinergic circuitry (Hashiguchi et al., 2015; Kim et al., 2016). Enormous research outcomes on the efficacy and safety of EGb against oxidative injury, neuronal damage, and regeneration of peripheral nerves are underexplored (Ekici-Günay, 2020).

4.5.2 *Ginkgo biloba* and Mitochondrial Protection

Amyloid-beta (Aβ) generates energy metabolism disruption as an aging index in oxidative stress linked with dementia (Machado et al., 2021). *G. biloba* enhances the phases of ATP synthesis that selectively influence electron transport chain enzymes, and bilobalide prevents mitochondrial permeability transition pore (mPTP). Hence, it decreases apoptosis, maintains neuronal integrity, and improves energy supply (Machado et al., 2021). *Ginkgo* inhibits caspase-3 activation in the mitochondria-dependent caspase pathway and maintains mitochondrial membrane potential through long-term regulatory effects for an effective, high-quality, and productive life (Tomino et al., 2021).

4.5.3 *Ginkgo biloba* and Neuroinflammation

Neuroinflammation is an immunological reaction to cues/stimuli common to neuropathology (Ozben & Ozben, 2019; Jia & Gao, 2021). Neuroinflammatory responses are regulated by glial cells for neuro-homeostasis (Asuku & Ogungbangbe, 2024). However, sustained inflammatory responses triggered by aggregates of protein deposits, pro-inflammatory cytokines, etc. result in physiological alteration of glial cells (Sochocka et al., 2017; Aramideh et al., 2022) and compromise the integrity of neurons and the upstream cascade events (Hashioka et al., 2021). Additionally, Sun et al. (2024) studied targeting the COX-2/PGE2 pathway; their findings show EGb 761 reduces neuroinflammation, which could explain the established clinical cognitive efficacy in AD, vascular, and mixed dementia.

4.5.4 The Antioxidant and Neuroprotective Nature of *Ginkgo biloba*

Antioxidants are molecules that detoxify ROS through enzymatic and non-enzymatic pathways (Zhu et al., 2019). The asymmetry connections between the production of ROS and the body's defense system result in neurodegenerative diseases, including lipid peroxidation, molecular disruption, misfolding of proteins, DNA mutations, fragmentation, mitochondrial dysfunction, and cell death (Morató et al., 2024a). Aggregates of Aβ are a major pathological factor that enhances ROS production, excessively damages neuronal cells, and promotes tau hyperphosphorylation, neurofibrillary tangles, neuronal energy disruption, and apoptosis due to mitochondrial dysfunction (Xie et al., 2014). ROS destroy the vascular endothelium, the blood–brain barrier, which causes ischemia and neuroinflammation (Kao et al., 2020; Morató et al., 2024b).

In dementia, oxidative stress biomarkers show increased levels of lipid peroxidation (MDA), 8-OHdG (oxidized DNA), and a drop in glutathione (GSH) and superoxide dismutase (SOD) levels (Nowak et al., 2021). Notably, oxidative injury is highly pronounced in the hippocampus and cortex, where heightening oxidative stress disrupts lipid metabolism and mitochondrial functionality (Akintoye et al., 2021; Morató et al., 2024b). Ultimately, ROS underpin dementia and are linked with pathological processes such as amyloid aggregation, tau pathology, mitochondrial dysfunction, and inflammation, which contribute to the onset and progression of brain-related diseases (Gaweł & Potulska-Chromik, 2015). Mitigating oxidative stress might decrease disease progression, enhance cognitive abilities, and promote neuronal survival through various antioxidant approaches (Asuku et al., 2023). Medicinal plants like *G. biloba* have the potential to be used for managing and treating dementia (Suárez-González et al., 2023). *Ginkgo* preserves neuronal integrity due to its phytochemical compositions, like flavonoids and terpenoids, such as quercetin, kaempferol, and ginkgolides, which neutralize free radicals (Shi et al., 2010; Suárez-González et al., 2023). Flavonoids are liable for chelation of pro-oxidative heavy metal ions, increasing the activity of SOD and GSH reductase (Shi et al., 2010). The activity of flavonoids across neuronal receptors, transcription factors, and proteins is associated with the learning and memory recovery process (Rendeiro et al., 2015).

In vascular dementia, it dilates cerebral vessels to a vasodilatory effect, improving oxygen and glucose delivery (Stoffregen et al., 2009; Tan et al., 2019). Terpene trilactones (diterpenoids and bilobalide) and diterpenoids (ginkgolides A, B, and C) inhibit platelet-activating factor by reducing inflammation, enhancing cognitive function, and enhancing cholinergic activity (Tripathi & Agarwal, 2024). *G. biloba* constituent bilobalide serves as a neuroprotective agent by inhibiting neuronal apoptosis, enhancing the physiology of mitochondria, and regulating GABAergic and glycinergic neuronal transmission (Van Beek & Montoro, 2009). Phenolic acids, proanthocyanidins, and lignans are strong free-radical scavengers, antioxidants, and anti-inflammatory properties actively involved in *G. biloba* (Tabassum et al., 2022). Alkylphenols (ginkgolic acids) and polyprenols are available at reduced levels and exhibit cytotoxic and antimicrobial potentials while supporting cellular activities (Tabassum et al., 2022).

4.6 METHODS OF USING *GINKGO BILOBA* FOR THE MANAGEMENT AND TREATMENT OF DEMENTIA

Traditionally, *G. biloba* has been an adjunct therapy in dementia management for years, with efficacy, safety, and tolerability for dementia patients (Rösler et al., 2019).

4.6.1 Decoction

In Chinese medicine, *G. biloba* leaves are properly washed and boiled for a few minutes; this procedure is called decoction. The water-soluble constituents supposedly contain flavonoids and phenolic compounds; their potency is poor and inefficient, and they possess a short shelf life.

4.6.2 Tinctures

It is a water-soluble and alcoholic (30–70% ethanol) extract whose process is by soaking *G. biloba* leaves for a few days or weeks; the alcoholic extract enhances the bioavailability of certain constituents such as flavonoids and terpene lactones and possesses a longer shelf life than decoctions but may not be suitable for children, the elderly, or patients with liver challenges due to the alcohol content and not being up to clinical standard measures (Orhan et al., 2021).

4.6.3 Powdered Form

G. biloba leaves are dried, pulverized, and consumed directly. It is believed that the powdery form contains the whole spectrum of the plants. These extensive raw phytochemical constituents reduce their bioavailability due to poor solubility and absorption of active constituents and can be incorporated into capsules or tablets. However, it may be toxic at high levels because the potency and dosage are not guaranteed according to the drug standard (Dziwenka & Coppock, 2021; Das et al., 2022).

4.7 CLINICAL APPLICATIONS OF *GINKGO BILOBA*

Gauthier et al. (2021) suggest that EGb 761 can be used as an adjunct concomitantly with conventional methods in dementia management and can enhance quality of life (Woelk & Schlierf, 2020). Some European states, like Germany and France, have legalized EGb 761 extract to be utilized in clinical settings (Nabavi et al., 2020). Adjunct therapy promotes retinal perfusion and decreases oxidative stress associated with age-related macular degeneration (AMD), retinal injury, and glaucoma (Chen et al., 2023). Savaskan and Hoerr's (2020) study reveals that cognitive disabilities can be managed using EGb 761 to improve neuroplasticity and neuroinflammation. EGb 761 penetrates the blood–brain barrier to aid ameliorative roles such as anti-cancer properties, anti-neurological activity, and anti-amyloidogenic roles, and it enhances synaptic transmission (Li et al., 2019; Kalai et al., 2022). It is also used as an adjunct concomitantly with conventional methods in managing and treating mild cognitive impairment (MCI) and early-stage dementia to enhance quality of life (Woelk & Schlierf, 2020).

The United States accedes to EGb 761 as a dietary supplement (Moriwaki et al., 2023; NCCIH, 2024). However, Nejabati & Roshangar (2022) suggest that due to ginkgo flavonoids' bioavailability limitation, Zheng et al. (2015) and Tao et al. (2022) employ an unstable liposome complex to improve ginkgo efficacy. Among several clinical studies, Savaskan et al. (2018) attest to EGb 761 efficacy in behavioral and psychological symptoms of dementia. Herrschaft et al. (2012) state that 240 mg once daily enhances cognitive abilities, improves psychopathological function, and savors existence. Le Bars' research focuses on the long-term EGb761 administration, which significantly strengthens mental performance in individuals struggling with dementia (Le Bars et al., 2019). In mild to moderate dementia, EGb761 showed no substantial difference compared with donepezil, elucidating its tolerance and efficacy (Mazza et al., 2006). An AChE inhibitor is an effective therapeutic medication that enhances memory in a disrupted neuronal system. Disruption of the cholinergic circuitry, comprising nicotinic and muscarinic receptors, vesicular transport system, and enzymatic activities, delays retrieval processes (Eldufani & Blaise, 2019). Also, in a combined therapy with medications like acetylcholinesterase inhibitors, mental and behavioral function is enhanced by EGb761 (García-Alberca et al., 2022).

Ginkgo demonstrates pharmacodynamic properties by halting hyperphosphorylated tau protein formation and amyloid protein aggregation (Bohlken et al., 2022; Morató et al., 2024b).

4.8 CHALLENGES AND FUTURE DIRECTIONS

Despite the wide acceptance of the standard dose of EGb 761, there is little variance and inconsistency across several studies, which may alter dosage and diagnosis procedures during clinical trials for dementia. Studies on the long-term evaluation of efficacy are minimal to support an effective dose duration of about six months. The authenticity of heterogeneous clinical bench work across several studies is unstable due to patients' age, dementia phase, research design, and outcome, which may prevent quality health. There is a need to combine adjunct therapy with conventional medication for efficacy. Further studies on the *G. biloba* mechanisms of action and the use of targeted nanocarriers, which can enhance bioavailability, can be carried out.

4.9 CONCLUSION

The receiving end, such as age-related cognitive impairment, neurosensory diseases, and dementia, occurs as a result of the sociodemographic transient associated with the aging census, which significantly disorients dementia patients, creeps out the caregiver, and poses a threat to the health sector and nation at large. *G. biloba* is a unique natural plant with potential remedies. The regulated form of *G. biloba* known as EGb761 has drawn distinguished researchers across the globe on account of its antioxidant, anti-inflammatory, neuroprotective, and vasoregulatory properties. Clinically, its efficacy and safety have been proven in managing and treating all stages of dementia. It is highly beneficial and economically used as an alternative or adjunct therapy, yet it should not be replaced with conventional therapies for dementia.

REFERENCES

Akintoye, O.O., Ajibare, A.J., Fabunmi, O.A., Asuku, A.O., & Tajudeen, R.A. (2021). Effect of Honey on Oxidative Stress in Brain Tissue of Sleep Deprived Rats. *Nigerian Journal of Basic and Applied Science*, 29(1): 83–89, http://dx.doi.org/10.4314/njbas.v29i1.10

Alzheimer's Disease International. (2020). *Dementia statistics*. Retrieved from https://www.alzint.org/about/dementia-facts-figures/dementia-statistics/

Aramideh, J.A., Maurel, C., Scherer, N.M., Don, E.K., Lee, A., Chung, R.S., Graeber, M.B., & Morsch, M. (2022). Microglia Morphophysiological Diversity and Its Implications for the CNS. *Frontiers in Immunology*, 13: 997786.

Asuku, A.O., Adeyemo, M.B., Adeyemo, R.O., Agoro, S.A., Odugbesan, O.A., & Ayoola, S.A. (2022). Applications of Stem Cell Therapy in Neurodegenerative Diseases. *European Journal of Biotechnology and Bioscience*, 10(2): 17–24.

Asuku, A.O., Ajibare, A.J., Ajagun, E.J., Babatunde, S.S., Olajide, T.S., Oyerinde, T.O., Ogungbangbe, G.O., & Odugbesan, O.A. (2024). Herbal Extracts in Anti-Aging Products. In: *Eco-Friendly Skin Solutions for Natural Cosmeceuticals*. IGI Publisher, http://doi.org/10.4018/979-8-3693-5473-5.ch012

Asuku, A.O., Ayinla, M.T., Olajide, T.S., Yakub, Z., & Owoyele, B.V. (2023). *Cola acuminata* Mitigates Cognitive Deficit and Oxidative Stress in Mercury Chloride-Induced Neurotoxicity in Male Wistar Rats. *Nigerian Journal of Neuroscience*, 14(2):33–40.

Asuku, A.O., & Ogungbangbe, G.O. (2024). Neuroinflammation and Binge Eating. In: *Neuroscientific Insights and Therapeutic Approaches to Eating Disorders* (pp. 147–170, Vol. 10). IGI Publisher, https://doi.org/10.4018/979-8-3693-3230-6.ch010

Belwal, T., Giri, L., Bahukhandi, A., Tariq, M., Kewlani, P., Bhatt, I.D., et al. (2019). *Ginkgo biloba*. In: S.M. Nabavi & A.S. Silva (Eds.), *Nonvitamin and Nonmineral Nutritional Supplements* (pp. 241–250).: Academic Press.

Bohlken, J., Peters, O., & Kostev, K. (2022). Association between *Ginkgo biloba* Extract Prescriptions and Dementia Incidence in Outpatients with Mild Cognitive Impairment in Germany: A Retrospective Cohort Study. *Journal of Alzheimer's Disease*, 86: 703–709.

Carroll, C.M., & Benca, R.M. (2024). Upsetting the Balance: How Modifiable Risk Factors Contribute to the Progression of Alzheimer's Disease. *Biomolecules*, 14: 274.

Chen, Y., Fu, C., Wu, Z., Xu, H., Liu, H., Schneider, H., & Lin, J. (2021). *Ginkgo biloba. Trends in Genetics*, 37(5): 488–489.

Chen, Y., Wu, H., & Zhang, H. (2023). *Ginkgo biloba* Extract EGb 761 in the Treatment of Ocular Disorders: A Systematic Review. *Frontiers in Pharmacology*, 14: 1170183.

Das, R., Lami, M.S., & Chakraborty, A.J. (2022). *Ginkgo biloba*: A Treasure of Functional Phytochemicals with Multimedicinal Applications. *Evidence-Based Complementary and Alternative Medicine*, 2022, 8288818.

Dower, J.I., Geleijnse, J.M., Gijsbers, L., Schalkwijk, C., Kromhout, D., & Hollman, P.C. (2015). Supplementation of the Pure Flavonoids Epicatechin and Quercetin Affects Some Biomarkers of Endothelial Dysfunction and Inflammation in (Pre)hypertensive Adults: a Randomized Double-Blind, Placebo-Controlled, Crossover Trial. *Journal of Nutrition*, 145: 1459–63.

Dziwenka, M., & Coppock, R. W. (2021). *Ginkgo biloba*. In: R. Gupta (Ed.), *Nutraceuticals* (pp. 715–732). Academic Press.

Ekici-Günay, N. (2020). *Ginkgo biloba* Extract as an Antioxidant in Nerve Regeneration. In *Phytochemicals as Lead Compounds for New Drug Discovery* (pp. 447–458). Academic Press.

Eldufani, J., & Blaise, G. (2019). The Role of Acetylcholinesterase Inhibitors Such as Neostigmine and Rivastigmine on Chronic Pain and Cognitive Function in Aging: A Review of Recent Clinical Applications. *Alzheimer's & Dementia: Translational Research & Clinical Interventions*, 5(1): 175–183.

Frank, J., Fukagawa, N.K., Bilia, A.R., et al. (2020). Terms and Nomenclature Used for Plant-Derived Components in Nutrition and Related Research: Efforts Toward Harmonization. *Nutrition Reviews*, 78(6): 451–458.

García-Alberca, J.M., Gris, E., & Mendoza, S. (2022). Combined Treatment with *Ginkgo biloba* Extract EGb 761 Plus Acetylcholinesterase Inhibitors Improved Cognitive Function and Neuropsychiatric Symptoms in Patients with Mild Cognitive Impairment. *Alzheimers Dement (N Y)*, 2;8(1):e12338, http://doi.org/10.1002/trc2.12338. PMID: 35929002; PMCID: PMC9345397.

Gauthier, S., Albert, M., Fox, N., Goedert, M., Kivipelto, M., Mestre, T.A., & Scheltens, P. (2021). Why We Need to Investigate the Efficacy of *Ginkgo biloba* in Early Alzheimer's Disease. *The Lancet Neurology*, 20(4): 297–306.

Gaweł, M., & Potulska-Chromik, A. (2015). Neurodegenerative Diseases: Alzheimer's and Parkinson's Disease. *Postępy Nauk Medycznych*, 7: 468–476.

Ge, W., Ren, C., Xing, L., Guan, L., Zhang, C., Sun, X., Wang, G., Niu, H., & Qun, S. (2021). *Ginkgo biloba* Extract Improves Cognitive Function and Increases Neurogenesis by Reducing Aβ Pathology in 5×FAD Mice. *American Journal of Translational Research*, 13(3): 1471–1482.

George-Carey, R., Adeloye, D., Chan, K.Y., Paul, A., Kolčić, I., Campbell, H., & Rudan, I. (2012). An Estimate of the Prevalence of Dementia in Africa: A Systematic Analysis. *Journal of Global Health*, 2(2):20401. http://doi.org/10.7189/jogh.02.020401.

Hashiguchi, M., Ohta, Y., Shimizu, M., Maruyama, J., & Mochizuki, M. (2015). Meta-Analysis of the Efficacy and Safety of Ginkgo biloba Extract for the Treatment of Dementia. *Journal of Pharmaceutical Health Care and Sciences*, 10(1):14. http://doi.org/10.1186/s40780-015-0014-7. PMID: 26819725; PMCID: PMC4729005.

Hashioka, S., Wu, Z., & Klegeris, A. (2021). Glia-Driven Neuroinflammation and Systemic Inflammation in Alzheimer's Disease. *Current Neuropharmacology*, 19(6): 908–924.

Hassan Mohammed, S.A., Tripathi, R., & Sreejith, K. (2020). Plant Metabolites: Methods for Isolation, Purification, and Characterization. In: *Plant Metabolites: Methods, Applications and Prospects (eds Sukumaran ST et al)*. Springer.

Herrschaft, H., Nacu, A., Likhachev, S., Sholomov, I., Hoerr, R., & Schlaefke, S. (2012). *Ginkgo biloba* Extract EGb 761® in Dementia with Neuropsychiatric Features: A Randomised, Placebo-Controlled Trial to Confirm the Efficacy and Safety of a Daily Dose of 240 mg. *Journal of Psychiatric Research*, 46(6):716–723. http://doi.org/10.1016/j.jpsychires.2012.03.003

Hori, T., Ridge, R.W., Tulecke, W., Del Tredici, P., Trémouillaux-Guiller, J., & Tobe, H. (2012). *Ginkgo biloba a Global Treasure: From Biology to Medicine*. Springer Science & Business Media.

Institute for Health Metrics and Evaluation. (2022). *Estimation of the global prevalence of dementia in 2019 and forecasted prevalence in 2050: An analysis for the Global Burden of Disease Study 2019.* Retrieved from https://www.healthdata.org/research-analysis/library/estimation-global-prevalence-dementia-2019-and-forecasted-prevalence-2050

Jia, X., & Gao, Z. (2021). Microglia in Depression: Current Perspectives. *Life Sciences*, 264: 118640.

Kalai, F.Z., Boulaaba, M., Ferdousi, F., & Isoda, H. (2022). Effects of Isorhamnetin on Diabetes and Its Associated Complications: A Review of In Vitro and In Vivo Studies and a Post Hoc Transcriptome Analysis of Involved Molecular Pathways. *International Journal of Molecular Science*, 23(2): 704.

Kao, Y.C., Ho, P.C., Tu, Y.K., Jou, I.M., & Tsai, K.J. (2020). Lipids and Alzheimer's Disease. *International Journal of Molecular Sciences*, 21(4): 1505, https://doi.org/10.3390/ijms21041505

Kim, Y.-J., Bostami, A.B.M.R., Islam, M.M., Mun, H.S., Ko, S.Y., & Yang, C.-J. (2016). Effect of Fermented Ginkgo biloba and Camelia Sinensis-Based Probiotics on Growth Performance, Immunity and Caecal Microbiology in Broilers. *International Journal of Poultry Science*, 15(2), 62–71.

Le Bars, P.L., Katz, M.M., Berman, N., Itil, T.M., Freedman, A.M., & Schatzberg, A.F. (2019). A Placebo-Controlled, Double-Blind, Randomized Trial of an Extract of *Ginkgo biloba* for Dementia. North American EGb Study Group. *JAMA*, 278(16): 1327–1332.

Lejri, I., Grimm, A., & Eckert, A. (2019). *Ginkgo biloba* Extract Increases Neurite Outgrowth and Activates the Akt/mTOR Pathway. *PLoS One*, 14(12): e0225761.

Li, M., Li, B., Xia, Z.M., Tian, Y., Zhang, D., Rui, W.J., Dong, J.X., & Xiao, F.J. (2019). Anticancer Effects of Five Biflavonoids from *Ginkgo biloba* L. Male Flowers in Vitro. *Molecules*, 24(8): 1496.

Li, Y., Zhu, X., Wang, K., Zhu, L., Murray, M., & Zhou, F. (2022). The Potential of *Ginkgo biloba* in the Treatment of Human Diseases and the Relationship to Nrf2–Mediated Antioxidant Protection. *Journal of Pharmacy and Pharmacology*, 74(12): 1689–1699.

Machado, M.M.F., Banin, R.M., Thomaz, F.M., de Andrade, I.S., Boldarine, V.T., Figueiredo, J.S., Hirata, B.K.S., Oyama, L.M., Lago, J.H.G., Ribeiro, E.B., & Telles, M.M. (2021). *Ginkgo biloba* Extract (GbE) Restores Serotonin and Leptin Receptor Levels and Plays an Antioxidative Role in the Hippocampus of Ovariectomized Rats. *Metabolic Brain Disease*, 36(3): 469–480.

Mazza, M., Capuano, A., Bria, P., & Mazza, S. (2006). *Ginkgo biloba* and Donepezil: A Comparison in the Treatment of Alzheimer's Dementia in a Randomized Placebo-Controlled Double-Blind Study. *European Journal Neurology*, 13(9): 981–985.

Montes, P., Ruiz-Sanchez, E., & Rojas, C. (2015). *Ginkgo biloba* Extract 761: A Review of Basic Studies and Potential Clinical Use in Psychiatric Disorders. *CNS & Neurological Disorders - Drug Targets*, 14(1): 132–143.

Morató, X., Tartari, J.P., Pytel, V., & Boada, M. (2024a). Neuroprotective and Antioxidant Effect of *Ginkgo biloba* Extract against Alzheimer's Disease and Other Neurological Disorders. *Journal of Alzheimer's Disease*, 101(S1): S285–S298.

Morató, X., Tartari, J.P., Pytel, V., & Boada, M. (2024b). Pharmacodynamic and Clinical Effects of *Ginkgo biloba* Extract EGb 761 and Its Phytochemical Components in Alzheimer's Disease. *Journal of Alzheimer's Disease*, 101: S285–S298.

Moriwaki, M., Tominaga, E., Kito, K., Nakagawa, R., Kapoor, M.P., Matsumiya, Y., Fukuhara, T., Kobayashi, J., Satomoto, K., Yamagata, H., & Kuroiwa, Y. (2023). Bioavailability of Flavonoids in Ginkgo biloba Extract-γ-Cyclodextrin Complex. *Natural Product Communications*, 18 (2023). https://doi.org/10.1177/1934578X231170221

Nabavi, S.M., Daglia, M., Braidy, N., Nabavi, S.F., & Loizzo, M.R. (2020). *Ginkgo biloba* as an Adjuvant Therapy for Neurological Disorders: A Review of Preclinical and Clinical Studies. *Current Medicinal Chemistry*, 27(36): 5982–5996.

NCCIH.(2024).*Dietarysupplementsandcognitivefunction,dementia,andAlzheimer'sdisease—Scienceandsafety overview*. National Center for Complementary and Integrative Health. https://www.nccih.nih.gov/health/providers/digest/dietary-supplements-and-cognitive-function-dementia-and-alzheimers-disease-science

Nejabati, H.R., & Roshangar, L. (2022). Kaempferol as a Potential Neuroprotector in Alzheimer's Disease. *Journal Food Biochemistry*. 46 (12)

Nguyen, V.T.T., Slotos, R.S., Guilherme, M.D.S., Nguyen, T.T., Weisenburger, S., Lehner, M.D., & Endres, K. (2025). *Ginkgo biloba* Extract EGb 761® Ameliorates Cognitive Impairment and Alleviates TNFα Response in 5xFAD Alzheimer's Disease Model Mice. *Phytomedicine*, 136: 156327.

Nowak, A., Kojder, K., Zielonka-Brzezicka, J., Wróbel, J., Bosiacki, M., Fabiańska, M., Wróbel, M., Sołek-Pastuszka, J., & Klimowicz, A. (2021). The Use of *Ginkgo biloba* L. as a Neuroprotective Agent in the Alzheimer's Disease. *Front Pharmaco*, 12: 775034.

Orhan, H., Arzuk, E., & Karaku, F. (2021). Toxicology of Herbal Medicines. In M. Shariati & D. Ekinci (Eds.), *Herbal Medicine* (pp. 221–240). CRC Press.

Ozben, T., & Ozben, S. (2019). Neuro-Inflammation and Anti-Inflammatory Treatment Options for Alzheimer's Disease. *Clinical Biochemistry*, 72: 87–89.

Patow, G., Escrichs, A., Ritter, P., & Deco, G. (2024). Whole-Brain Dynamics Disruptions in the Progression of Alzheimer's Disease: Understanding the Influence of Amyloid-Beta and Tau. *bioRxiv*. https://doi.org/10.1101/2024.03.29.587333

Pendarvis, M., Crawley, P., & John, L. (2018). *Exploring Biology in the Laboratory, 3e*. Morton Publishing Company.

Rauf, A., Imran, M., Patel, S., Khan, I.A., & Mubarak, M.S. (2022). Neuroprotective Effects of Plant-Derived Bioactive Compounds: A Review. *Frontiers in Pharmacology*, 13: 822336.

Rendeiro, C., Rhodes, J.S., & Spencer, J.P.E. (2015). The Mechanisms of Action of Flavonoids in the Brain: Direct Versus Indirect Effects. *Neurochemistry International*, 89: 126–139.

Rojas, C., Rojas-Castañeda, J., & Rojas, P. (2016). Antioxidant Properties of a *Ginkgo biloba* Leaf Extract (EGb 761) in Animal Models of Alzheimer's and Parkinson's Diseases. *Current Topics in Nutraceutical Research*, 14(1): 1–16.

Rösler, M., Thiemann, S., & Dimpfel, W. (2019). Safety Aspects of *Ginkgo biloba* Extract EGb 761. *Pharmacopsychiatry*, *52*(2), 78–87.

Savaskan, E., & Hoerr, R. (2020). *Ginkgo biloba* Extract in the Treatment of Dementia: Current Evidence and Potential Mechanisms. *Pharmacopsychiatry*, 53(1): 24–30.

Savaskan, E., Mueller, Heiko, Hoerr, R., von Gunten, Armin, & Gauthier, S. (2018). Treatment Effects of *Ginkgo biloba* Extract EGb 761® on the Spectrum of Behavioral and Psychological Symptoms of Dementia: Meta-Analysis of Randomized Controlled Trials. *International Psychogeriatrics*, 30(3): 285–293.

Shi, C., Liu, J., Wu, F., & Yew, D.T. (2010). *Ginkgo biloba* Extract in Alzheimer's Disease: from Action Mechanisms to Medical Practice. *International Journal of Molecular Sciences*, 11(1): 107–123.

Shi, C., Zhao, L., Zhu, B., Li, Q., Yew, D.T., Yao, Z., & Xu, J. (2009). Protective Effects of *Ginkgo biloba* Extract (EGb761) and Its Constituents Quercetin and Ginkgolide B against Beta-Amyloid Peptide-Induced Toxicity in SH-SY5Y Cells. *Chemico-Biological Interactions*, 181: 115–123.

Singh, S.K., Barreto, G.E., Aliev, G., & Echeverria, V. (2017). *Ginkgo biloba* as an Alternative Medicine in the Treatment of Anxiety in Dementia and Other Psychiatric Disorders. *Current Drug Metabolism*, 18(2): 112–119.

Sochocka, M., Diniz, B.S., & Leszek, J. (2017). Inflammatory Response in the CNS: Friend or Foe? *Molecular Neurobiology*, 54(10): 8071–8089.

Stoffregen, P., Banks, S.J., Sander, M., Wack, D.S., & Lowe, M.J. (2009). Effects of *Ginkgo biloba* Extract on Cerebral Blood Flow Measured by DSC-MRI in Older Adults. *Journal of Magnetic Resonance Imaging*, 30(2): 443–449.

Suárez-González, E., Sandoval-Ramírez, J., Flores-Hernández, J., & Carrasco-Carballo, A. (2023). *Ginkgo biloba*: Antioxidant Activity and in Silico Central Nervous System Potential. *Current Issues in Molecular Biology*, 45(12): 9674–9691.

Sun, L., Apweiler, M., Tirkey, A., Klett, D., Normann, C., Dietz, G.P.H., Lehner, M.D., & Fiebich, B.L. (2024). Anti-Neuroinflammatory Effects of *Ginkgo biloba* Extract EGb 761 in LPS-Activated BV2 Microglial Cells. *International Journal of Molecular Sciences*, 25(15): 8108.

Tabassum, N.-E., Das, R., Lami, M.S., Chakraborty, A.J., Mitra, S., Tallei, T.E., Idroes, R., Mohamed, A.A., Hossain, M.J., Dhama, K., Mostafa-Hedeab, G., & Emran, T.B. (2022). *Ginkgo biloba*: A Treasure of Functional Phytochemicals with Multimedicinal Applications. *Evidence-Based Complement Alternate Medicine*. 28(2022):8288818. https://doi.org/10.1155/2022/8288818. PMID: 35265150; PMCID: PMC8901348.

Tan, M.S., Yu, J.T., Tan, C.C., Wang, H.F., Meng, X.F., Wang, C., & Tan, L. (2019). Efficacy and Adverse Effects of *Ginkgo biloba* for Cognitive Impairment and Dementia: A Systematic Review and Meta-Analysis. *Journal of Alzheimer's Disease*, 63(2): 513–527.

Tan, R.H., Kril, J.J., Fatima, M., McGeachie, A., McCann, H., Shepherd, C., Forrest, S.L., Affleck, A., Kwok, J.B., Hodges, J.R., & Halliday, G.M. (2015). TDP-43 Proteinopathies: Pathological Identification of Brain Regions Differentiating Clinical Phenotypes. *Brain*, 138(10): 3110–3122.

Tao, Y., Zhu, F., Pan, M., Liu, Q., & Wang, P. (2022). Pharmacokinetic, Metabolism, and Metabolomic Strategies Provide Deep Insight into the Underlying Mechanism of *Ginkgo biloba* Flavonoids in the Treatment of Cardiovascular Disease. *Front Nutrition*, 9: 857370.

Tomino, C., Ilari, S., Solfrizzi, V., Malafoglia, V., Zilio, G., Russo, P., Proietti, S., Marcolongo, F., Scapagnini, G., Muscoli, C., & Rossini, P.M. (2021). Mild Cognitive Impairment and Mild Dementia: The Role of *Ginkgo biloba* (EGb 761®). *Pharmaceuticals*, 14(4): 305.

Tripathi, P., & Agarwal, V. (2024). Research Progress on the Types, Functions, Biosynthesis, and Metabolic Regulation of Ginkgo Terpenoids. *Plant Physiology and Biochemistry*, 212, Article 108754.

Van Beek, T.A., & Montoro, P. (2009). Chemical Analysis and Quality Control of *Ginkgo biloba* Leaves, Extracts, and Phytopharmaceuticals. *Journal of Chromatography A*, 1216(11): 2002–2032.

Venkat, P., Chopp, M., Zacharek, A., Cui, C., Zhang, L., Li, Q., Lu, M., Zhang, T., Liu, A., & Chen, J. (2017). White Matter Damage and Glymphatic Dysfunction in a Model of Vascular Dementia in Rats With No Prior Vascular Pathologies. *Neurobiology of Aging*, 50: 96–106.

Wang, D.P., Liu, K.J., Kasper, G., Lin, Q., & Hai, J. (2017). Inhibition of SENP3 by URB597 Ameliorates Neurovascular Unit Dysfunction in Rats with Chronic Cerebral Hypoperfusion. *Biomedicine & Pharmacotherapy*, 91: 872–879.

Wang, J., Mao, S., Wang, J., Jiao, Y., Zhang, X., Zhang, Y., Wang, Z., Zhang, Q., & Li, S. (2018). Effect of Ginkgo biloba Extract on Cognitive Function and Neurotransmitter Levels in Rats with Vascular Dementia. *Indian Journal of Animal Research*, 52(8), 1141–1145. https://doi.org/10.18805/ijar.v0iOF.8467

Wang, L., Liu, J., Liu, J., Wei, H., Fang, Y., Wang, D., Chen, R., & Gu, W. (2023). Revealing the Long-Term Trend of the Global-Scale *Ginkgo biloba* Distribution and the Impact of Future Climate Change Based on the Ensemble Modeling. *Biodiversity and Conservation*, 32(6): 2077–2100.

Woelk, H., & Schlierf, G. (2020). Efficacy of *Ginkgo biloba* Extract EGb 761 in Patients with Mild to Moderate Anxiety Disorder: A Randomized, Placebo-Controlled Trial. *Journal of Psychiatric Research*, 129: 236–243.

World Health Organization. (2023). *Dementia*.

Xie, H., Wang, J.-R., Yau, L.-F., Liu, Y., Liu, L., Han, Q. B., Zhao, Z., & Jiang, Z.-H. (2014). Catechins and Procyanidins of *Ginkgo biloba* Inhibit β-Amyloid Peptide Aggregation. *Molecules*, 19(4): 5119–5134.

Yin, Y., Ren, Y., Wu, W., et al. (2013). Protective Effects of Bilobalide on Aβ(25–35) Induced Learning and Memory Impairments in Male Rats. *Pharmacology Biochemistry and Behavior*, 106: 77–84.

Zheng, B., Teng, L., Xing, G., Bi, Y., Yang, S., Hao, F., Yan, G., Wang, X., Lee, R.J., Teng, L., & Xie, J. (2015). Proliposomes Containing a Bile Salt for Oral Delivery of *Ginkgo biloba* Extract: Formulation Optimization, Characterization, Oral Bioavailability and Tissue Distribution in Rats. *European Journal of Pharmaceutical Sciences*, 77: 254–264.

Zhu, J., Arsovska, B., & Kozovska, K. (2019). Alzheimer's Disease Treatment with Traditional Chinese Medicine. *International Journal of Research – Granthaalayah*, 7(10): 448–451.

5 *Bacopa monnieri* for Managing and Treating Dementia

Gideon Opeyemi Ayilara, Patrick Oluwole Abolarin, Abosede Oreoluwa Bolaji, Oluwafemi Abiola Onaolapo, and Olutayo Folajimi Olaseinde

5.1 INTRODUCTION

Dementia is a progressive neurodegenerative condition that is characterized by decline in cognitive ability and memory loss. The most prevalent forms of dementia are Alzheimer disease (AD) and vascular dementia with other forms being Lewy body dementia, frontotemporal dementia, and mixed dementia (Asuku et al., 2024, 2025). Although there are available conventional therapies, they have several limitations impacting their efficacy and medication adherence among patients (Arvanitakis et al., 2019). This has necessitated the need for further research into alternative therapies, with herbal medicine being in the frontline. Indeed, significant progress has been made in herbal medicine as alternative approach for the treatment and management of some neurological diseases like AD, schizophrenia, and Parkinson's disease, among others (Puri et al., 2022). *Bacopa monnieri* (*BM*) (mostly referred to as Brahmi) is gaining attention for its neuroprotective and cognitive-enhancing properties (Neto et al., 2024; Ayilara & Owoyele, 2025).

BM is a plant that belongs to the *Scrophulariaceae* family and has found usefulness in the traditional Indian medicine for thousands of years. This plant has a unique obovate foliage and little white blossoms. Traditionally, it has been used to improve memory (Gościniak et al., 2025) and regulate blood glucose levels (Mittal et al., 2025), sleeplessness, epilepsy, and as an anxiolytic (Kolape et al., 2025). The extensive applications of this plant in numerous ailments are due to its rich phytochemical substances like saponins and alkaloids, of which bacosides A is the most prominent (Manap et al., 2019).

5.2 PHYTOCHEMICAL PROFILE OF *BACOPA MONNIERI*

BM has several metabolites such as saponins, flavonoids, alkaloids, sterol glycosides, sugars, brahmin, hydrocotyline, and cucurbitacins (Fatima et al., 2022). There are also specific amino acids like alanine, aspartic acid, serine, and glutamic acid found in the extracts of *BM*. The major compounds that are responsible for the biological effects of this plant are the Bacopasides (Sekhar et al., 2019). They are present in all parts of the plant and are specifically known for the roles they play in neuronal health (Paul et al., 2025). Among these saponins (Bacopasides), bacoside A is the major fraction and they account for about 38% of the plant's methanolic extract (Sairam et al., 2002).

Brahmine is also a primary alkaloid found in *BM* and is believed to play significant role in the neuroprotective effects of the plant (Dubey et al., 2024). Herpestine and monnieri are other alkaloids believed to be responsible for the cognitive effects of *BM*. They achieve these effects by influencing the neurotransmitter systems (Nishanth et al., 2023). Cucurbitacins are a category of highly oxygenated tetracyclic triterpenoids recognized for their bitter flavor and medicinal qualities. The cucurbitacins include cucurbitacin B and cucurbitacin E and they enhance the plant's biological activities. The inhibitory effects of cucurbitacins on the proliferation of breast, lung, colon, and central nervous system cancer cell lines were also recently reported (Gościniak et al., 2025).

 DOI: 10.1201/9781003657385-5

Considering these chemical complexities, *BM*'s traditional use in cognitive and neuroprotective therapies may therefore be justifiable.

5.3 BIOLOGICAL ACTIVITY OF *BACOPA MONNIERI* IN DEMENTIA MANAGEMENT

The effects of the bioactive components on different neurological processes make *BM* to be effective in the management of dementia. *Bacopa* has been found to upregulate cholinergic tone by inhibiting acetylcholinesterase (AChE), thus prolonging availability of acetylcholine (Ach) at the synapses (Gościniak et al., 2025). This potentially makes *Bacopa* beneficial in dementia such as AD since studies have established that the cognitive decline in AD is largely due to a deficit in cholinergic neurotransmission (Chen et al., 2022). Likewise, the loss of cholinergic neurons and reduced levels of ACh are strongly correlated with memory impairment (Akintoye et al., 2023). *In silico* and *in vitro* research show dose-dependent AChE inhibition activity of *Bacopa* (Shoukat et al., 2023), while randomized controlled trials (RCT) in older adults reported reduced peripheral AChE activity alongside attention/memory gains during supplementation (typically 300–600 mg/day of standardized extract) (Peth-Nui et al., 2012; Eraiah et al., 2024).

The antioxidative property of *BM* may also be responsible for its beneficial effect in dementia. Oxidative stress is an early driver of neurodegeneration seen in dementia (Korovesis et al., 2023): *Bacopa* reduces lipid peroxidation and boosts enzymatic antioxidants in hippocampus and cortex (Bhattacharya et al., 2000; Simpson et al., 2015). *BM* achieves these due to its compounds like bacosides A and B, quercetin, betulinic acid, and asiatic acid that scavenge free radicals, reduce mitochondrial dysfunction, and enhance endogenous antioxidant enzymes (Shinomol & Muralidhara, 2011). These antioxidant effects help to slow dementia progression by preserving neuronal integrity and improving synaptic plasticity. Likewise, the anti-inflammatory property of *Bacopa* has been exploited. It is found to downregulate pro-inflammatory mediators, such as the interleukins, tumor necrosis alpha, and inducible enzymes (COX-2, iNOS), largely via suppression of NF-κB pathway activity (Gościniak et al., 2025). By minimizing glial cells' activation, *Bacopa* may protect synapses from cytokine-mediated injury common in Alzheimer's and mixed dementias (Al-Ghraiybah et al., 2022). In vitro studies that used neuronal cell lines primed with TNF-α and IFN-γ found that *Bacopa* significantly reduce the expression of these proteins and mitigates cytotoxic damage Costantini et al. (2023).

Accumulation of Aβ peptides is a hallmark of AD, leading to abnormalities such as neuronal death, oxidative stress, and synaptic dysfunction (Asuku et al., 2022). Preclinical studies indicate that *Bacopa* can mitigate Aβ-driven cognitive deficits and oxidative injury in the hippocampus (Neto et al., 2024). In silico study suggests that bacosides in *BM* may interact with β-secretase, a rate-limiting enzyme in Aβ production, thus slowing the progression of AD (Sahu, 2024). *Bacopa*, according Dubey et al. (2023), can inhibit tau aggregation in vitro, reducing the formation of neurofibrillary tangles that are toxic to neurons. In addition, it lowers phosphorylation of glycogen synthase kinase-3β (GSK-3β), a kinase responsible for hyperphosphorylating tau proteins (Dubey et al., 2023). Studies also showed that *Bacopa* can upregulate molecular components such as BDNF, TrkB, and NMDA receptor involved in long-term potentiation (Neto et al., 2024). It was also found to increase hippocampal cell proliferation and neuroblast differentiation and thus support structural and functional plasticity (Srivastava & Jhelum, 2024).

Beyond its effect on the cholinergic pathway, *Bacopa* helps normalize stress-altered monoamines (5-HT, NA, dopamine) and corticosterone in animal models (Bhattacharya et al., 2008). It may therefore potentially improve mood and cognitive efficiency in prodromal stages of dementia. These cognitive effects are caused by the triterpenoid saponins (bacosides A/B) with contributions from flavonoids and alkaloids.

Table 5.1 and Figure 5.1 summarized the biological activities of *BM* in dementia management.

TABLE 5.1
Biological Activities of *Bacopa monnieri* in Dementia Management

Mechanism of Action	Active Compounds	Observed/Proposed Effects	References
Cholinergic modulation	Bacosides A and B, herpestine	Inhibits acetylcholinesterase (AChE); increases acetylcholine levels; improves memory and attention	Gościniak et al. (2025), Shoukat et al. (2023)
Antioxidant activity	Bacosides A/B, quercetin, asiatic acid	Scavenges free radicals; reduces lipid peroxidation; preserves neuronal integrity	Shinomol and Muralidhara (2011), Simpson et al. (2015)
Anti-inflammatory activity	Saponins, alkaloids	Suppresses NF-κB pathway; reduces pro-inflammatory cytokines.	Al-Ghraiybah et al. (2022), Costantini et al. (2023)
Anti-amyloidogenic effect	Bacosides, flavonoids	Inhibits β-secretase activity; reduces amyloid-β accumulation; protects synapses	Neto et al. (2024), Sahu (2024)
Anti-tau effects	Bacosides, cucurbitacins	Inhibits tau aggregation; reduces GSK-3β hyperphosphorylation	Dubey et al. (2023)
Neurotrophic and synaptic support	Bacosides, triterpenoids	Upregulates BDNF, TrkB, NMDA receptors; enhances long-term potentiation and neurogenesis	Srivastava and Jhelum (2024), Neto et al. (2024)
Mood and stress regulation	Bacosides, alkaloids	Normalizes monoamines (serotonin, i.e. 5-HT; noradrenaline, i.e. NA; dopamine, i.e. DA) and corticosterone; improves mood and cognition	Bhattacharya et al. (2008)

5.4 CLINICAL EVIDENCE OF *BACOPA MONNIERI* ROLE IN DIAGNOSED DEMENTIA

Asides preclinical studies on the potential therapeutic role of *BM* in dementia, several clinical studies have explored potential therapeutic role of *BM* in dementia. Evidences gathered so far are preliminary relative to gold standard pharmacotherapies. However, these evidences are pivotal to generating robust data validating efficacy of *BM*. Of note is the standardized bacosides contained in the *BM* extract, which ameliorate cognitive and memory dysfunction in dementia individuals.

FIGURE 5.1 Mechanisms of action of *Bacopa monnieri* in dementia management.

Basheer et al. (2022) showed that RCT and quasi-RCT provide the most consistent insights about *BM*, its extract or bioactive compound, compared with placebo or a cholinesterase inhibitor among individuals with dementia-associated AD and those with AD-associated mild cognitive impairment. Authors observed that *BM* supplementation improved memory in the aspect of attention, processing, and retention in the dementia patients. They also reported no major safety issues in the included trials as well as no significant difference between *BM,* placebo, or donepezil in AD patients. Prabhakar et al. (2020) reported in a 52-week, randomized, double-blind parallel-group phase 2b trial involving 48 patients that, based on PGI (postgraduate institute) memory scale and AD Assessment Scale–Cognitive Subscale (ADAS-Cog scores), there was no significant difference between *BM* and donepezil in cognitive function and quality of life measures of AD patients after one year. The therapeutic benefits of *BM* are attributed to its cholinergic modulatory properties and anti-inflammatory and antioxidant potentials. This assertion was also reinforced by Zanotta et al. (2014) in a multicenter prospective cohort trial. Authors recruited 104 subjects with mild cognitive impairment and tested a dietary supplement containing *BM,* vitamin E, phosphatidylserine, and astaxanthin daily for 60 days. Treatment over this period was associated with enhanced Mini-Mental State Examination (MMSE), ADAS-Cog scores, and clock drawing test scores, with memory tasks showing the greatest improvements and minimal adverse effects and supporting further controlled, long-term reports.

In another randomized controlled trail (n=62), Delfan et al. (2024) revealed that *BM* (160 mg/day for 2 months) at baseline, 1 month, and 2 months of cognitive assessment (Montreal Cognitive Assessment) and sleep evaluation (Pittsburgh Sleep Quality Index) significantly improved verbal fluency (month 2: p=0.003), attention (month 1: p=0.033; month 2: p=0.004), and total cognitive score at study end (p=0.029) compared to placebo. Authors notably concluded on the domain-specific beneficial effects of *BM* on cognition but not sleep, as quality of sleep did not show improvement through the experimental time course. Furthermore, reports from observational and open-label studies have shown that *BM* may mitigate neuropsychiatric symptoms, including restlessness, irritability, and fatigue thereby promoting dementia patients' quality of life. According to Neto et al. (2024), 22 clinical trials revealed that *BM* enhances mood, sleep, emotional regulation, cognitive functions, memory, attention, and mood through its synaptic-restorative, neuroprotective, antioxidant, and anti-apoptotic properties. However, these findings are inconsistent due to methodological limitations, including small sample size, variability in extract formulations, and short follow-up duration.

Largely, current beneficial clinical reports on *BM*'s effects on dementia are still modest in terms of cognitive and behavioral function. Nonetheless, there are insufficient data to reinforce conclusive therapeutic efficacy. As such, robust, multicenter RCT with consistent formulations and extensive follow-up durations are required before *BM* can be validated as a clinical candidate in the management of dementia.

5.5 THE SAFETY PROFILE OF *BACOPA MONNIERI*

As the interest in natural cognitive enhancers increases, *BM* (Brahmi) has transitioned from the realms of Ayurvedic medicine to the forefront of contemporary neuroscience. Renowned for its memory-boosting and adaptogenic characteristics, its effectiveness is progressively corroborated by clinical studies (Aguiar & Borowski, 2013). Nevertheless, a pivotal question persists for clinicians, researchers, and consumers alike: what is its safety for modern application? The literature suggests that while *BM* demonstrates a positive safety profile, its application requires careful consideration of dosage, context, and individual patient characteristics.

The most consistent observation across human trials is the exceptional tolerability of standardized *BM* extracts. A meta-analysis conducted by Kumar et al. (2016) aggregated data from various randomized controlled trials. The study concluded that adverse events were rare, mild, and temporary. The side effects most frequently reported are gastrointestinal in nature, including nausea, cramping, and diarrhea. These effects are often dose-dependent and can typically be alleviated if the supplement is taken with food or by starting therapy with a lower dose that is gradually increased (Kean et al., 2016). This implies that for the majority of healthy adults, *BM* is well-tolerated over durations generally ranging from 12 to 24 weeks, which is the standard length of most studies (Kean et al., 2016; Kumar et al., 2016).

The safety of any bioactive substance is inherently associated with its dosage. Contemporary clinical trials usually employ standardized extracts, often containing 50% bacosides (the presumed active ingredients), at doses varying from 300 to 450 mg/day (Kumar et al., 2016).

At these levels, no significant systemic toxicity is observed. Chronic toxicity studies conducted on animal support this conclusion; researches by Singh et al. (2018) and Shinomol and Muralidhara (2011) revealed that even at doses considerably exceeding the human equivalent, *BM* extract did not cause notable adverse alterations in blood biochemistry, hematological parameters, or the histology of essential organs in rodent models. This suggests a high therapeutic index, indicating a substantial margin between an effective dose and a potentially harmful one.

Nevertheless, safety encompasses more than just the absence of acute toxicity. There is need to consider polypharmacy, given the chances of possible interactions with pharmaceutical drugs since patients with dementia are most likely placed on two or more medications. An in vitro study indicated that *BM* may affect the activity of cytochrome P450 enzymes, a group of liver enzymes that metabolize a wide range of medications (Srivastava et al., 2020). This introduces a theoretical risk that it could modify the blood levels of medications such as anticoagulants (e.g., warfarin) or specific antidepressants. Although clinically significant interactions are infrequently reported in human trials, this highlights the importance for patients on long-term medications to seek advice from a healthcare provider prior to initiating supplementation (Singh, 2013). Caution is the wisest approach.

Safety information for vulnerable groups is limited, with a lack of clinical trials on *BM* for pregnant or breastfeeding women, leading to advisories against its use due to potential risks to fetal and infant development (Dutta & Basu, 2019). While traditional uses in children exist, modern clinical evidence is scarce, so pediatric use should be supervised by professionals. For older adults, *BM* appears safe for cognitive enhancement, but caution is advised due to polypharmacy and metabolic changes with age.

Product quality is crucial; the label "*Bacopa monnieri*" does not guarantee a standardized or safe product. Differences in cultivation and processing can lead to variations in potency and phytochemical profiles, with potential contaminants like heavy metals. Consumers and healthcare providers should choose products from reputable manufacturers that provide proof of standardization and third-party testing for purity (Smith & Oliver, 2021). Clinical safety is linked to high-quality extracts.

To summarize, *BM* is generally safe and well-tolerated as an herbal nootropic for healthy adults when used correctly, with mild and manageable side effects.

5.6 CONCLUSION

Dementia remains a challenging global health burden with limited effective therapeutic options. *BM* is a traditional Ayurvedic herb showing promising results in the management and treatment of dementia due to its neuroprotective and cognitive-enhancing properties. The plant is renowned for its phytochemical richness, especially in bacosides, alkaloids, flavonoids, and cucurbitacins. Both preclinical and clinical findings supported the benefits in dementia management through multiple mechanisms such as inhibition of amyloid-β and tau pathology, modulation of synaptic plasticity, neurogenesis, cholinergic modulation, and antioxidant and anti-inflammatory activities. The safety profile of *BM* is encouraging although caution is advised in cases of polypharmacy. Overall, *BM* represents a promising adjunctive or alternative candidate for dementia management. However, large-scale RCT are required to validate their therapeutic role.

REFERENCES

Aguiar, S., & Borowski, T. (2013). Neuropharmacological review of the nootropic herb *Bacopa monnieri*. *Rejuvenation Research*, *16*(4), 313–326.

Akintoye, O.O., Ajibare, A.J., Folawiyo, M.A., Asuku, A.O., Jimoh Abdulghaffar, H.O., Akintoye, A.O., Babalola, K.T., Idowu, O.O., & Oyiza, Y.G. (2023). Virgin coconut oil-supplemented diet reverses behavioural phenotypes of sodium benzoate-model of acetylcholinesterase dysfunction and cognitive impairment: Role of Nrf2/NfKb signaling. *Nigerian Journal of Physiological Sciences*, *37*, 225–233.

Al-Ghraiybah, N.F., Wang, J., Alkhalifa, A.E., Roberts, A.B., Raj, R., Yang, E., & Kaddoumi, A. (2022). Glial cell-mediated neuroinflammation in Alzheimer's disease. *International Journal of Molecular Sciences*, *23*(18), 10572. https://doi.org/10.3390/ijms231810572

Arvanitakis, Z., Shah, R.C., & Bennett, D.A. (2019). Diagnosis and management of dementia: Review. *JAMA*, *322*(16), 1589–1599. https://doi.org/10.1001/jama.2019.4782

Asuku, A.O., Adeyemo, M.B., Adeyemo, R.O., Agoro, S.A., Odugbesan, O.A., & Ayoola, S.A. (2022). Applications of stem cell therapy in neurodegenerative diseases. *European Journal of Biotechnology and Bioscience*, *10*(2): 17–24, 2321–9122.

Asuku, A.O., Ayinla, M.T., Gbonjubola, O.O., Babatunde, S.S., Olajide, T.S., & Oyerinde, T.O. (2024). Neurocognitive Aspects of Dementia. In: *Nutrition in Brain Aging and Dementia. Nutritional Neurosciences*, Springer, Singapore.

Asuku, A.O., Ogungbangbe, G.O., & Olajide, T.S. (2025). Effect of Happiness and Positive Emotions on Dementia. In: *Neuroscience of Feelings. Progress in Brain Research*, Elsevier, https://doi.org/10.1016/bs.pbr.2025.03.001

Ayilara, G.O., & Owoyele, B.V. (2025). Effectiveness of *Bacopa Monnieri* (Brahmi) in the management of schizophrenia: A systematic review. *Nutritional Neuroscience*, *28*(7), 788–795. https://doi.org/10.1080/1028415X.2024.2421782

Basheer, A., Agarwal, A., Mishra, B., Gupta, A., Srivastava, M.V.P., Kirubakaran, R., & Vishnu, V. (2022). Use of *Bacopa monnieri* in the treatment of dementia due to Alzheimer disease: Systematic review of randomized controlled trials. *Interactive Journal of Medical Research*, *11*(2), e38542.

Bhattacharya, S.K., Bhattacharya, A., Kumar, A., & Ghosal, S. (2000) Antioxidant activity of *Bacopa monniera* in rat frontal cortex, striatum and hippocampus. *Phytotherapy Research*, 14(3):174–179. https://doi.org/10.1002/(sici)1099-1573(200005)14:3<174::aid-ptr624>3.0.co;2-o

Bhattacharya, S.K., Bhattacharya, D., Kumar, A., & Ghosal, S. (2008). Effect of *Bacopa monniera* on stress-induced changes in plasma corticosterone and brain monoamines in rats. *Journal of Ethnopharmacology*, *71*(3), 287–291.

Chen, Z.R., Huang, J.B., Yang, S.L., & Hong, F.F. (2022). Role of cholinergic signaling in Alzheimer's disease. *Molecules (Basel, Switzerland)*, *27*(6), 1816. https://doi.org/10.3390/molecules27061816

Costantini, E., Jarlapoodi, S., Serra, F., Aielli, L., Khan, H., Belwal, T., Falasca, K., & Reale, M. (2023). Neuroprotective potential of *Bacopa monnieri*: Modulation of inflammatory signals. *CNS & Neurological Disorders – Drug Targets*, *22*(3), 441–451.

Delfan, M., Kordestani-Moghaddam, P., Gholami, M., Kazemi, K., & Mohammadi, R. (2024). Evaluating the effects of *Bacopa monnieri* on cognitive performance and sleep quality of patients with mild cognitive impairment: A triple-blinded, randomized, placebo-controlled trial. *EXPLORE*, *20*(5), 102990.

Dubey, P., Upadhyaya, R., & Prakash, A. (2024). Phytochemicals activity of *Bacopa monnieri* (L.) Pennell: A brief review. *International Journal of Biotechnology and Microbiology*, *6*(2), 5–8.

Dubey, T., Kushwaha, P., Thulasiram, H.V., Chandrashekar, M., & Chinnathambi, S. (2023). *Bacopa monnieri* reduces Tau aggregation and Tau-mediated toxicity in cells. *International Journal of Biological Macromolecules*, *234*, 123171. https://doi.org/10.1016/j.ijbiomac.2023.123171

Dutta, S., & Basu, U.P. (2019). Therapeutic potential of *Bacopa monnieri*: A review. *International Journal of Research in Pharmaceutical Sciences*, *10*(3), 1823–1828.

Eraiah, M.M., Shekhar, H.C., Joshua, L., & Thomas, J.V. (2024). Effect of *Bacopa monnieri* extract on memory and cognitive skills in adult humans: A randomized, double-blind, placebo-controlled study. *Journal of Psychiatry and Cognitive Behaviour*, *8*, 168.

Fatima, U., Roy, S., Ahmad, S., Ali, S., Elkady, W.M., Khan, I., Alsaffar, R.M., Adnan, M., Islam, A., & Hassan, M.I. (2022). Pharmacological attributes of *Bacopa monnieri* extract: Current updates and clinical manifestation. *Frontiers in Nutrition*, *9*, Article 972379.

Gościniak, A., Stasiłowicz-Krzemień, A., Szeląg, M., Pawlak, J., Skiera, I., Kwiatkowska, H., Nowak, N., Bernady, K., Trzaskoma, P., Zimak-Krótkopad, O., & Cielecka-Piontek, J. (2025). *Bacopa monnieri*: Preclinical and clinical evidence of neuroactive effects, safety of use and the search for improved bioavailability. *Nutrients*, *17*(11), 1939. https://doi.org/10.3390/nu17111939

Kean, J.D., Downey, L.A., & Stough, C. (2016). A systematic review of the Ayurvedic herb *Bacopa monnieri* in child and adolescent populations. *Complementary Therapies in Medicine*, *29*(1), 56–62.

Kolape, S.B., Ghodake, S.R., & Kamble, H.V. (2025). Pharmacological insights into *Bacopa monnieri* as a potential antiepileptic agent. *International Journal of Research Publication and Reviews*, *6*(8), 1426–1432.

Korovesis, D., Rubio-Tomás, T., & Tavernarakis, N. (2023). Oxidative stress in age-related neurodegenerative diseases: An overview of recent tools and Findings. *Antioxidants (Basel, Switzerland)*, *12*(1), 131. https://doi.org/10.3390/antiox12010131

Kumar, N., Abichandani, L.G., & Thawani, V. (2016). Efficacy of standardized extract of *Bacopa monnieri* (Bacognize®) on cognitive functions of medical students: A six-week, randomized placebo-controlled trial. *Evidence-Based Complementary and Alternative Medicine*, *2*(1), 410–423.

Manap, A.S.A., Vijayabalan, S., Madhavan, P., Chia, Y.Y., Arya, A., Wong, E.H., Rizwan, F., Bindal, U., & Koshy, S. (2019). *Bacopa monnieri*, a neuroprotective lead in Alzheimer disease: A review on its properties, mechanisms of action, and preclinical and clinical studies. *Drug Target Insights*, *13*, 1–13.

Mittal, C., Mondal, M., & Chatterjee, S. (2025). A review: Phytochemistry and medicinal activities of *Bacopa monnieri*. *International Journal of Advanced Biochemistry Research*, *9*(3), 550–555.

Neto, L.J.V., de Araujo, M.R., Junior, R.C.M., Machado, N.M., Joshi, R.K., dos Santos Buglio, D., & Barbalho, S.M. (2024). Investigating the neuroprotective and cognitive-enhancing effects of *Bacopa monnieri:* A systematic review focused on inflammation, oxidative stress, mitochondrial dysfunction, and apoptosis. *Antioxidants*, *13*(4), 393.

Nishanth, B.J., Vijayababu, P., & Kurian, N.K. (2023). *Bacopa monnieri* extract as a neuroprotective and cognitive enhancement agent. *International Journal of Drug Discovery and Pharmacology*, *3*(4), 1–12.

Paul, P., Majumdar, S., Basu, A., Jha, S., & Halder, M. (2025). Bacosides and Neuroprotection. In *Natural Products* (pp. 1–30). Springer.

Peth-Nui, T., Wattanathorn, J., Muchimapura, S., Tong-Un, T., Piyavhatkul, N., Rangseekajee, P., Ingkaninan, K., & Vittaya-Areekul, S. (2012). Effects of 12-week *Bacopa monnieri* consumption on attention, cognitive processing, working memory, and functions of both cholinergic and monoaminergic systems in healthy elderly volunteers. *Evidence-Based Complementary and Alternative Medicine: eCAM*, *2012*, 606424. https://doi.org/10.1155/2012/606424

Prabhakar, S., Vishnu, V.Y., Modi, M., Mohanty, M., Sharma, A., Medhi, B., & Avasthi, A. (2020). Efficacy of *Bacopa monnieri* (Brahmi) and donepezil in Alzheimer's disease and mild cognitive impairment: A randomized double-blind parallel Phase 2bstudy. *Annals of Indian Academy of Neurology*, *23*(6), 767–773.

Puri, V., Kanojia, N., Sharma, A., Huanbutta, K., Dheer, D., & Sangnim, T. (2022). Natural product-based pharmacological studies for neurological disorders. *Frontiers in Pharmacology*, *13*, 1011740. https://doi.org/10.3389/fphar.2022.1011740

Sahu, A. (2024). In silico analysis and docking study of the active phytocompounds of *Bacopa monnieri* against Alzheimer disease. Research & reviews. *Journal of Computational Biology*, *13*(2), 1–13.

Sairam, K., Dorababu, M., Goel, R.K., & Bhattacharya, S.K. (2002). Antidepressant activity of standardized extract of Bacopa monniera in experimental models of depression in rats. *Phytomedicine: International Journal of Phytotherapy and Phytopharmacology*, *9*(3), 207–211. https://doi.org/10.1078/0944-7113-00116

Sekhar, V.C., Viswanathan, G., & Baby, S. (2019). Insights into the molecular aspects of neuroprotective bacoside A and bacopaside I. *Current Neuropharmacology*, *17*(5), 438–446.

Shinomol, G.K., & Muralidhara (2011). *Bacopa monnieri* modulates endogenous cytoplasmic and mitochondrial oxidative markers in prepubertal mice brain. *Phytomedicine: International Journal of Phytotherapy and Phytopharmacology*, *18*(4), 317–326. https://doi.org/10.1016/j.phymed.2010.08.005

Shoukat, S., Zia, M.A., Uzair, M., Attia, K.A., Abushady, A.M., Fiaz, S., Ali, S., Yang, S.H., & Ali, G.M. (2023). *Bacopa monnieri*: A promising herbal approach for neurodegenerative disease treatment supported by *in silico* and *in vitro* research. *Heliyon*, *9*(11), e21161. https://doi.org/10.1016/j.heliyon.2023.e21161

Simpson, T., Pase, M., & Stough, C. (2015). *Bacopa monnieri* as an antioxidant therapy to reduce oxidative stress in the aging brain. *Evidence-Based Complementary and Alternative Medicine: eCAM*, *2015*, 615384. https://doi.org/10.1155/2015/615384

Singh, A., Singh, S.K., & Verma, A. (2018). Neuroprotective potential of *Bacopa monnieri* in Parkinson's disease: A review. *International Journal of Nutrition, Pharmacology, Neurological Diseases*, *8*(1), 11–21.

Singh, H.K. (2013). Brain enhancing ingredients from Ayurvedic medicine: Quintessential example of *Bacopa monniera*, a narrative review. *Nutrients*, *5*(2), 478–497.

Smith, A., & Oliver, G. (2021). Quality control of herbal supplements: A case study with *Bacopa monnieri*. *Journal of Dietary Supplements*, *18*(2), 113–125.

Srivastava, P., Yadav, R.S., & Chandravanshi, L. (2020). *Bacopa monnieri* modulatesantioxidant responses and protects against cadmium-induced oxidative stress in rat brain. *Biological Trace Element Research*, *19*(5), 483–492.

Srivastava, S., & Jhelum, P. (2024). Restoration of hippocampal adult neurogenesis by CDRI-08 (*Bacopa monnieri* extract) relates with the recovery of BDNF-TrkB levels in male rats with moderate-grade hepatic encephalopathy. *Neurochemistry International*, *172*, 105406.

Zanotta, D., Puricelli, S., & Bonoldi, G. (2014). Cognitive effects of a dietary supplement made from extract of *Bacopa monnieri,* astaxanthin, phosphatidylserine, and vitamin E in subjects with mild cognitive impairment: A noncomparative, exploratory clinical study. *Neuropsychiatric Disease and Treatment*, *10*, 225–230.

6 Curcumin for Managing and Treating Dementia

Ruqayyah Yetunde Ibiyeye, Abdullahi Mohammed, Fatimo Ajoke Sulaimon, Ambali Olanrewaju Danwahab, Olabisi Elizabeth Ayoola, and Olugbemi Tope Olaniyan

6.1 INTRODUCTION

Bioresources are highly renewable and biodegradable materials that occur naturally. They include crops, agricultural and industrial wastes and marine resources such as fishes and aquatic crustaceans. Bioactive compounds gotten from plants are referred to phytochemicals; valuable bioresources endowed with a number of health benefits such as mitigation and treatment of diabetes, hyperlipidemia, cancer, neurodegenerative diseases and so on (Kumar et al., 2023). Researchers have reported that phytochemicals filled with antioxidants and anti-inflammatory effects such as curcumin may cause age-related cognitive decline and susceptibility of emerging various diseases (Porro & Panaro, 2023).

Curcuma longa (also known as turmeric) is a rhizome, non-woody perennial plant that is of family Zingiberaceae. The bright yellow orange color of turmeric is known as curcuminoids, which is a fat-soluble polyphenolic pigment. Curcuminoids constitute about 2–9% of the turmeric plant, which is a mixture of three main compounds, curcumin being the highest constituent, accounting for 77% of the entire curcuminoid. Other compounds are desmethoxycurcumin (17%) and bisdesmethoxycurcumin (3%). Curcumin is in enolate form which makes it to be in alkaline condition predominantly (Iweala et al., 2023). Curcumin has IUPAC nomenclature of (1E,6E)-1,7-bis(4-hydroxy-3-methoxyphenyl)-1,6-heptadiene-3,5-dione, with chemical formula ($C_{21}H_{20}O_6$) and its molecular weight is 368.38 g/mol. Curcumin acquired most of its biological and therapeutic properties because of its chemical structural makeup such as phenolic hydroxyl groups, the central bis-α, β-unsaturated β-diketone dual conjugated bonds, and methoxy groups (Porro & Panaro, 2023).

The first time curcumin was isolated was in 1815 and got purified to a crystal-like substance in 1870. It was referred to as 'Wonder drug of life'. Curcumin is a known spice, food dye and addictive, which was also used in ancient times for a broad spectrum of diseases, such as the treatment of liver and gastrointestinal organ inflammation, and wound healing. It has been employed by vast number of communities, mostly in the Asian countries as far back as 2500 years (Sharifi-Rad et al., 2020). Advanced medical technology has proved curcumin to have wide range of biological and biomedical applications such as anti-inflammatory, antioxidant, anticancerous, immunomodulatory, brain-protective, antiproliferative, antibacterial and chemoprotective properties due to its ability to regulate many signaling molecules that modulates biological activities. Importantly, curcumin has now been considered a drug candidate for neurological diseases such as Parkinson's disease (PD), Huntington's disease (HD), amyotrophic lateral sclerosis (ALS), multiple sclerosis (MS), Alzheimer's disease (AD) and prion disease (Genchi et al., 2024).

Dementia is fast becoming a significant growing crisis in modern medicine, with its financial implications skyrocketing. The most prevalent form is AD, then vascular dementia (VaD), Lewy body dementia (LBD) and PD (Asuku et al., 2022, 2024). Mixed pathologies are more predominant. There are lots of pathogenesis interplaying in the progression of dementia, which includes aging, brain ischemia, experimental toxins and oxidative stress (Asuku et al., 2025). However, there are limited approved medications for dementia, particularly cholinesterase inhibitors and memantine.

 DOI: 10.1201/9781003657385-6

Additionally, other drugs such as antipsychotics, antidepressants, tranquillizers and hypnotics are often prescribed (Alzobaidi et al., 2021).

6.2 MECHANISM OF ACTION OF CURCUMIN IN TREATING DEMENTIA

Curcumin exhibits its therapeutic potential in mitigating dementia by targeting key pathological hallmark such as impediment of buildup of amyloid-beta (Aβ) and tau proteins. Aβ accumulation is linked to compromised synaptic function, reduced neurite outgrowth, and cerebral atrophy, all of which are associated with dementia, hence its mitigation will help in dementia cases (Abdullahi & Oluwole, 2022). Furthermore, curcumin exhibits a range of mechanisms that helps the modulation of oxidative stress, inflammation, brain-derived neurotrophic factor levels (BDNF) and various signaling pathways (Sarraf et al., 2019).

Some of those mechanisms are highlighted below:

6.2.1 Anti-Inflammatory Activities of Curcumin in Dementia

Nuclear factor kappa-light-chain-enhancer of activated B cells (NF-κB) typically responds to cellular stress and enhances the activities of pro-inflammatory cytokines (Akintoye et al., 2023). Curcumin has proved its ability to decrease the activation of NF-κB, thereby mitigating amount of inflammatory cytokines such as tumor necrosis factor alpha (TNF-α) and interleukins (IL-1, 6) as shown in Table 6.1. Curcumin also modulates the activities of cyclooxygenase-2 (COX-2) as well as inducible nitric oxide synthase (iNOS). Neurodegenerative diseases such as dementia lead in these cytokine activities (Maugeri et al., 2023; Rosińska-Lewandoska et al., 2024).

6.2.2 Curcumin Reduces Oxidative Stress

Curcumin functions as a potent antioxidant, by neutralizing reactive oxygen (or nitrogen) species (ROS/RNS) (Table 6.1). It enhances activity of detoxifying enzymes, such as superoxide dismutase (SOD), glutathione peroxidase (GPX) and so on, thus maintaining redox homeostasis in neuronal cells, which in turn contributes to the treatment of neuronal damage and decreased cognition in dementia (Dhankhar et al., 2023; Kothawade et al., 2023).

Additionally, curcumin activates the pathway responsible for Nrf2-antioxidant response element (ARE) signaling, that is critical to cellular defense against oxidative damage, Usually, Keap1 protein binds to Nrf2, and leading to its breakdown. However, in response to oxidative stress, curcumin disrupts this interaction, which leads to the accumulation of Nrf2, translocation to the nucleus, and initiates the transcription of various cytoprotective genetic factor, including those encoding enzymes that combat oxidative stress (Shahcheraghi et al., 2021).

6.2.3 Curcumin Enhances Energy Metabolism

Curcumin assists in the activation of AMP-activated protein kinase (AMPK) signaling pathway, that is fundamental to cellular energy homeostasis and metabolic regulation. It promotes catabolic processes while inhibiting the anabolic pathways, leading to improved energy metabolism and reduced neuronal damage associated with metabolic dysregulation in dementia (Xue et al., 2025).

Curcumin enhances glucose metabolism by modulating insulin pathways, scavenging ROS and inhibit lipid peroxidation, thus augmenting the capacity of neurons to utilize glucose more efficiently (Kesharwani et al., 2025; Xue et al., 2025). Also, curcumin activates phosphoinositide 3-kinase/Akt pathway and elevates BDNF which bolster neuronal health by enhancing glucose uptake and energy production, which helps in the enhancement of cognitive function and helps in the treatment of dementia (Ahmad et al., 2019; Abdullahi & Oluwole, 2022; Aditi et al., 2022).

TABLE 6.1
Summary of the Mechanisms of Action of Curcumin in the Treatment and Management of Dementia, Showing Molecular Targets, Signaling Pathways and the Impact on Hallmark Features of Dementia

Mechanism	Molecular/Cellular Target	Pathways Involved	Effect/Outcome	Relevance to Dementia
Anti-inflammatory	Microglia, astrocytes, cytokines (TNF-α, IL-6, IL-1β)	NF-κB inhibition	Decreased neuroinflammation	Reduces neuronal injury and progression of neurodegeneration
Antioxidant	ROS, RNS, mitochondrial enzymes	Nrf2-ARE pathway activation	Scavenging free radicals; enhancing antioxidant enzyme expression	Protects neurons from oxidative damage.
Anti-amyloidogenic	Aβ peptides, amyloid plaques	Inhibition of β-secretase and aggregation	Prevents Aβ plaque formation and enhances disaggregation	Slows or reverses Alzheimer's disease pathology
Tau protein modulation	Hyperphosphorylated tau	GSK-3β inhibition	Reduces tau clusters and formation of neurofibrillary tangle	Targets Alzheimer's pathology
Neurogenesis and synaptic plasticity	BDNF, synaptic proteins	ERK/MAPK pathway, CREB activation	Enhances neurogenesis and synaptic strength	Improves learning, memory, and cognition
Metal ion chelation	Iron, copper, zinc	Direct chelation by phenolic groups	Prevents metal-induced Aβ aggregation and oxidative stress	Decreases toxicity associated with metal imbalance in AD brains
Mitochondrial protection	Mitochondrial membrane	Mitochondrial biogenesis via PGC-1α	Stabilizes mitochondrial function and reduces apoptosis	Sustains energy supply critical for cognitive functions
Epigenetic modulation	Histone acetylation, DNA methylation	HDAC inhibition, DNMT modulation	Alters gene expression favorably	May influence memory and learning via chromatin remodeling
Cholinesterase inhibition	Acetylcholinesterase (AChE), butyrylcholinesterase	Enzyme binding	Increases acetylcholine levels in synaptic cleft	Enhances cholinergic neurotransmission
Regulation of apoptosis and necrosis	Caspases (caspase-3, -9), Bcl-2 family proteins, cytochrome c, RIPK1/RIPK3	PI3K/Akt, JNK, p53, mitochondrial apoptotic pathway	Inhibits pro-apoptotic signals; stabilizes mitochondria; prevents necroptosis	Reduces neuronal loss by preventing programmed and unregulated cell death in the brain

6.2.4 Curcumin Modulates Apoptosis

Curcumin has been reported to increase the expression of Bcl-2 (an anti-apoptotic protein) and decrease Bax and caspase-3 (which are pro-apoptotic proteins). This may promote cell survival by inhibiting the apoptotic signaling cascade, hence protecting neuronal cells from death under stress conditions (Mahnashi et al., 2024). Moreover, curcumin exhibits powerful neuroprotective effects against oxidative stress, a significant contributor to neuronal cell death. It has been shown earlier in this chapter to counteract ROS production. This is linked to both apoptosis and necrosis. By reducing oxidative stress, curcumin helps maintain neuronal integrity and functionality (Nagu et al., 2022).

Curcumin attenuates excitotoxicity, characterized by excessive stimulation of glutamate receptors leading to neuronal death. It modulates glutamate receptors and prevents calcium influx into cells, thus reducing excitotoxic cell death. This action is significant in conditions like stroke, dementia and traumatic brain injury, where excitotoxicity is prevalent. Curcumin inhibits necroptosis, a form of programmed necrosis. Furthermore, curcumin protects against iron-induced neurotoxicity by reducing necroptotic cell death in primary cortical neurons. Hence, curcumin offers neuroprotection by targeting multiple forms of cell death (Magdaleno Roman & Chapa González, 2024).

6.3 CHALLENGES OF USING CURCUMIN FOR TREATING DEMENTIA

Nutraceutical bioavailability is firmly reliant on the transformation that occurs in the GIT and bioaccessibility. Curcumin, a distinguished polyphenol, has numerous bioactive properties but a poor pharmacokinetic profile, including a low absorption rate in the GIT, wide-ranging reductive conjugative metabolism in the liver, and a high rate of elimination through the gall bladder and clearance. It also exhibits reduced solubility in H_2O and unstable chemical structure, which compromise therapeutic potential, even at high doses. A reductase in the hepatocytes and enterocytes reduces the double bond in curcumin to dihydrocurcumin, tetrahydrocurcumin, hexahydrocurcumin, and octahydrocurcumin. After oral ingestion, curcumin reaches its serum peak 1–2 hours afterward and declines after 12 hours, hence it has a low half-life (Lopresti, 2018; Agrawal & Jaiswal, 2022). Research has helped improved bioavailability, permeability, circulation, half-life and basic metabolic process through the establishment of curcumin formulation (Figure 6.1). The formulations accept chemical curcumin derivatives and analogs that have metabolic adjuvants, micelles, liposomes, nanoparticles, nanostructured lipid carriers and phospholipid complexes (El-Saadony et al., 2023).

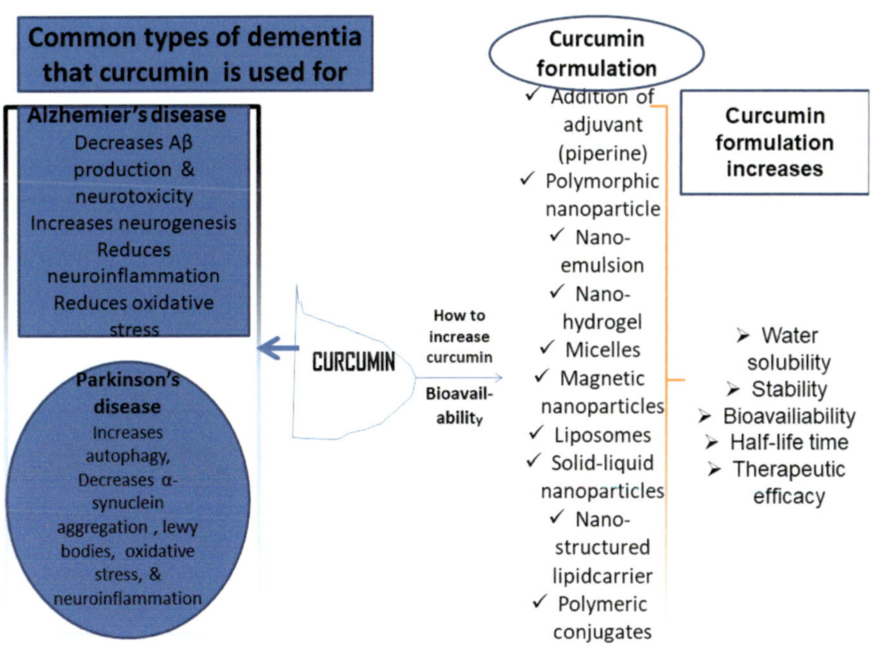

FIGURE 6.1 Chart showing the use of curcumin and its formulation for treating dementia.

6.3.1 Factors Responsible for Challenges from the Use of Curcumin in Mitigating Dementia and Solutions

6.3.1.1 Poor Bioavailability

Curcumin therapeutic activities are dependent on its bioavailability. Bioavailability is strictly reliant on the change that occurs in the digestive system. Factors responsible for poor bioavailability include poor adsorption, fast metabolism, structural instability and systemic elimination (Lopresti, 2018). Bioavailability of curcumin can be increased by the use of adjuvant such as administration of curcumin and piperine simultaneously; these increase the solubility by stimulating bile secretion; piperine is an effective inhibitor of cytochrome P450, hence there is reduction in the rate of metabolism, and facilitate membrane permeability. Also, combining curcumin with dietary lipids such as lecithin-rich ingredients like eggs, vegetable oil, yogurt, buttermilk helps to increase its bioaccessibility (Agrawal & Jaiswal, 2022), although, buttermilk is more effective than yogurt. Hence, the addition of adjuvant such as piperine, yogurt and buttermilk helps to inhibit metabolism, enhance its solubility, increase half-life and improve its cellular uptake. Furthermore, formulations such as nanoparticle-based curcumin, liposomal curcumin and xanthan–chitosan nanofibers increased the bioavailability of curcumin by about 70 times, 8 times and 4 times more, respectively, compared to free curcumin (Pratti et al., 2024). Organic phospholipid bilayer is used for microcapsule of curcumin, the organic phospholipid bilayer is synonymous to cell membrane which has hydrophilic, hydrophilic and amphiphilic molecules embedded in them, rate of absorption is increased (Rinkunaite et al., 2021). Furthermore, esterification of phenyl hydroxyl group, which changes analog of curcumin, helps to protect curcumin's phase II metabolism and makes it more stable hence it makes the curcumin derivative has greater activity and more bioavailable (Agrawal & Jaiswal, 2022).

Blood–brain barrier (BBB) penetration: The limited capability of curcumin to pass through the BBB poses a significant challenge for transporting curcumin to the brain after ingestion. Melt-crystallization of curcumin can enhance its solubility by 2627-fold compared to pure curcumin, potentially increasing its ability to cross the BBB. Combining low-intensity-focused ultrasound with curcumin nanobubbles to encapsulate curcumin can also open the BBB, enabling curcumin delivery to the brain and achieving a cumulative release rate of 30% within 6 hours. The advantages of using LIFU-induced BBB opening include its non-invasive nature, its capacity for indigenous and proficient drug delivery and its suitability for both small and large molecular-sized drugs (Yan et al., 2021). Liposomal curcumin and organic phospholipid bilayer used for microcapsule formulation of curcumin lead to increased rate of absorption and the ability of curcumin to pass through BBB (Rinkunaite et al., 2021; Pratti et al., 2024). Surfactant micelles are colloidal particles with hydrophobic interior, skillful of solubilizing poorly polarized molecules such as curcumin, as well as a water-loving exterior that is readily dispersed in water. They have thermodynamically favorable systems, which makes the formulation stable indefinitely. Micelles (Tween 80 and Surfynol 465) are used to cover and stabilize curcumin (Ryu et al., 2021).

6.4 EVIDENCE OF CURCUMIN INTERVENTION FROM PRECLINICAL STUDIES

Preclinical studies have established a variety of mechanisms through which curcumin can exhibit beneficial properties on cognition, particularly in cases of neurodegenerative diseases such as dementia. Animal models are used to investigate the biochemical and physiological impacts of curcumin supplementation on cognition and brain health (Heidari et al., 2023).

Studies stated that curcumin can reduce oxidative stress and inflammation in the brains of animals as observed in animal models, which in turn influences cognition, thereby mitigating the pathogenesis of dementia. By modulating neuroinflammatory responses, curcumin could potentially mitigate neuronal damage and promote cognitive health (Kotha & Luthria, 2019; Kothawade et al., 2023).

In preclinical models, hormetic effect has been attributed to curcumin, where low doses provide significant cognition while increased doses lead to diminished effects or give side effects. This non-linear dose-response association underscores the importance of dosage in curcumin's efficacy, suggesting that optimal dosing is crucial for maximizing its cognitive-enhancing properties (Das et al., 2023).

Emerging evidence points to the gut–brain axis as a critical pathway through which curcumin may exert its effects on cognition. Curcumin may enhance gut health, which in turn can positively influence brain function. These indicate that nutritional interventions show a major role in mitigating cognitive decline (Mukherjee et al., 2021).

Behavioral indices in animal models have shown improvement in cognition in animal models treated with curcumin. These improvements include enhanced learning and memory capabilities, often assessed through various cognitive tests such as maze navigation and recognition tasks, as hypothesized by some studies that curcumin supplementation can improve cognitive outcomes (Abdullahi & Oluwole, 2022).

A comparative analysis of curcumin formulations for dementia shows different delivery methods and formulations have varying efficacy. Some studies reported cognitive improvements with high doses of free curcumin administered over extended periods, others could not substantiate same, indicating the necessity for standardized methodologies and larger sample sizes to draw conclusive evidence (Far et al., 2024). Curcumin formulations, such as curcumin-glucoside-methoxylated (CGM) supplements, exhibit ability to enhance cognitive performance as measured by the Mini-Mental State Examination scores. Notably, a study using 400 mg CGM for six months indicated significant benefits over placebo controls (Stohs et al., 2020).

Use of standard curcumin-complex as a control provides a clearer understanding of the potential advantages offered by bio-enhanced formulations in terms of bioavailability and clinical outcomes (Han et al., 2024). Literature also emphasizes that while various proprietary formulations like Meriva®, Longvida® and Theracumin® demonstrate improved bioavailability compared to standard curcumin, the clinical significance of these enhancements remains debated. For instance, trials using Longvida® at a dosage of 0.4 g/day reported improved working memory, suggesting that lower doses of bio-enhanced curcumin can mitigate gastrointestinal issues often associated with higher dosages of traditional curcumin (Mukherjee et al., 2021; Das et al., 2023).

Furthermore, several experiments indicate the lack of consistency in measuring biochemical markers, which could elucidate the molecular tools behind the effect of curcumin on cognition. Absence of standardized biomarker assessment could obscure the potential benefits and mechanisms of action related to curcumin supplementation in dementia patients (Mukherjee et al., 2021; Heidari et al., 2023).

6.5 CLINICAL TRIALS

Curcumin has gained attention as a possible beneficial agent in the management of dementia because of its anti-inflammatory, antioxidant and Aβ-lowering properties as discussed earlier. However, there are disappointments in the level of efficacy of curcumin when subjected to preclinical and clinical trials. But, recent studies utilizing improved formulations and targeted cohorts have begun to yield promising results.

Recent advancements in curcumin formulations help to enhance its pharmacokinetic properties, including bioavailability and water solubility, while curcumin analogs and nanoencapsulation techniques facilitate better drug delivery (Awad et al., 2025). For instance, injectable polymeric micro-particles demonstrated sustained curcumin levels in the bloodstream and improved BBB penetration in animal studies (Ahmad et al., 2019). BCM-95 and Longvida have emerged as the most promising curcumin formulation. Data from human trials also supports this (Keservani & Lohani, 2025).

Timing of intervention is a major influence in the efficacy of curcumin in treating dementia. Most studies focus on treating individuals already diagnosed with dementia, in which significant neuronal loss would have occurred. However, emerging research suggests that using curcumin for individuals in the preclinical or prodromal phases may yield better outcomes, as interventions introduced earlier could help slow down pathogenic mechanisms before substantial synaptic and neuronal loss occurs (Aquino, 2021).

Given the dynamic nature of epigenetic alterations associated with dementia, curcumin's potential to induce epigenetic changes could offer additional avenues for disease prevention (Kesharwani et al., 2025).

Clinical trials indicate that curcumin is generally safe for consumption, with adverse events typically reported as mild and comparable to placebo. It is also worthy to note that some individuals experience gastrointestinal discomfort, which resolve without the need for dose adjustment. A dose-escalation study involving children and adolescents with inflammatory bowel disease (IBD) administered curcumin of about 4000 mg/day for 21 days without serious reactions. Only two patients reported mild increased gassiness, and no laboratory values fell outside normal ranges during the study. Furthermore, curcumin is safe at a daily dose of 30 mg/kg of turmeric powder for children aged 7 years and older, and 500 mg of curcumin per day for adolescents for ten weeks (Ancona et al.,2023).

Curcumin can interact with various medications, including anti-diabetic and blood-thinning drugs. The potential for cumulative effects with specific medications demands caution when co-administering curcumin with the following classes of drugs: alkylating agents, anti-tumor antibiotics, and topoisomerase I inhibitors (Wiggers et al., 2017).

6.6 CURRENT STATUS AND FUTURE DIRECTIONS

Curcumin has garnered interest for its possible beneficial effects in mitigating dementia, particularly AD. Current research suggests that curcumin can exert beneficial effects through multiple mechanisms, such as regulation of BDNF and anti-inflammatory markers (Kothawade et al., 2023). Specifically, curcumin has been shown to reduce pro-inflammatory markers and reduce levels of Aβ (Aβ42) and tau; they are implicated in cognition and neurodegeneration associated with AD (Su et al., 2022; Sadeghi et al., 2023).

Various degrees of efficacy were ascertained by human trials. Curcumin could have multitargeted mode of action in improving cognitive and locomotive function, although challenges remain due to limited human studies and the complexity of conducting trials with elderly populations (Hegde et al., 2023).

Furthermore, the use of ultrasound to bypass BBB presents a promising non-invasive method for delivering curcumin, particularly in neurodegenerative disorders; the practicality and security of such techniques are still under investigation. The method allows for targeted drug delivery and could enhance the therapeutic effects of curcumin in combating conditions like PD (Yan et al., 2021).

Hence, comprehensive human trials are needed to confirm the safety and efficacy of curcumin for dementia treatment conclusively. Further investigation should focus on explaining the precise operating principle, improving dosing strategies, and discovering combination therapies which is likely to enhance curcumin's therapeutic potential while minimizing adverse effects. Overall, while curcumin shows promise as a treatment for dementia, ongoing investigations are critical to overcoming existing challenges and fully realizing its benefits.

REFERENCES

Abdullahi, A.M., & Oluwole, B.A. (2022). The Effects of Flavonoids in Curcumin on Neurobehavioral Deficits in Insulin-Resistant Rats. *The Journal of Neurobehavioral Sciences*, *9*(2), 51–57. https://doi.org/10.4103/jnbs.jnbs_17_22

Aditi, K., Singh, A., Shakarad, M.N., & Agrawal, N. (2022). Management of Altered Metabolic Activity in Drosophila model of Huntington's Disease by Curcumin. *Experimental Biology and Medicine (Maywood, N.J.)*, *247*(2), 152–164.

Agrawal, N., & Jaiswal, M. (2022). Bioavailability Enhancement of Curcumin via Esterification Processes: A Review. *European Journal of Medicinal Chemistry Reports*, *6*, 100081. https://doi.org/10.1016/j.ejmcr.2022.100081

Ahmad, S.S., Khan, H., Danish Rizvi, S.M., Ansari, S.A., Ullah, R., Rastrelli, L., Mahmood, H.M., & Siddiqui, M.H. (2019). Computational Study of Natural Compounds for the Clearance of Amyloid-Beta: A Potential Therapeutic Management Strategy for Alzheimer's Disease. *Molecules (Basel, Switzerland)*, *24*(18), 3233. https://doi.org/10.3390/molecules24183233

Akintoye, O.O., Ajibare, A.J., Folawiyo, M.A., Asuku, A.O., Abdulghaffar, J., Akintoye, H.O., Babalola, A.O., Idowu, K.T., & Oyiza, O.O. (2023). Virgin Coconut Oil-Supplemented Diet Reverses Behavioural Phenotypes of Sodium Benzoate-Model of Acetylcholinesterase Dysfunction and Cognitive Impairment: Role of Nrf2/NfKb Signaling. *Nigerian Journal of Physiological Sciences*, *37*, 225–233. https://doi.org/10.54548/njps.v37i2.10

Alzobaidi, N., Quasimi, H., Emad, N.A., Alhalmi, A., & Naqvi, M. (2021). Bioactive Compounds and Traditional Herbal Medicine: Promising Approaches for the Treatment of Dementia. *Degenerative Neurological and Neuromuscular Disease*, *11*, 1–14. https://doi.org/10.2147/DNND.S299589

Ancona, S., Signa, S., Longo, C., Cangemi, G., Carfora, R., Drago, E., La Rosa, A., Crocco, M., Chiaro, A., Gandullia, P., & Arrigo, S. (2023). Dose escalation of Adalimumab as a Strategy to Overcome Anti-Drug Antibodies: A Case Report of Infantile-Onset Inflammatory Bowel Disease. *World Journal of Gastroenterology*, *29*(38), 5428–5434. https://doi.org/10.3748/wjg.v29.i38.5428

Asuku, A.O., Adeyemo, M.B., Adeyemo, R.O., Agoro, S.A., Odugbesan, O.A., & Ayoola, S.A. (2022). Applications of Stem Cell Therapy in Neurodegenerative Diseases. *European Journal of Biotechnology and Bioscience*, *10*(2), 17–24. 2321–9122.

Asuku, A.O., Ayinla, M.T., Gbonjubola, O.O., Babatunde, S.S., Olajide, T.S., & Oyerinde, T.O. (2024). Neurocognitive Aspects of Dementia. In *Nutrition in Brain Aging and Dementia: Nutritional Neurosciences*. Springer.

Asuku, A.O., Ogungbangbe, G.O., & Olajide, T.S. (2025). Effect of Happiness and Positive Emotions on Dementia. In *Neuroscience of Feelings: Progress in Brain Research*. Elsevier. https://doi.org/10.1016/bs.pbr.2025.03.001

Awad, S.M., Attia, Y.A., ElSayed, H., Abdelhafez, S.H., Keshta, A.T., Rashad, E., Khalil, H.M.A., & Fathy, A.T. (2025). Efficacy of Curcumin-Selenium Nanoemulsion in Alleviating Oxidative Damage Induced by Aluminum Chloride in a Rat Model of Alzheimer's Disease. *Journal of Molecular Histology*, *56*(2), 122. https://doi.org/10.1007/s10735-025-10406-6

Das, S.S., Gopal, P.M., Thomas, J.V., Mohan, M.C., Thomas, S.C., Maliakel, B.P., Krishnakumar, I.M., & Pulikkaparambil Sasidharan, B.C. (2023). Influence of CurQfen®-Curcumin on Cognitive Impairment: A Randomized, Double-Blinded, Placebo-Controlled, 3-Arm, 3-Sequence Comparative Study. *Frontiers in Dementia*, *2*. https://doi.org/10.3389/frdem.2023.1222708

de Aquino, C.H. (2021). Methodological Issues in Randomized Clinical Trials for Prodromal Alzheimer's and Parkinson's Disease. *Frontiers in Neurology*, *12*. https://doi.org/10.3389/fneur.2021.694329

Dhankhar, J., Shrivastava, A., & Agrawal, N. (2023). Amendment of Altered Immune Response by Curcumin in Drosophila Model of Huntington's Disease. *Journal of Huntington's Disease*, *12*(4), 335–354. https://doi.org/10.3233/JHD-230595

El-Saadony, M.T., Yang, T., Korma, S.A., Sitohy, M., Abd El-Mageed, T.A., Selim, S., Al Jaouni, S.K., Salem, H.M., Mahmmod, Y., Soliman, S.M., Mo'men, S.A.A., Mosa, W.F.A., El-Wafai, N.A., Abou-Aly, H.E., Sitohy, B., Abd El-Hack, M.E., El-Tarabily, K.A., & Saad, A.M. (2023). Impacts of Turmeric and Its Principal Bioactive Curcumin on Human Health: Pharmaceutical, Medicinal, and Food Applications: A Comprehensive Review. *Frontiers in Nutrition*, *9*, 1040259. https://doi.org/10.3389/fnut.2022.1040259

Far, B.F., Safaei, M., Pourmolaei, A., Adibamini, S., Shirdel, S., Shirdel, S., Emadi, R., & Kaushik, A.K. (2024). Exploring Curcumin-Loaded Lipid-Based Nanomedicine as Efficient Targeted Therapy for Alzheimer's Diseases. *ACS Applied Bio Materials*, *7*(6), 3535–3555. https://doi.org/10.1021/acsabm.4c00112

Genchi, G., Lauria, G., Catalano, A., Carocci, A., & Sinicropi, M.S. (2024). Neuroprotective Effects of Curcumin in Neurodegenerative Diseases. *Foods*, *13*(11), 1774. https://doi.org/10.3390/foods13111774

Han, Y., Fu, S., Yang, X., Wang, X., Zhao, H., & Yang, X. (2024). Recent Nanotechnology Improvements in Curcumin Bioavailability and Related Applications. *Food Bioscience*, *61*, 104660. https://doi.org/10.1016/j.fbio.2024.104660

Hegde, M., Girisa, S., Bharathwaj Chetty, B., Vishwa, R., & Kunnumakkara, A.B. (2023). Curcumin Formulations for Better Bioavailability: What We Learned from Clinical Trials Thus Far? *ACS Omega*, *8*(12), 10713–10746. https://doi.org/10.1021/acsomega.2c07326

Heidari, H., Bagherniya, M., Majeed, M., Sathyapalan, T., Jamialahmadi, T., & Sahebkar, A. (2023). Curcumin-Piperine Co-Supplementation and Human Health: A Comprehensive Review of Preclinical and Clinical Studies. *Phytotherapy Research: PTR*, *37*(4), 1462–1487. https://doi.org/10.1002/ptr.7737

Iweala, E.J., Uche, M.E., Dike, E.D., Etumnu, L.R., Dokunmu, T.M., Oluwapelumi, A.E., Okoro, B.C., Dania, O.E., Adebayo, A.H., & Ugbogu, E.A. (2023). *Curcuma longa* (Turmeric): Ethnomedicinal Uses, Phytochemistry, Pharmacological Activities and Toxicity Profiles—A Review. *Pharmacological Research – Modern Chinese Medicine*, *6*, 100222. https://doi.org/10.1016/j.prmcm.2023.100222

Keservani, R.K., & Lohani, A. (Eds.). (2025). *Nutraceuticals in Obesity Management and Control*. Apple Academic Press. https://doi.org/10.1201/9781003557883

Kesharwani, A., Sree, B.K., Singh, N., Gajbhiye, R.L., Murti, K., Peraman, R., Pandey, K., Limoli, C.L., Velayutham, R., & Parihar, V.K. (2025). Dehydrozingerone Improves Mood and Memory in Diabetic Mice via Modulating Core Neuroimmune Genes and Their Associated Proteins. *ACS Pharmacology & Translational Science*, *8*(6), 1694–1710. https://doi.org/10.1021/acsptsci.5c00046

Kotha, R.R., & Luthria, D.L. (2019). Curcumin: Biological, Pharmaceutical, Nutraceutical, and Analytical Aspects. *Molecules (Basel, Switzerland)*, *24*(16), 2930. https://doi.org/10.3390/molecules24162930

Kothawade, S.M., Buttar, H.S., Tuli, H.S., & Kaur, G. (2023). Therapeutic Potential of Flavonoids in the Management of Obesity-Induced Alzheimer's Disease: An Overview of Preclinical and Clinical Studies. *Naunyn-Schmiedeberg's Archives of Pharmacology*, *396*(11), 2813–2830. https://doi.org/10.1007/s00210-023-02529-y

Kumar, A., Nirmal, P., Kumar, M., Jose, A., Tomer, V., Oz, E., Proestos, C., Zeng, M., Elobeid, T., Sneha, K., & Oz, F. (2023). Major Phytochemicals: Recent Advances in Health Benefits and Extraction Method. *Molecules*, *28*(2), 887. https://doi.org/10.3390/molecules28020887

Lopresti, A.L. (2018). The Problem of Curcumin and Its Bioavailability: Could Its Gastrointestinal Influence Contribute to Its Overall Health-Enhancing Effects? *Advances in Nutrition*, *9*(1), 41–50. https://doi.org/10.1093/advances/nmx011

Magdaleno Roman, J.Y., & Chapa González, C. (2024). Glutamate and Excitotoxicity in Central Nervous System Disorders: Ionotropic Glutamate Receptors as a Target for Neuroprotection. *Neuroprotection*, *2*(2), 137–150. https://doi.org/10.1002/nep3.46

Mahnashi, M.H., Ayaz, M., Ghufran, M., Almazni, I.A., Alqahtani, O., Alyami, B.A., Alqahtani, Y.S., Khan, H.A., Sadiq, A., & Waqas, M. (2024). Phytochemicals-Based β-Amyloid Cleaving Enzyme-1 and MAO-B Inhibitors for the Treatment of Alzheimer's Disease: Molecular Simulations-Based Predictions. *Journal of Biomolecular Structure & Dynamics*, *42*(16), 8359–8371. https://doi.org/10.1080/07391102.2023.2265494

Maugeri, A., Russo, C., Patanè, G.T., Barreca, D., Mandalari, G., & Navarra, M. (2023). The Inhibition of Mitogen-Activated Protein Kinases (MAPKs) and NF-κB Underlies the Neuroprotective Capacity of a Cinnamon/Curcumin/Turmeric Spice Blend in Aβ-Exposed THP-1 Cells. *Molecules (Basel, Switzerland)*, *28*(24), 7949. https://doi.org/10.3390/molecules28247949

Mukherjee, S., Mishra, A.K., Peer, G.D.G., Bagabir, S.A., Haque, S., Pandey, R.P., Raj, V.S., Jain, N., Pandey, A., & Kar, S.K. (2021). The Interplay of the Unfolded Protein Response in Neurodegenerative Diseases: A Therapeutic Role of Curcumin. *Frontiers in Aging Neuroscience*, *13*. https://doi.org/10.3389/fnagi.2021.767493

Nagu, P., Sharma, V., Behl, T., Pathan, A.K.A., & Mehta, V. (2022). Molecular Insights to the Wnt Signaling during Alzheimer's Disorder: A Potential Target for Therapeutic Interventions. *Journal of Molecular Neuroscience: MN*, *72*(4), 679–690. https://doi.org/10.1007/s12031-021-01940-5

Porro, C., & Panaro, M.A. (2023). Recent Progress in Understanding the Health Benefits of Curcumin. *Molecules*, *28*(5), 2418. https://doi.org/10.3390/molecules28052418

Pratti, V.L., Thomas, M., Bhoite, R., & Satyavrat, V. (2024). Investigating Bioavailability of Curcumin and Piperine Combination in Comparison to Turmeric Rhizomes: An in vitro Study. *Journal of Experimental Pharmacology*, *16*, 37–47. https://doi.org/10.2147/JEP.S427818

Rinkunaite, I., Simoliunas, E., Alksne, M., Dapkute, D., & Bukelskiene, V. (2021). Anti-Inflammatory Effect of Different Curcumin Preparations on Adjuvant-Induced Arthritis in Rats. *BMC Complementary Medicine and Therapies*, *21*(1), 39. https://doi.org/10.1186/s12906-021-03207-3

Rosińska-Lewandoska, D., Lewandowska, D., Morawska, M., Jarosz, K., Wolff, A., Kożuchowska, K., Gruszka, S., Szewczyk, M., Pilarz, D., & Ufnal, J. (2024). The Impact of Curcumin Supplementation on Inflammatory Processes and Recovery in Athletes. *Journal of Education, Health and Sport*, *76*, 56583. https://doi.org/10.12775/JEHS.2024.76.56583

Ryu, V., Ruiz-Ramirez, S., Chuesiang, P., McLandsborough, L.A., McClements, D.J., & Corradini, M.G. (2021). Use of Micellar Delivery Systems to Enhance Curcumin's Stability and Microbial Photoinactivation Capacity. *Foods*, *10*(8), 1777. https://doi.org/10.3390/foods10081777

Sadeghi, M., Dehnavi, S., Asadirad, A., Xu, S., Majeed, M., Jamialahmadi, T., Johnston, T.P., & Sahebkar, A. (2023). Curcumin and Chemokines: Mechanism of Action and Therapeutic Potential in Inflammatory Diseases. *Inflammopharmacology*, *31*(3), 1069–1093. https://doi.org/10.1007/s10787-023-01136-w

Sarraf, P., Parohan, M., Javanbakht, M.H., Ranji-Burachaloo, S., & Djalali, M. (2019). Short-Term Curcumin Supplementation Enhances Serum Brain-Derived Neurotrophic Factor in Adult Men and Women: A Systematic Review and Dose-Response Meta-Analysis of Randomized Controlled Trials. *Nutrition Research (New York, N.Y.)*, *69*, 1–8. https://doi.org/10.1016/j.nutres.2019.05.001

Shahcheraghi, S.H., Salemi, F., Peirovi, N., Ayatollahi, J., Alam, W., Khan, H., & Saso, L. (2021). Nrf2 Regulation by Curcumin: Molecular Aspects for Therapeutic Prospects. *Molecules*, *27*(1), 167. https://doi.org/10.3390/molecules27010167

Sharifi-Rad, J., Rayess, Y.E., Rizk, A.A., Sadaka, C., Zgheib, R., Zam, W., Sestito, S., Rapposelli, S., Neffe-Skocińska, K., Zielińska, D., Salehi, B., Setzer, W.N., Dosoky, N.S., Taheri, Y., El Beyrouthy, M., Martorell, M., Ostrander, E.A., Suleria, H.A.R., Cho, W.C., …, & Martins, N. (2020). Turmeric and Its Major Compound Curcumin on Health: Bioactive Effects and Safety Profiles for Food, Pharmaceutical, Biotechnological and Medicinal Applications. *Frontiers in Pharmacology*, *11*, 01021. https://doi.org/10.3389/fphar.2020.01021

Stohs, S.J., Chen, O., Ray, S.D., Ji, J., Bucci, L.R., & Preuss, H.G. (2020). Highly Bioavailable Forms of Curcumin and Promising Avenues for Curcumin-Based Research and Application: A Review. *Molecules*, *25*(6), 1397. https://doi.org/10.3390/molecules25061397

Su, I.-J., Hsu, C.-Y., Shen, S., Chao, P.-K., Hsu, J.T.-A., Hsueh, J.-T., Liang, J.-J., Hsu, Y.-T., & Shie, F.-S. (2022). The Beneficial Effects of Combining Anti-Aβ Antibody NP106 and Curcumin Analog TML-6 on the Treatment of Alzheimer's Disease in APP/PS1 Mice. *International Journal of Molecular Sciences*, *23*(1), 556. https://doi.org/10.3390/ijms23010556

Wiggers, H.J., Zaioncz, S., Cheleski, J., Mainardes, R.M., & Khalil, N.M. (2017). Chapter 7 – Curcumin, a Multitarget Phytochemical: Challenges and Perspectives. In Atta-ur-Rahman (Ed.), *Studies in Natural Products Chemistry* (Vol. 53, pp. 243–276). Elsevier. https://doi.org/10.1016/B978-0-444-63930-1.00007-7

Xue, C., Yan, Z., Cheng, W., Zhang, D., Zhang, R., Duan, H., Zhang, L., Ma, X., Hu, J., Kang, J., & Ma, X. (2025). Curcumin Ameliorates Aging-Induced Blood-Testis Barrier Disruption by Regulating AMPK/mTOR Mediated Autophagy. *PLoS One*, *20*(4), e0321752. https://doi.org/10.1371/journal.pone.0321752

Yan, Y., Chen, Y., Liu, Z., Cai, F., Niu, W., Song, L., Liang, H., Su, Z., Yu, B., & Yan, F. (2021). Brain Delivery of Curcumin through Low-Intensity Ultrasound-Induced Blood–Brain Barrier Opening via Lipid-PLGA Nanobubbles. *International Journal of Nanomedicine*, *16*, 7433–7447. https://doi.org/10.2147/IJN.S327737

7 *Huperzine A* for Managing and Treating Dementia

Olatoun Motunrayo Olutunde, Kehinde Oluwaseyi Adeniji, Abosede Mary Ayoola, Anne Fiyinfoluwa Sanwo, David Temidayo Ovioke, Victor Bassey Archibong, Rhoda Mama Kolo, Kelechi Roselyn Onyema, Moyosoluwa Mary Dada, and Temitope Oluwatobi Adeniji

7.1 INTRODUCTION

Dementia is a progressive neurological disease. It is characterized by mental deterioration and memory loss, with major variants including Alzheimer's disease (AD), vascular dementia (VaD), Lewy Body, and frontotemporal dementia (Hermann & Zerr, 2022; Asuku et al., 2025).

Globally, about 4 million adults aged 30–60 live with dementia, with 370,000 recorded annually (Hendriks et al., 2023). Dementia results from a cholinergic deficit (Ayinla et al., 2020), as shown in Figure 7.1. Available pharmaceutical drugs (donepezil, galantamine, rivastigmine, and memantine) only treat symptoms and have side effects such as gastrointestinal distress, dizziness, and bradycardia (Hyde et al., 2013). None profoundly alters disease progression; therefore, there is an urgent need

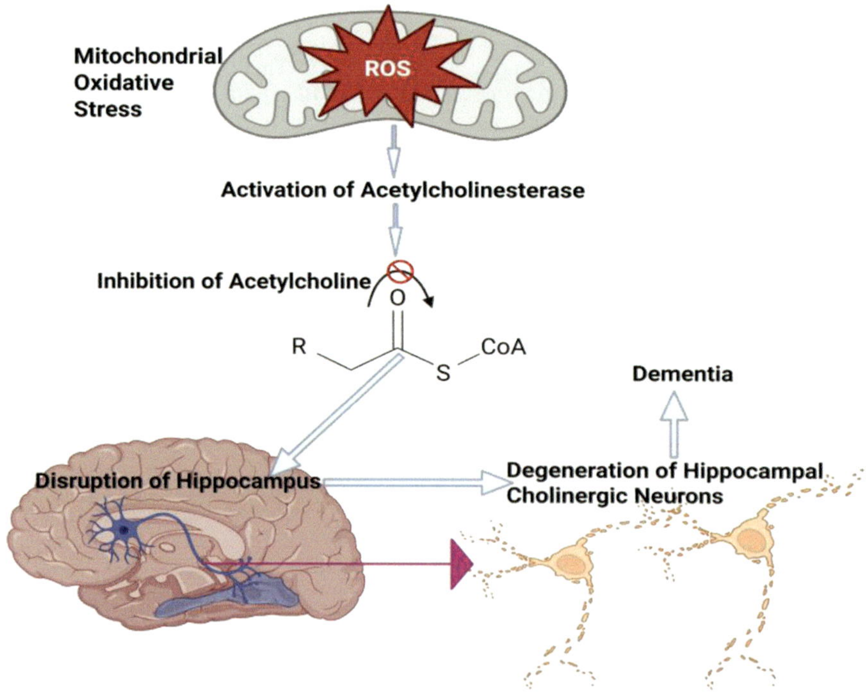

FIGURE 7.1 Schematic illustration of cholinergic deficit in dementia.

DOI: 10.1201/9781003657385-7

for safer, more effective treatments. This unmet need has intensified the search for compounds from natural bioresources. Medicinal plants that synthesize secondary metabolites (alkaloids, tannins, flavonoids) have documented efficacy and safety in treating neurological, neurodegenerative, and infectious diseases (Sundarrajan & Bhagtaney, 2023; Ayinla et al., 2024). Huperzine A (HupA) is one such compound. It is distributed in Asia and North America, and for centuries, has been used in traditional Chinese medicine (Ferreira et al., 2016). HupA demonstrates broad therapeutic potential for neuropathic pain, neurotoxicity, and mental health disorders via anti-inflammatory and antioxidative mechanisms (Friedli & Inestrosa, 2021).

Primarily, HupA is a potent AChE inhibitor that prolongs acetylcholine availability and attenuates amyloid-β (Aβ) aggregation (Mak et al., 2021) in acetylcholine-deficient dementias like AD (caused by disruption of cholinergic signaling (often from hippocampal neuron inhibition) and activation of acetylcholinesterase (AChE) by oxidative stress).

It is tolerable, has multi-target benefits, and favorable pharmacokinetics, making it a compelling candidate.

7.2 *HUPERZINE A*: SOURCE AND CHEMISTRY

HupA is a *Lycopodium* alkaloid isolated from *Huperzia serrata* (Quian Ceng Ta). It is used traditionally in China to treat inflammation, schizophrenia, contusions, and memory disorders (Kumbhar et al., 2020; Ghassab-Abdollahi et al., 2021).

Chemically, *HupA* is a molecule with a three-ring system that includes a tetrahydroquinolinone moiety, a three-carbon bridge, an exocyclic ethylidene group, and a primary amino group (Figure 7.2). Its empirical formula is $C_{15}H_{18}N_2O$ (molecular weight: 242.32 g/mol), and its IUPAC name (1R,9S,13E)-1-amino-13-ethylidene-11-methyl-6-azatricyclo[7.3.1.02,7]trideca-2(7), 3,10-trien-5-one] (National Center for Biotechnology Information, 2025). It is soluble in methanol and ethanol and insoluble in water. The compound is relatively stable but degrades with prolonged light/heat exposure. It exists as two stereoisomers: the natural (–)-*Huperzine A*, which is the biologically active form, with higher AChE inhibitory potency than the synthetic (+)-*Huperzine A* form (Suvaiv et al., 2025).

7.2.1 Extraction and Isolation

Solvents and chromatographic separations are used to extract HupA from *H. serrata*. High-Performance Liquid Chromatography (HPLC) on C18 columns with methanol/acetonitrile mobile phases is the standard for quantifying yield and purity.

Enhanced sensitivity and specificity for structural validation and quantification are via Liquid Chromatography–Mass Spectrometry (LC–MS) (Kumbhar et al., 2020).

Endophytic microorganisms associated with *H. serrata* are being investigated as alternative biosynthetic sources to overcome plant supply limitations.

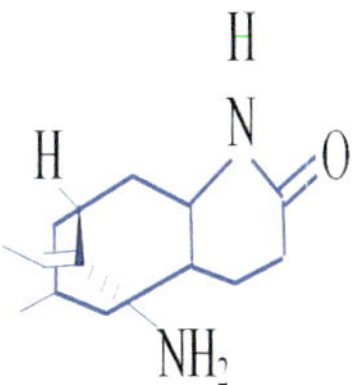

FIGURE 7.2 Structural formula of Huperzine.

7.3 PHARMACOKINETICS OF *HUPERZINE A*

HupA, being a small lipophilic alkaloid (Wang et al., 2011), diffuses easily across lipid membranes and is absorbed rapidly following oral administration with high bioavailability (Figure 7.3). Owing to its lipophilicity, small size, and ability to permeate the blood–brain barrier (BBB), it is widely distributed in the body more effectively than donepezil or tacrine (Wang et al., 2011).

HupA binds human serum albumin, which reduces the free unbound drug, creating a reservoir for prolonged action (Shamsi et al., 2022; Stielow et al., 2023).

In humans, *HupA* is primarily excreted unchanged in urine with little to no hepatic metabolism (Lin et al., 2016), and about 2.4% is recovered in feces (Ma et al., 2003). However, it is metabolized in rats by CYP1A2 and possibly CYP3A1/2 (Ma et al., 2003). Its elimination half-life (T1/2) is approximately 4–5 hours; at least 4–17 times longer than tacrine or physostigmine, supporting less frequent dosing (Friedli & Inestrosa, 2021). Table 7.1 shows the different factors affecting the pharmacokinetics of *Huperzine A*

This schematic illustrates the absorption, distribution, metabolism, and excretion of orally administered *Huperzine A*.

7.4 PHARMACOLOGICAL MECHANISMS OF ACTION OF *HUPERZINE A*

In China, (–)-HupA is clinically used to treat schizophrenia and memory impairment, while in the United States, it is used as a dietary supplement (Yan et al., 2022). Its primary mechanism is reversible AChEI, which hydrolyzes acetylcholine to terminate neuronal signaling. In AD, AChE aggregates with Aβ in plaques, accelerating fibrillar neurotoxicity.

A rationale for the clinical use of *HupA* in AD therapy is that it prolongs acetylcholine availability, disrupting AChE–Aβ interactions (Bartolini et al., 2003; Friedli & Inestrosa, 2021).

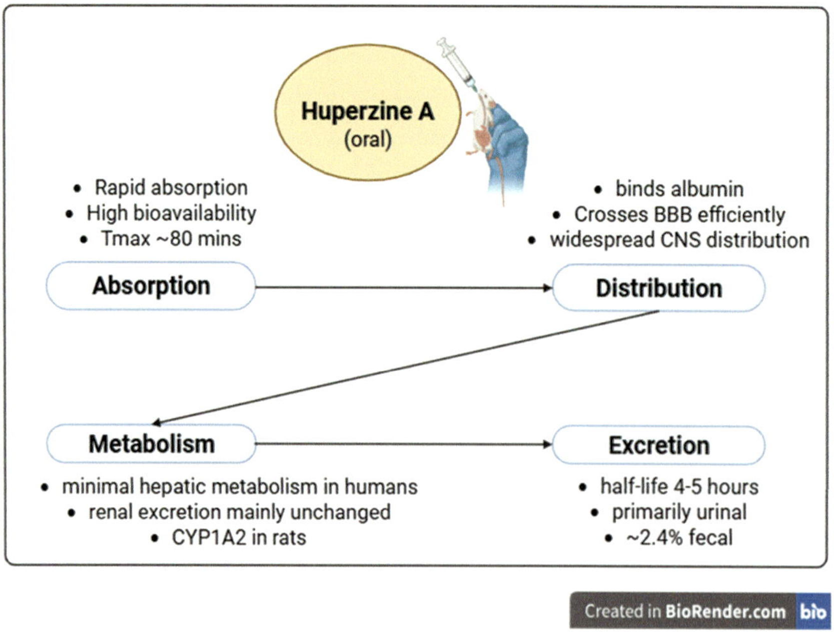

FIGURE 7.3 Pharmacokinetics of *Huperzine A*.

TABLE 7.1
Factors Affecting Pharmacokinetics *of Huperzine A*

Phase	Factors Affecting Pharmacokinetics of *Huperzine A*
Absorption	Route of administration Physicochemical properties of the drug Gastric pH and motility.
Distribution	Plasma protein binding, blood flow to tissue Tissue binding, Blood–brain barrier permeability.
Metabolism	Genetics, age Liver impairment, first-pass metabolism.
Excretion	Kidney function, urinary pH
Patient-related factors	Age, gender, and pregnancy Genetics, nutrition, disease state

7.4.1 Inhibition of Acetylcholinesterase

HupA is a potent, selective, reversible AChE inhibitor; within 30 minutes of administration, it inhibits AChE across brain regions (cortex, hippocampus, striatum) in a dose-dependent manner (Friedli & Inestrosa, 2021). This inhibitory activity is 24-fold and 180-fold greater than that of donepezil and tacrine, respectively, on a molar basis (Tun & Herzon, 2012). It decreases peripheral cholinergic adverse effects (AEs) and enhances memory deficits in older adults (Zhang & Wang, 2024).

HupA is highly selective for AChE over butyrylcholinesterase (BuChE), reducing peripheral cholinergic side effects associated with excessive BuChE inhibition (Tun & Herzon, 2012). Its reversible binding was confirmed through enzyme activity recovery after metabolite extraction (Tun & Herzon, 2012). X-ray crystallography shows the (–)-isomer binds strongly to aromatic residues (Trp84, Phe330) in AChE's active site.

Due to its bioavailability and favorable safety profile, HupA provides neuroprotection in animal models without significant abnormalities, unlike pyridostigmine (which does not cross the BBB) or physostigmine (which has severe side effects).

This dual advantage (therapeutic efficacy in AD and neuroprotection against chemical warfare agents) highlights the unique translational potential of *HupA* in both clinical and defense settings, but its clinical application remains underexplored globally, with limited research evaluating its long-term efficacy and safety (Gul et al., 2019; Suvaiv et al., 2025).

7.4.2 Neuroprotection by *Huperzine A* (*HupA*)

Neuroprotection by HupA is via multiple pathways:

1. *Cognitive enhancement*: longer-lasting memory than standard AChE inhibitors demonstrated in animal studies (Ratia et al., 2013).
2. It protects against Aβ-induced apoptosis, oxidative injury, and mitochondrial dysfunction (Qian & Ke, 2014).
3. It restores Krebs cycle activity and respiratory chain function (complexes II–IV) by preventing Aβ penetration into mitochondria (Yang et al., 2012).
4. Its antioxidant effects elevate glutathione peroxidase and catalase activity while reducing lipid peroxidation (Xiao et al., 2000).

5. Glutamate-induced excitotoxicity is reduced by suppressing calcium mobilization, reactive oxygen species (ROS) overproduction, and caspase-3 activity (Xiao et al., 2002; Yan et al., 2022).
6. It activates the PI3K/Akt/mTOR pathway, increasing anti-apoptotic Bcl-2 and reducing neuronal death (Mao et al., 2016).
7. It modulates amyloid precursor protein (APP) processing toward the non-amyloidogenic pathway via Wnt/β-catenin signaling, reducing plaque formation (De Ferrari et al., 2003).

These findings suggest that *HupA* combines potent and selective AChE inhibition with mitochondrial stabilization, antioxidant activity, anti-apoptotic signaling, and modulation of amyloid metabolism. This constellation of effects underlines its promise as a therapeutic agent for AD and related neurodegenerative disorders.

7.4.3 Iron Regulation, Anticonvulsant Potential, and Amyloid Clearance

Iron overload in the brain disrupts tau phosphorylation and reduces the ferroxidase activity of APP, contributing to dementia in AD (Lei et al., 2012).

Emerging evidence shows *HupA* modulates iron homeostasis:

1. Decreases cortical/hippocampal iron levels in transgenic mice, downregulates transferrin receptor 1, and reduces Aβ and phosphorylated tau (Huang et al., 2014), suggesting that *HupA* restores metal homeostasis to mitigate amyloid and tau pathology (Huang et al., 2014).
2. It exhibits anticonvulsant potential by enhancing GABA-mediated inhibition and suppressing NF-κB and IL-1β pathways (Gersner et al., 2015). *HupA* may modulate neuroinflammation to stabilize the excitatory–inhibitory balance in the cortex.
3. It lowers Aβ-positive neurons, enhances peptide clearance, and reduces fibrillar amyloid deposits (Huang et al., 2009; Wang et al., 2011; Yan et al., 2022).

These multifaceted actions position *HupA* as a broad-spectrum neuroprotective agent.

7.5 CLINICAL EVIDENCE FOR EFFICACY OF *HUPERZINE A* IN DEMENTIA MANAGEMENT

7.5.1 Alzheimer's Disease (AD)

The clinical evidence for *HupA*'s efficacy in AD and VaD is summarized in Table 7.2. The current evidence primarily addresses AD, and to a lesser extent, VaD, with *HupA* showing modest benefits in both conditions. No study has evaluated *HupA* for mild cognitive impairment (MCI) or its preventive potential (Yue et al., 2012). A school-based trial had some short-term memory improvement, and a quasi-randomized study in healthy military personnel had no significant benefits (Morasch et al., 2015). These trials were far from MCI contexts.

Small randomized controlled trials (RCTs) were aggregated, the effect size ratio was estimated, and longer treatment correlated with better outcomes (Qian & Ke, 2014; Damar et al., 2017). However, most studies were small, short-term, and mainly from Chinese cohorts, limiting generalizability. Long-term research and direct comparisons with standard AChEIs are sparse (Chow, 2008). There was high heterogeneity, risk of bias, overall low rigor of underlying studies, and minimal safety or long-term follow-up data (Wang et al., 2009; Xing et al., 2014). Table 7.2 shows the RCT and meta-analyses on the use of *HupA* for AD and VaD management.

TABLE 7.2
Summary of RCTs and Meta-Analyses on *HupA* for AD and VaD Management

Study (Author, Year)	Design	Population (Diagnosis, n)	Intervention (Dose, Duration)	Comparator	Primary Outcomes	Key Limitation
NIA 2008 (ICH GCP, 2008)	DB-RCT (Phase II: NCT00083590)	AD, n~150	200 or 400 µg BID 16 weeks	Placebo	ADAS-Cog improvement; ADCS-CGIC, ADL positive; Tolerable	Small, single population; Short-term
Wang et al. (2009)	Meta-analysis	AD, 4 RCTs	300–500 µg/day 8–24 weeks	Placebo	Significant improvement in MMSE and ADL; Larger effect sizes with longer treatment; mild cholinergic AEs only.	Small studies; All Chinese; Short-term
Rafii et al. (2011)	DB-RCT multicenter (Phase II)	Mild-to-moderate AD N = 210 (~177 completed)	200 or 400 µg BID 16 weeks	Placebo	200 µg BID: no ADAS-Cog benefit; 400 µg BID: Improvements at weeks 11 (2.27 points; p=0.001) and 16 (1.92 points; p=0.07); No significant effect on ADCS-CGIC, ADL, or NPI	Small sample, Class III evidence Secondary outcomes positive; Short duration Interpret with caution
Xu et al. (2012)	DB-RCT	Mild-to-moderate VD, n=78	100 µg BID 12 weeks	Placebo	Significant improvement in MMSE, CDR, and ADL scores (p<0.01) scores; No AEs reported	Small sample Short-term Single trial
Yang et al. (2013)	Meta-analysis	AD, 20 RCTs, n=1823	Mixed doses, duration	Placebo/control	Improved MMSE, HDS, ADL, WMS; some global benefits; No severe AEs	Poor quality; Risk of bias
Xing et al. (2014)	Meta-analysis	AD n=733 VaD n=92	Mixed	Placebo	Improved MMSE and ADL; Better tolerability in VaD	Small; Heterogeneity
NCT01282619 (ICH GCP, 2011).	Phase II DB-RCT Multicenter	Mild-to-moderate AD N=390	Sustained-release 400 µg QD vs. 200 µg BID 6 months after a 2-week run-in	Placebo	Primary measure: ADAS-Cog; Secondary: CDIC, ADL, NPI; Data unavailable	Results unavailable; Status unclear

TABLE 7.3
Common Adverse Effects of *HupA*

Adverse Events	Typical Incidence (%)	Severity (Usually Mild/Moderate)
Diarrhea	5	Mild
Somnolence	5	Mild
Abdominal pain	10	Mild
Nausea	35	Mild-to-moderate
Constipation	15	Mild-to-moderate
Vomiting	20	Mild-to-moderate
Nasal obstruction	5	Mild
Edema	5	Mild
Hyperactivity	5	Mild
Sweating	10	Mild
Dizziness	20	Mild
Thirst	10	Mild
Excitability	10	Mild
Insomnia	15	Mild
Anorexia	25	Mild
Bradycardia	10	Mild

7.6 SAFETY AND TOLERABILITY PROFILE OF *HUPERZINE A*

7.6.1 Common Adverse Effects (AEs)

HupA is generally well-tolerated, but common AEs are mild, transient, and cholinergic; these include nausea, vomiting, constipation, dizziness, and gastrointestinal discomfort (Yang et al., 2013; Ghassab-Abdollahi et al., 2021) as shown in Table 7.3. Some studies report mild, dose-dependent bradycardia (Zhang, 2012). Serious AEs are rare.

Dose of 200–400 μg BID for up to 24 weeks is well-tolerated without severe AEs (Suvaiv et al., 2025).

Contraindications include pre-existing gastrointestinal conditions and cardiovascular disease due to bradycardia risk (Zhang, 2012; Ghassab-Abdollahi et al., 2021). *HupA* may interact with other anticholinergics and beta-blockers, potentiating cholinergic effects (Behere et al., 2019). Synergistic actions with memantine have been noted (Zagami & Alabboud, 2023). No overdose cases have been reported in clinical trials using doses up to 400 μg BID (Zagami & Alabboud, 2023).

7.7 CURRENT USE, FORMULATION, AND DOSING

7.7.1 Regulatory Status

- *China*: *HupA* was approved for AD and MCI (Zhang, 2012).
- *United States*: It is a dietary supplement under the Dietary Supplement Health and Education Act (DSHEA), bypassing FDA pre-market approval (Crawford et al., 2020).
- *Europe*: It is sold only as a food supplement, as no *HupA*-based medicine has qualified for clinical use under the Traditional Herbal Medicinal Products Directive (THMPD). Herbal medicines must demonstrate 30 years of traditional use (15 within the EU) (Jędrejko et al., 2023).
- *Japan*: Local trials are required before approval
- *India*: Over-the-counter sales are done as a nutraceutical with minimal oversight.

Regulatory disparities and purity variations (60–70% in supplements vs. ≥98% pharmaceutical-grade) create a global "gray zone" for its use (Crawford et al., 2020).

7.7.2 Available Formulations

Standard oral tablets/capsules require two to three daily doses (Zhang, 2012). However, research focuses on novel delivery systems to enhance bioavailability and BBB penetration.

- Intranasal formulations (polylactide-co-glycoside (PLGA) nanoparticles, nanoemulsions, in situ gels) (Zhang et al., 2013a; Chen et al., 2019).
- Self-microemulsifying systems for oral delivery (Li et al., 2017).
- Transdermal methods (microneedle patches and ethosomal gels) (Yan et al., 2020; Wu et al., 2024).

Together, these emerging systems optimize therapeutic outcomes by improving drug stability, brain penetration, and dosing convenience.

7.7.3 Dosing Recommendations

The recommended clinical dose in China is 150–250 µg twice daily (Zhang, 2012). Clinical trials commonly use 200–800 µg/day for 8–24 weeks, with 400 µg/day being the best-tolerated (Yang et al., 2013; Xing et al., 2014).

An NIH-funded, multicenter Phase II trial evaluated 200 and 400 µg twice daily (total 400–800 µg/day) in patients with mild-to-moderate AD; there were significant cognitive benefits compared with placebo (Rafii et al., 2011). The study employed a stepwise titration beginning at 100 µg/day and escalating to 600 µg/day before reaching the target dose, ensuring tolerability.

Standardization of content and purity is essential for reliable dosing but remains challenging (Li et al., 2025).

7.7.4 Practical Guidance for Clinicians

Start with 100 µg daily, titrate to 200–400 µg twice daily based on tolerance and response (Rafii et al., 2011). Ensure to monitor cognitive outcomes, cholinergic AEs, and cardiovascular parameters (heart rate, blood pressure). Use third-party certified products to ensure quality. It is generally safe for up to six months with regular review. Exercise caution in older adults due to polypharmacy risks (Zhang et al., 2013b; Chu et al., 2024).

7.7.5 *HUPA'S* Place in Therapy

HupA only shows modest short-term cognitive benefits vs. placebo, but it's not a first-line option compared to established AChEIs due to insufficient evidence from large, high-quality trials (Rafii et al., 2011; Yang et al., 2013; Crawford et al., 2020).

Adjunctive evidence from Chinese studies suggests combining *HupA* with memantine may improve outcomes more than memantine with standard AChEIs, but data are limited and heterogeneous (Shao, 2015). Current data support use only when standard AChEIs are not tolerated.

7.8 CHALLENGES, FUTURE DIRECTIONS, AND INTEGRATION OF *HUPA* AS A BIORESOURCE FOR MANAGING DEMENTIA

Integrating *HupA*'s into global dementia therapeutics faces significant challenges. Scientific and clinical challenges include small, short-term, heterogeneous trials predominantly from China, limiting translational potential and generalizability (Suvaiv et al., 2025). Large, long-term, multicenter RCTs outside China must be done to confirm efficacy, long-term safety, and effects across dementia stages.

Direct comparisons with standard AChEIs and studies on drug interactions are also needed.

Supply chain and sustainability should be looked into as *H. serrata* grows slowly, and predominantly in Asia, creating risks of overharvesting, ecological damage, and supply instability (Yumkham et al., 2022). Alternatives like microbial biosynthesis or plant tissue culture are being explored but face cost barriers. Variable supplement quality and purity undermine research reproducibility and clinical safety.

Regulatory disparities between regions (e.g., drug vs. supplement status) hinder international recognition and clinical adoption; A standardized global regulatory framework is needed to enforce consistent quality control, extraction protocols, and dosing.

Ultimately, it is important to overcome the scientific, ecological, and regulatory challenges through coordinated global effort, sustainable production, and rigorous clinical validation before HupA can be part of a bioresource-based strategy for dementia. Its distinct multi-target mechanism, combining potent AChE inhibition with broad neuroprotection, offers a unique advantage.

7.9 CONCLUSION

Huperzine A is a unique natural compound with a multi-target mechanism of action that extends far beyond simple AChEI; it also confers broad neuroprotection through antioxidant, anti-apoptotic, anti-amyloid, and metal-chelating activities, as reviewed in this chapter. This multifaceted profile distinguishes it from many single-target synthetic pharmaceuticals. Clinically, *HupA* has shown modest but consistent efficacy in improving cognitive and functional outcomes in patients with AD and VaD, primarily in studies conducted in China via its pharmacokinetic properties (high oral bioavailability, efficient BBB penetration, and a favorable half-life).

Furthermore, it is suitable for patients who cannot tolerate standard AChE inhibitors as it exhibits a benign safety and tolerability profile.

Therefore, *Huperzine A* is an up-and-coming complementary and alternative agent. Its rich pharmacological profile and historical use provide a strong foundation for further development.

It can be integrated into alleviating dementia by addressing the highlighted challenges above through a concerted global effort.

REFERENCES

Asuku, A. O., Ogungbangbe, G. O., & Olajide, T. S. (2025). Effect of Happiness and Positive Emotions on Dementia. In: Neuroscience of Feelings: Progress in Brain Research. Elsevier. https://doi.org/10.1016/bs.pbr.2025.03.001

Ayinla, M. T., Asuku, A. O., Bayo-Olugbami, A., Ayeni, A. O., Abiola, A. A., & Owoyele, B. V. (2020). Melatonin and vitamin C modulate cholinergic neurotransmission and oxidative stress in scopolamine–induced rat model of memory impairment. *Journal of African Association of Physiological Sciences*, 8(1), 20.

Ayinla, M. T., Yakub, Z. M., Asuku, A. O., & Muzzamil, S. S. (2024). Ameliorative effect of aqueous leaf extract of aqueous leaf extract of *Cola acuminata* on mercury chloride – Induced neurotoxicity in male Wistar rat. *Bulletin of Faculty of Pharmacy, Cairo University*, 62(11), 18, https://doi.org/10.54634/2090-9101.1049

Bartolini, M., Bertucci, C., Cavrini, V., & Andrisano, V. (2003). Amyloid aggregation induced by human acetylcholinesterase: Inhibition studies. *Biochemical Pharmacology*, 65, 407–416.

Behere, P. B., Das, A., & Behere, A. P. (2019). Drugs Used in Dementia. In: Clinical Psychopharmacology. Springer, Singapore. https://doi.org/10.1007/978-981-13-2092-7_7

Chen, Y., Cheng, G., Hu, R., Chen, S., Lu, W., Gao, S., Xia, H., Wang, B., Sun, C., Nie, X., Shen, Q., & Fang, W. (2019). A nasal temperature and pH dual-responsive in situ gel delivery system based on microemulsion of huperzine A: Formulation, evaluation, and in vivo pharmacokinetic study. *AAPS PharmSciTech*, 20(7), 301.

Chow, T. W. (2008). Review: Insufficient evidence on Huperzine A for Alzheimer's. *Evidence Based Mental Health*, 11(4), 112.

Chu, Z., Sun, Q., Mao, M., Wu, Y., Yu, L., Xu, J., Liu, K., Qin, L., & Zhu, B. (2024). Pharmacology, phytochemistry, and traditional uses of *Huperzia serrata* (Thunb. ex Murray) Trev. *Fitoterapia*, 180, 106304.

Crawford, C., Wang, Y. H., Avula, B., Bae, J. Y., Khan, I. A., & Deuster, P. A. (2020). The scoop on brain health dietary supplement products containing Huperzine A. *Clinical Toxicology*, 58(10), 991–996.

Damar, U., Gersner, R., Johnstone, J. T., Schachter, S., & Rotenberg, A. (2017). Huperzine A: A promising anticonvulsant, disease modifying, and memory enhancing treatment option in Alzheimer's disease. *Medical Hypotheses*, 99, 57–62.

De Ferrari, G. V., Chacon, M. A., & Barria, M. I. (2003). Activation of Wnt signaling rescues neurodegeneration and behavioral impairments induced by β-amyloid fibrils. *Molecular Psychiatry*, 8:195–208.

Ferreira, A., Rodrigues, M., Fortuna, A., Falcão, A., & Alves, G. (2016). *Huperzine A* from *Huperzia serrata:* A review of its sources, chemistry, pharmacology and toxicology. *Phytochemistry Reviews*, 15(Issue 1), 51–85.

Friedli, M. J., & Inestrosa, N. C. (2021). Huperzine A and its neuroprotective molecular signaling in Alzheimer's disease. *Molecules*, 26(21), 6531.

Gersner, R., Ekstein, D., Dhamne, S. C., Schachter, S. C., & Rotenberg, A. (2015). *Huperzine A* prophylaxis against pentylenetetrazole-induced seizures in rats is associated with increased cortical inhibition. *Epilepsy Research*, 117, 97–103.

Ghassab-Abdollahi, N., Mobasseri, K., Dehghani Ahmadabad, A., Nadrian, H., & Mirghafourvand, M. (2021). The effects of Huperzine A on dementia and mild cognitive impairment: An overview of systematic reviews. *Phytotherapy Research*, 35(9), 4971–4987.

Gul, A., Bakht, J., & Mehmood, F. (2019). Huperzine-A response to cognitive impairment and task switching deficits in patients with Alzheimer's disease. *Journal of the Chinese Medical Association*, 82(1), 40–43.

Hendriks, S., Peetoom, K., Bakker, C., Koopmans, R., van der Flier, W., Papma, J., & Verhey, F. (2023). Global incidence of young-onset dementia: A systematic review and meta-analysis. *Alzheimer's Dement*, 19(3), 831–843.

Hermann, P., & Zerr, I. (2022). Rapidly progressive dementias—Aetiologies, diagnosis and management. *Nature Reviews Neurology*, 18(6), 363–376.

Huang, X.-T., Qian, Z.-M., & He, X. (2014). Reducing iron in the brain: A novel pharmacologic mechanism of Huperzine A in the treatment of Alzheimer's disease. *Neurobiol Aging*, 35, 1045–1054.

Huang, Z. S., Zhang, S. J., Xie, H. Y., Lin, X., Jiang W. Z., & Huang, R. B. (2009). Key gene and protein changes in the beta-amyloid pathway following Longyanshen poly-saccharides treatment in a mouse model of Alzheimer's disease. *Neural Regeneration Research*, 4, 756–762.

Hyde, C., Peters, J., Bond, M., Rogers, G., Hoyle, M., Anderson, R., Jeffreys, M., Davis, S., Thokala, P., & Moxham, T. (2013). Evolution of the evidence on the effectiveness and cost-effectiveness of acetylcholinesterase inhibitors and memantine for Alzheimer's disease: Systematic review and economic model. *Age and Ageing*, 42(1), 14–20.

ICH GCP. (2008). Huperzine A in Alzheimer Disease – Clinical Trials Registry – ICH GCP. ICH GDP. https://ichgcp.net/clinical-trials-registry/NCT00083590

ICH GCP. (2011). Study of Huperzine A Sustained-Release Tablets in Patients With Alzheimer's Disease. ICH GCP. https://ichgcp.net/clinical-trials-registry/NCT01282619

Jędrejko, K., Catlin, O., Stewart, T., Anderson, A., Muszyńska, B., & Catlin, D. H. (2023). Unauthorized ingredients in "nootropic" dietary supplements: A review of the history, pharmacology, prevalence, international regulations, and potential as doping agents. *Drug Testing and Analysis*, 15(7), 506–519.

Kumbhar, S. A., Hangargekar, C. B., & Joshi, A. A. (2020). Huperzine A from *Huperzia serrata* – A systematic review. *Journal of Advanced Scientific Research*, 11(3), 22–29.

Lei, P., Ayton, S., Finkelstein, D. I., et al. (2012). Tau deficiency induces parkinsonism with dementia by impairing APP-mediated iron export. *Nature Medicine*, 18, 291–295.

Li, F., Hu, R., Wang, B., Gui, Y., Cheng, G., Gao, S., Ye, L., & Tang, J. (2017). Self-microemulsifying drug delivery system for improving the bioavailability of *Huperzine A* by lymphatic uptake. *Acta Pharmaceutica Sinica B*, 7(3), 353–360.

Li, W., Wang, Z., Zhu, Q., & Tian, P. (2025). Systematic investigations of the *Huperzine A* – Producing endophytic fungi of *Huperzia serrata* in China and fermentation optimization using OSMAC strategy. *Frontiers in Microbiology*, 16, 1549272.

Lin, P., Li, X., Yuan, F., Chen, W., Yang, M., & Xu, H. (2016). Evaluation of the in vitro and in vivo metabolic pathway and cytochrome P450 inhibition/induction profile of *Huperzine A*. *Biochemical and Biophysical Research Communtications*, 480(2), 248–253.

Ma, X., Wang, H., Xin, J., Zhang, T., & Tu, Z. (2003). Identification of cytochrome P450 1A2 as enzyme involved in the microsomal metabolism of Huperzine A. *European Journal of Pharmacology*, 461(2–3), 89–92.

Mak, S., Li, W., Fu, H., Luo, J., Cui, W., Hu, S., & Han, Y. (2021). Promising tacrine/huperzine A-based dimeric acetylcholinesterase inhibitors for neurodegenerative disorders: From relieving symptoms to modifying diseases through multitarget. *Journal of Neurochemistry*, 158(6), 1381–1393.

Mao, X. Y., Zhou, H. H., Li, X., & Liu, Z. Q. (2016). Huperzine A alleviates oxidative glutamate toxicity in hippocampal HT22 cells via activating BDNF/TrkB-dependentPI3K/Akt/mTOR signaling pathway. *Cellular and Molecular Neurobiology*, 36, 915–925.

Morasch, K. C., Aaron, C. L., Moon, J. E., & Gordon, R. K. (2015). Physiological and neurobehavioral effects of cholinesterase inhibition in healthy adults. *Physiology and Behavior*, 138, 165–172.

National Center for Biotechnology Information. (2025). PubChem Compound Summary for CID854026, Huperzine A. Retrieved August 29, 2025, from https://pubchem.ncbi.nlm.nih.gov/compound/huperazine-A

Qian, Z. M., & Ke, Y. (2014). Huperzine A: Is it an effective disease-modifying drug for Alzheimer's disease? *Frontiers in Aging Neuroscience*, 6, 216.

Rafii, M. S., Walsh, S., Little, J. T., Behan, K., Reynolds, B., Ward, C., Jin, S., Thomas, R., & Aisen, P. S. (2011). A phase II trial of *Huperzine A* in mild to moderate Alzheimer disease. *Neurology*, 76(16), 1389–1394.

Ratia, M., Giménez-Llort, L., & Camps, P. (2013). *Huprine X* and *Huperzine A* improve cognition and regulate some neurochemical processes related with Alzheimer's disease in triple transgenic mice (3xTg-AD). *Neurodegenerative Diseases*, 11, 129–140.

Shamsi, A., Shahwan, M., Khan, M. S., Alhumaydhi, F. A., Alsagaby, S. A., Al Abdulmonem, W., Abdullaev, B., & Yadav, D. K. (2022). Mechanistic insight into binding of *Huperzine A* with human serum albumin: Computational and spectroscopic approaches. *Molecules*, 27(3), 797.

Shao, Z. Q. (2015). Comparison of the efficacy of four cholinesterase inhibitors in combination with memantine for the treatment of Alzheimer's disease. *International Journal of Clinical and Experimental Medicine*, 8(2), 2944–2948.

Stielow, M., Witczyńska, A., Kubryń, N., Fijałkowski, Ł., Nowaczyk, J., & Nowaczyk, A. (2023). The bioavailability of drugs—The current state of knowledge. *Molecules*, 28(24), 8038.

Sundarrajan, P., & Bhagtaney, L. (2023). Traditional Medicinal Plants as Bioresources in Health Security. In: Ethnic Knowledge and Perspectives of Medicinal Plants (pp. 53–75). Apple Academic Press, Palm Bay, FL.

Suvaiv, Singh, K., Hasan, S. M., Kumar, A., Khan, A., Shahanawaz, M., Hasan Zaidi, S. M., & Verma, K. (2025). *Huperzine A:* A natural acetylcholinesterase inhibitor with multifunctional neuroprotective effects. *Beni-Suef University Journal of Basic and Applied Sciences*, 14(82), 1–14.

Tun, M. K., & Herzon, S. B. (2012). The pharmacology and therapeutic potential of (–)-Huperzine A. *Journal of Experimental Pharmacology*, 4, 113–123.

Wang, B. S., Wang, H., Wei, Z. H., Song, Y. Y., Zhang, L., & Chen, H. Z. (2009). Efficacy and safety of natural acetylcholinesterase inhibitor Huperzine A in the treatment of Alzheimer's disease: An updated meta-analysis. *Journal of Neural Transmission*, 116(4), 457–465.

Wang, C. Y., Zheng, W., Wang, T., Xie, J. W., Wang, S. L., Zhao, B. L., Teng, W. P., & Wang, Z. Y. (2011). *Huperzine A* activates Wnt/β-catenin signaling and enhances the nonamyloidogenic pathway in an Alzheimer transgenic mouse model. *Neuropsychopharmacology*, 36(5), 1073–1089.

Wu, J., Xu, R., Xu, X., Ye, S., & Huang, A. (2024). Preparation and evaluation of transdermal permeation of Huperzine A ethosomes gel in vitro. *BMC Pharmacology and Toxicology*, 25(1), 21.

Xiao, X. Q., Wang, R., & Tang, X. C. (2000). *Huperzine A* and tacrine attenuate β-amyloid peptide-induced oxidative injury. *Journal of Neuroscience Research*, 61, 564–569.

Xiao, X. Q., Zhang, H. Y., & Tang, X. C. (2002). *Huperzine A* attenuates amyloid β-peptide fragment 25-35-induced apoptosis in rat cortical neurons via inhibiting reactive oxygen species formation and caspase-3 activation. *Journal of Neuroscience Research*, 67, 30–36.

Xing, S. H., Zhu, C. X., Zhang, R., & An, L. (2014). *Huperzine A* in the treatment of Alzheimer's disease and vascular dementia: A meta-analysis. *Evidence-Based Complementary and Alternative Medicine, Volume 2014*, 363985.

Xu, Z. Q., Liang, X. M., Juan-WuZhang, Y. F., Zhu, C. X., & Jiang, X. J. (2012). Treatment with *Huperzine A* improves cognition in vascular dementia patients. *Cell Biochemistry and Biophysics*, 62(1), 55–58.

Yan, Q., Wang, W., Weng, J., Zhang, Z., Yin, L., Yang, Q., Guo, F., Wang, X., Chen, F., & Yang, G. (2020). Dissolving microneedles for transdermal delivery of huperzine A for the treatment of Alzheimer's disease. *Drug Delivery*, 27(1), 1147–1155.

Yan, Y. P., Chen, J. Y., & Lu, J. H. (2022). Disease-modifying activity of *Huperzine A* on Alzheimer's disease: Evidence from preclinical studies on rodent models. *International Journal of Molecular Sciences*, 23(23), 15238.

Yang, G., Wang, Y., Tian, J., & Liu, J. P. (2013). Huperzine A for Alzheimer's disease: A systematic review and meta-analysis of randomized clinical trials. *PLoS One*, 8(9), e74916.

Yang, L., Ye, C.-Y., Huang, X.-T., Tang, X.-C., & Zhang, H.-Y. (2012). Decreased accumulation of subcellular amyloid-β with improved mitochondrial function mediates the neuroprotective effect of Huperzine A. *Journal of Alzheimer's Disease*, 31, 131–142.

Yue, J., Dong, B. R., Lin, X., Yang, M., Wu, H. M., & Wu, T. (2012). Huperzine A for mild cognitive impairment. *Cochrane Database of Systematic Reviews*, 12.

Yumkham, S. D., Khomdram, S. D., & Yaiphaba, N. (2022). Clubmosses (*Huperzia* Bernh.) of North East India: Genetic Resources, Utilization and Sustainability. In: Mahdi, S. S., & Singh, R. (eds) Innovative Approaches for Sustainable Development. Springer, Cham.

Zagami, F., & Alabboud, M. (2023). Myasthenia gravis symptom response to *Huperzine A,* pyridostigmine bromide, and an immunomodulatory incorporated regimen: A multi-case study. *DYSONA – Life Science*, 4(2), 36–49.

Zhang, H. Y. (2012). New insights into Huperzine A for the treatment of Alzheimer's disease. *Acta Pharmacologica Sinica*, 3;33(9), 1170–1175

Zhang, K., & Wang, H. (2024). Development and validation of a highly sensitive UPLC–MS/MS method for the determination of Huperzine A in rat plasma. *Biomedical Chromatography*, 38(11), e6011.

Zhang, L., Han, L., Qin, J., Lu, W., & Wang, J. (2013a). The use of borneol as an enhancer for targeting aprotinin-conjugated PEG–PLGA nanoparticles to the brain. *Pharmaceutical Research*, 30(10), 2560–2572.

Zhang, L., Song, Y., Lu, C., Zhang, J., Yuan, J., Wang, T., & Fu, F. (2013b). The effects of Huperzine A on gastrointestinal acetylcholinesterase activity and motility after single and multiple dosing in mice. *Experimental and Therapeutic Medicine*, 5(3), 793–796.

8 *Withania somnifera* for Managing and Treating Dementia

Tolulope Oluwatimilehin Alabi, Favour Ayomide Akinpelu, Oluwapelumi Lydia Apanisile, Azeezat Dagbo Yakub, and Faith Ewaoluwa Ajayi

8.1 INTRODUCTION

Natural products and plants have been a focus for researchers, especially for the treatment of different medical conditions as well as new drug discovery. Extracts from plants are especially of concern as possible medications and parts of complementary therapy that will lower the patient's dosage of the medication. Also, there is a multifunctional activity of the compounds of plants, because of which they can be used for preventive treatment, and one can obtain proper homeostasis in the body. Furthermore, they can be included in standard treatment, for which adequate proof of efficacy is needed (Wiciński et al., 2024).

Withania somnifera (WS), often referred to as Ashwagandha, is a valuable medicinal plant with a history of application in Ayurvedic medicine for over 3 millennia. It is a woody, evergreen shrub of the Solanaceae family used by many names with different meanings, e.g., "Indian winter cherry" or "Indian ginseng" in English, "Punir" or "Asgandh" in Hindi, and "Asgand" in Urdu. The herb has been designated as "somnifera", i.e., "sleep-inducer" in Latin, due to its noteworthy anti-stress activity; "Ashwagandha" (horse smells is "ashwa" and smell is "gandha") as the roots possess a character "wet horse" smell; and Indian ginseng due to its similarity to the pharmacologic effect and conventional uses of Korean ginseng tea (Paul et al., 2021).

Leaves, stems, roots, and flowers of the plant carry medicinal values, and 29 common metabolites from the root and leaf extracts (Dutta et al., 2019). The plant root includes 0.13–0.31% alkaloids, volatile oils, starch, amino acids, glycosides, reducing sugars, and steroids. Additionally, the root and leaves are used in the treatment of fever and swellings, flowers as astringents and diuretics, fruits for ulcers of the skin, tumors, and carbuncles, and seeds for increasing sperm count and enlarging testicles (Paul et al., 2021). This therapeutic plant has been reported to exert anti-epileptic, anti-inflammatory, anti-arthritic, anti-depressant, anti-coagulant, anti-oxidant, anti-diabetic, and anti-pyretic activities, besides palliative effects such as analgesia, rejuvenation, regeneration, and growth-stimulating activities (Figure 8.1).

8.1.1 Botanical Description and Scientific Classification

WS is a grey, stellate-tomentose erect undershrub with long tuberous roots, attaining a height of up to 30–150 cm and covered with white-brownish woolly pubescence. Leaves ovate, about 10 cm long and 2.5–5 cm wide, margins alternately disposed. Flowers are typically yellow or green, and axillary fascicles have bright red berries (Santhanu & Senthil, 2021). The stems are brown in color, ranging from prostrate to erect. The calyx is 5-lobed, about 5 mm long, and the fruit is 20 mm long, spherical, membrane-bound, and 5–10 ribbed. The corolla is slightly campanulate, five-lobed, 5–8 mm long, bright yellow to yellow-green. The five stamens are

DOI: 10.1201/9781003657385-8

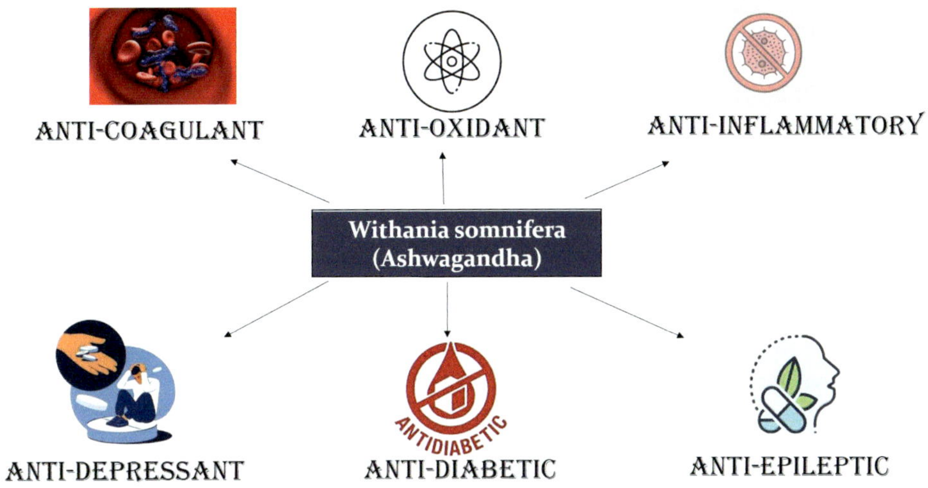

FIGURE 8.1 The health benefits of *Withania somnifera*.

protruding and yellow-orange. The fruit is a globose, hairless berry with an inflated, membrane-persistent calyx that is reddish-orange to crimson as it reaches maturity, measuring 5–8 mm in diameter. Fruit contains many pale brown, kidney-shaped seeds that are crushed to a coarse netted surface. Roots are meaty, thick, and possess secondary branches that resemble fibers that develop from the primary root. They taste harsh and caustic, and they smell awful (Paul et al., 2021).

Scientifically, WS is a member of the Solanaceae family (Table 8.1), which has over 90 genera and 3000 species globally, predominantly in the tropical regions. The genus *Withania* was established by Pauquy in 1825, and it remains so even today; about 26 species are found around the globe. Only two species have been identified to occur in India, viz., WS and *Withania coagulans* (WC) (Kumar et al., 2023). The scientific classification of WS is as follows:

8.1.2 Distribution and Folk Use

WS is the most geographically distributed species in the genus, which is mainly distributed in Africa, Europe, Australia, and Asia. It is grown heavily in both cultivated and wild farms in India

TABLE 8.1
Scientific Classification of *Withania somnifera* Dunal Species

Kingdom	Plantae
Sub-kingdom	Viridiplantae
Division	Angiosperms
Class	Dicotyledons
Order	Tubiflorae
Family	Solanaceae
Genus	*Withania*
Species	*Somnifera*
Botanical name	*Withania somnifera* (L.) Dunal

(Ratnaparkhi et al., 2023). It is widely grown in India, South Australia, and Europe for medicinal and industrial uses. In India, it is grown in the subtropical parts of Rajasthan, Haryana, Maharashtra, Punjab, Gujarat, Madhya Pradesh, and Uttar Pradesh. WS has been discovered in numerous countries of Africa, including South Africa, Egypt, Tanzania, Mali, Lesotho, Sudan, Djibouti, Swaziland, South Sudan, Nigeria, Liberia, and the Congo. WS is said to be native to South Africa and is employed widely as a hypnotic and sedative, and has been used effectively against a variety of disorders in Southern Africa, including Lesotho (Ratnaparkhi et al., 2023). However, the main reason it is planted as a medicinal crop in India is because of its meaty roots, which contain an infinite number of different bioactive compounds with a range of pharmacological benefits (Paul et al., 2021).

The root of the plant is utilized as a regular supplement in conventional medicine and is said to reduce nervousness and pressure. In addition to its high antioxidant content, eating the roots helps people avoid baldness, strengthen their heart muscles, lower blood pressure and edema, regulate cholesterol, and promote cardiovascular health (Afewerky et al., 2021). Furthermore, the roots have been used to treat cattle in veterinary medicine. A decoction of roots and leaves heated in water and given to sheep, cows, and buffalo improves milk production and serves as a sexual tonic and antipyretic (Aziz et al., 2018).

8.2 PHYTOCHEMISTRY AND BIOACTIVE CONSTITUENTS IN *WITHANIA SOMNIFERA*

Plant secondary metabolites isolation and identification are critical for discovering new drugs for different health conditions. Several phytochemicals found in WS have been isolated and characterized using chromatographic and spectroscopic analytical methods, including column chromatography, gas chromatography-mass spectroscopy (GC–MS), liquid chromatography-mass spectroscopy (LC–MS), nuclear magnetic resonance (NMR), and X-ray diffraction studies (Bashir et al., 2023). The phytochemical composition shows that bioactive chemicals are present all over the leaves, roots, fruits, and stem bark of the plant (Santhanu & Senthil, 2021). WS contains physiologically active chemical constituents such as alkaloids (isopertierine, anaferine, quahygrin, anahygrine, etc.), steroidal lactones (withanolides, withaferin), and saponins. Ashwagandha consists of sitoindoside and acylsteryl glucoside, which are anti-stress agents. Ashwagandha active compounds, including cytoindoside VII–X and withaferin-A, have been found to exhibit potent anti-stress activity in acute stress models (Ratnaparkhi et al., 2023).

The roots of WS contain amino acids, alkaloids, dulcitol, glycosides, hentriacontane, starch, steroids, reducing sugar, volatile oil, and withaniol. The prominent alkaloids of the roots are anahygrine, anaferine, cuscohygrine, isopelletierine, pseudotropine, pseudo-withanine, tropine, somnine, somniferine, somniferinine, withananinnie, and withananine. The plant leaves have 12 different types of withanolides and unidentified alkaloids. Withaferin A is the major withanolide, and free amino acids, condensed tannins, chlorogenic acid, flavonoids, glucose, and glycosides are present in the leaves (Santhanu & Senthil, 2021). In vitro, studies indicated that withanolide A and withanosides IV and VI are responsible for axon and dendritic elongation. Withanolide A can also redesign neural networks. Besides, withanolides can inhibit acetylcholinesterase (AChE) and induce the growth of dendrites in human neuroblastoma cells (Paul et al., 2021).

However, studies have shown that ethanolic extract of WS stem bark mainly comprises withanolides, which include somniferanolide, withasomnilide, somniferawithanolide, somniwithanolide, and withasomniferanolide, whereas unsaturated and saturated fatty acids such as palmitic acid, linoleic acid, elaidic acid, oleic acid, and tetracosanoic acid are present in oil extracts of fresh WS berries. Methanolic fruit extract of the plant contains withanamides A-I and 6,7α-epoxy-1α,3β, 5α-trihydroxy-witha-24enolide (Afewerky et al., 2021).

8.3 POTENTIAL THERAPEUTIC ACTION OF *WITHANIA SOMNIFERA*

8.3.1 Antioxidant Activity (Scavanges Free Radicals)

WS is an Ayurvedic plant that is widely used to treat memory loss, stress, anxiety, depression, and insomnia (Speers et al., 2021). Because of the antioxidant activity of the extract (H_2O_2 scavenging, inhibition of lipid peroxidation), it is neuroprotective and could reduce neuroinflammation by scavenging free radicals and guard against oxidative stress (Joshi, 2024). Oral treatment of aqueous extract of WS (at doses of 0, 0.5, or 2.5 mg/mL for four weeks) was found to be effective in managing dementia like symptoms in the 5xFAD mouse model of β-amyloid (Aβ) buildup (Gladen-Kolarsky et al., 2024).

8.3.2 Anti-inflammatory Effects (modulated NF-κB, TNF-Alpha)

WS has been identified as having excellent anti-inflammatory activities by improving cognitive function in a variety of rodent models of neurodegeneration (Khalil et al., 2021). Different research on WS have indicated that it possesses anti-inflammatory properties, which strengthen its neuroprotective properties since it inhibits the activity of certain enzymes and transcription factors implicated in inflammation, such as NF-κB, a key mediator of inflammatory processes, as well as pro-inflammatory cytokines IL-1β, IL-6, and TNF-α (Saha et al., 2024).

8.3.3 Modulation of Amyloid Beta and Tau Protein

Aβ deposition constitutes the hallmark of dementia (Asuku et al., 2024b; Asuku et al., 2025), and Withanolide A compound of WS possesses therapeutic significance through inhibition of Aβ secretion, aggregation, and neurotoxicity (Bai et al., 2022). It improves cognitive function through augmentation of neuroprotection, reduction of tau aggregation, and regulation of microglial cells, preventing NF-κB inflammatory pathway and iNOS, thus lowering Aβ-induced neurotoxicity proteins (Barua et al., 2020). Studies by Syed et al. show that WS, rich in withanolide and withanoside, not only corrected behavioral impairments but also plaque pathology, β-amyloid peptide (Aβ), and oligomer deposition in the brains of middle-aged transgenic mice with Alzheimer's disease (AD) (Syed et al., 2021).

8.3.4 Acetylcholinesterase (AChE) Inhibition

WS possesses therapeutic potential as an AD treatment by restoring cholinergic indicators, blocking AChE and butyrylcholinesterase (BuChE), and increasing acetylcholine levels (Dar & Ahmad, 2020). WS is reported to facilitate cholinergic neurotransmission and memory retention without disturbing other receptors. Studies by Raziya et al. demonstrated that WS at 50 mg/kg causes neuroprotection by increasing in dopamine, potentially reversing motor defects (Raziya Banu et al., 2020).

8.4 *WITHANIA SOMNIFERA'S* PHARMACOLOGICAL EFFECTS ON THE BRAIN WELLNESS

8.4.1 Neuroprotective Properties

8.4.1.1 Impact on Oxidative Damage

Neurodegenerative disease has been correlated with compromised antioxidant defense mechanisms and heightened free radical generation (Asuku et al., 2024a). WS has profound antioxidant activity by modulating the prominent enzymatic and non-enzymatic defense mechanisms

(Gladen-Kolarsky et al., 2024). Through elevated activity of pertinent antioxidant enzymes like glutathione peroxidase (GPx), catalase (CAT), and superoxide dismutase (SOD) within the striatum and major nigra, it acts as a neuroprotectant by suppressing oxidative stress. By decreasing nigrostriatal oxidative stress, augmenting tyrosine hydroxylase-positive cells in the substantia nigra, and enhancing motor coordination, an ethanolic root extract of WS is neuroprotective against maneb-paraquat-induced dopaminergic neurodegeneration in Parkinson's disease (PD) mice (Lerose et al., 2024). Lerose et al. also made the same observations in PD models developed using MPTP. These studies demonstrated that WS (roots/leaf extracts) decreased lipid peroxidation (MDA) in the midbrain striatum, normalized the levels of antioxidant enzymes (SOD, CAT, GPx, and GSH), and countered the overproduction of ROS. In behavioral testing, treatment improved the motor deficits induced by MPTP, demonstrating neuroprotection of susceptible substantia nigra neurons against oxidative injury. These results depict the antioxidant activity of WS.

8.4.1.2 Blocks the Formation of AB Fibrils and Tau Phosphorylation

AD causes neurodegeneration that is marked by tau hyperphosphorylation, neuroinflammation, and Aβ fibril deposition. Aβ fibrils cause oxidative stress and suppress the uptake of glucose by neurons, whereas hyperphosphorylated tau destabilizes microtubules and causes cell death (Tiwari et al., 2019; Asuku et al., 2025). However, WS has been shown to have neuroprotective properties by mitigating Aβ toxicity and neuroinflammation. Withaferin A (WA), a bioactive component present in WS, inhibits Aβ aggregation, tau deposition, and oxidative stress, and regulates heat shock proteins (HSPs) and NF-κB-mediated neuroinflammation (Das et al., 2021). WA's neuroprotection for the release of Aβ in SH-APP cells, a human neuroblastoma cell line, was recently validated. Aβ (1–40) was reduced significantly at 2 μM without toxicity. Atluri et al. also employed Congo red staining to validate that withaferin A treatment reduced Aβ aggregation compared to controls (Atluri et al., 2020).

8.4.1.3 Enhancement of Mitochondrial Activity

Neurodegenerative disease is linked with mitochondrial dysfunction. WS enhances mitochondrial biogenesis and ATP synthesis by enhancing the efficiency of the electron transport chain and stimulating the peroxisome proliferator-activated receptor gamma coactivator 1-alpha (PGC-1α) (Kuboyama et al., 2023). Its action to inhibit cytochrome c release and thus prevent apoptosis of neurons exposed to ischemia or hypoxic conditions further supports its neuroprotective function (Singh et al., 2023).

8.4.2 Cholinergic Modulation

Lerose et al. illustrate that WS has considerable protective effects against scopolamine-induced cholinergic impairment through various mechanisms. Scopolamine, a non-specific muscarinic receptor antagonist, interferes with memory and learning and replicates significant clinical characteristics of dementia by interfering with cholinergic neurotransmissions. Treatment with the alcoholic extract of *WS* or its active component, withaone, reverses these by redressing neural markers and decreasing oxidative stress (Lerose et al., 2024).

8.4.3 Stress and Cortisol Regulation

Chronic stress enhances neurodegeneration by interfering with the hypothalamic-pituitary-adrenal (HPA) axis (Ayinla & Asuku, 2025). Lopresti et al. showed in a controlled study (N=60) that 300 mg of WS root extract twice daily for eight weeks reduced serum cortisol by 27.9% ($p<0.001$ vs. placebo) and stress scores (44% reduction, DASS-42; $p=0.003$), and cognitive fatigue and sleep quality (Lopresti et al., 2019). Withanolides have the ability to normalize the HPA axis through inhibition of hypothalamic corticotropin-releasing hormone release and adrenal 11 βm-hydroxylase, resulting in reduced cortisol production (Zhang et al., 2024).

8.4.3.1 Support for Hippocampus Neurogenesis

WS triggers hippocampus neurogenesis through brain-derived neurotrophic factor (BDNF) or cAMP response element binding protein (BDNF/CREB) signaling. WS root extract elevated the level of BDNF by 50% and triggered CREB phosphorylation ($p<0.01$ vs. controls) in the chronic unpredictable mild stress (CUMS) animal model, reversing dendritic arborization and synaptic plasticity (Bhattacharya & Muruganandam, 2023). Kuboyama et al. found reduced pro-inflammatory cytokines (IL-6, TBF-α; $p<0.05$) and oxidative stress markers (MDA; 30%↓) created a favorable environment for neurogenesis (Kuboyama et al., 2023). Stressed individuals exhibit enhanced memory consolidation, perhaps via cortisol-modulated neurogenesis and mitochondrial biogenesis (Lopresti et al., 2023).

8.5 PRECLINICAL AND CLINICAL EVIDENCE FOR *WITHANIA SOMNIFERA* IN DEMENTIA CARE

8.5.1 Preclinical Evidence

WS is gifted with diverse pharmacological activities through its bioactive compounds.

Preclinical experiments show several evidence for the neuroprotective activity of WS in different models of dementia, especially AD (Paul et al., 2021; Polumackanycz et al., 2023; Tancreda et al., 2025) (Figure 8.2). Its bioactive constituents, which are primarily withanolides, withanosides, and

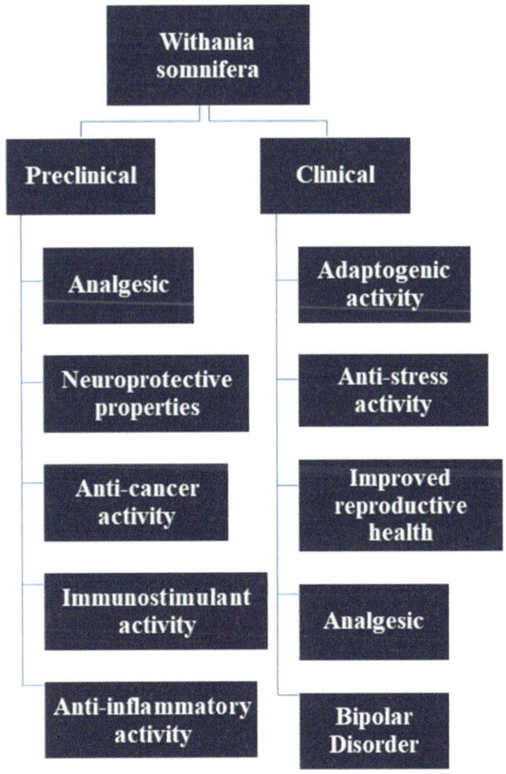

FIGURE 8.2 The clinical and preclinical activity of *Withania somnifera*.

withanone, possess diverse therapeutic effects by antioxidant, anti-inflammatory, anti-apoptotic, and neuroregenerative pathways, through the attenuation of oxidative stress, mitochondrial injury, and neuronal apoptosis (Paul et al., 2021; Lerose et al., 2024).

In 5xFAD Alzheimer's model mice, Gladen-Kolarsky et al. showed that aqueous WS root extract improved spatial memory and reduced anxiety-like behavior, and significantly reduced hippocampal β-amyloid deposition. The study also established normalization of microglial and astrocyte function to validate the anti-inflammatory and neuroprotective action of the extract (Gladen-Kolarsky et al., 2024).

Similarly, in a phenolic-rich extract evaluation, Polumackanycz et al. showed that WS exhibited strong AChE and AChE inhibition in vitro, alongside potent free radical scavenging ability. These results suggest the extract's potential for protecting cholinergic neurons and mitigating oxidative stress in neurodegenerative disorders (Polumackanycz et al., 2023).

WS mitigates NMDA-induced excitotoxicity in Neuro2a cells effectively by reducing malondialdehyde levels and preventing DNA damage as measured by Poly (ADP-ribose) polymerase-1 cleavage (Dar & Ahmad, 2020). Scopolamine-induced amnesia is also reversed by extracts of Ashwagandha with oxidative stress, which suggests therapeutic effectiveness in neurological disorders (Er et al., 2025).

Gregory et al. further detailed the anti-inflammatory effects of WS, demonstrating downregulation of pro-inflammatory cytokines (IL-1β, TNF-α) and inhibition of NF-κB signaling in rat models. The extract also supported the survival of hippocampal neurons and inhibited apoptosis by regulating the Bax/Bcl-2 ratio and preventing caspase-3 activation (Gregory et al., 2021).

Furthermore, the neurotrophic activity of withanolides has been ascribed to increased dendritic arborization and synaptic plasticity in hippocampal neurons (Lerose et al., 2024). This has been caused by increased expression of BDNF and glial fibrillary acidic protein (GFAP) regulation, as observed in sleep-deprivation models and aging-induced neurodegeneration. The active constituents (Withanolides and Withanosides) also ensured better cognitive behavior in aged Alzheimer's models (Lerose et al., 2024).

8.5.2 Clinical Evidence

Several clinical trials show WS's therapeutic effect in several diseases. For adaptogenic effectiveness, 3 g/day root powder increased hemoglobin level and physical working capacity in healthy adults. Herb's anti-stress effect was reported at the doses of 300 mg, reducing cortisol levels and anxiety scores vs. placebo in various studies. Insomnia patients reported improved sleeping quality following 300 mg of its root extract (Salve et al., 2019). In a clinical trial, 600 mg/day of root extract for eight weeks effectively normalized thyroid hormone levels in hypothyroid patients (Sharma et al., 2018).

Ng et al. identified five randomized, placebo-controlled, double-blind studies involving older adults with mild cognitive impairment (MCI) and patients with psychiatric disorders. The review concluded that WS supplementation consistently improved cognitive tasks, executive function, attention, and reaction time, with high tolerability and minimal side effects (Ng et al., 2020).

In 50 participants with MCI, administration of 300 mg of root extract twice daily for eight weeks resulted in notable enhancements in immediate and general memory, executive function, sustained attention, and information-processing speed in comparison to a placebo (Naik et al., 2024).

In a placebo-controlled trial of 59 healthy adults, acute dosing (225 mg) and 30-day liposomal WS supplementation led to marked enhancements in word recall, choice reaction time, picture recognition, digit vigilance, Stroop testing, and reduced tension and fatigue—all while maintaining a favorable safety profile (Leonard et al., 2024).

8.6 CONSIDERATION OF *WITHANIA SOMNIFERA* USE FOR DEMENTIA TREATMENT

8.6.1 Dosage and Preparations

8.6.1.1 Standardized Extracts

The roots standardized to 5–10% withanolides (the primary bioactive component) are frequently used in clinical trials. Dosages range from 300 to 600 mg/day, either divided twice a day (e.g., 300 mg twice a day). Cognitive enhancement was robust in a randomized controlled trial of the elderly with MCIs after taking 300 mg/day of an extract for more than eight weeks (Ng et al., 2020). They have been investigated at increased doses, but additional long-term safety studies are required. Higher withanolide concentrations, such as KSM-66® or Sensoril®, are better suited for stable concentrations of active compounds (withaferin A and withanolide A) that modulate Aβ clearance and neuroinflammation (Bhat et al., 2022).

8.6.1.2 Traditional Preparations

Traditionally, 3–6 g of the root powder is consumed per day in divided doses, frequently with milk or ghee. Phytochemical variability due to growing conditions and processing, however, restricts its therapeutic use. Ayurvedic preparations such as Brahmi Ghrita, a medicated ghee, use decoctions and water extracts. These have the potential to increase the bioavailability of hydrophilic substances but are below modern standards in comparison to modern extracts (Farooqui et al., 2018).

8.6.2 Safety and Interactions

8.6.2.1 General Safety

The National Center for Complementary and Integrative Health (National Center for Complementary and Integrative Health (NCCIH), 2023) reported that Ashwagandha is safe and tolerable for short-term use (up to three months), but there is little information about its long-term effects. Some people develop stomach discomfort, drowsiness, or headaches from Ashwagandha use. Ashwagandha has been contraindicated in the following conditions:

- The serum should not be used during pregnancy since increased dosage of Ashwagandha root extract may cause miscarriage (Brar & Malhotra, 2021).
- While Ashwagandha root extract is used to treat male infertility, it is contraindicated in hormone-sensitive prostate cancer patients. Polumackanycz et al. found that the herb elevates testosterone levels, which can promote disease progression (Połumackanycz et al., 2020).
- Ashwagandha should be used with caution in hyperthyroidism. Though it reduces symptoms such as anxiety, fatigue, hand trembling, and palpitations, it raises serum T3/T4 levels, worsening thyroid dysfunction (Kamal et al., 2022).

8.6.2.2 Drug Interactions

Ashwagandha can potentiate the central nervous system (CNS) depressant effect of benzodiazepines, barbiturates, and alcohol through action on GABA receptors. This can lead to enhanced sedation, motor impairment, and inhibition of respiration. Ashwagandha can inhibit CYP450 enzymes, thus impacting drug metabolism and action (Mikulska et al., 2023).

8.7 CONCLUSION

WS (Ashwagandha) shows significant therapeutic potential in the treatment of dementia by demonstrating several neuroprotective properties. The phytochemical constituents in the plant demonstrate robust antioxidant activity, prevention of oxidative stress and mitochondrial damage, and its anti-inflammatory potential promotes counterbalancing neuroinflammation—a major element of dementia pathophysiology. In addition, Ashwagandha's cholinergic modulation, including AChE inhibition and enhancement of acetylcholine levels, also contributes to improved cognitive function, memory, and executive function.

Preclinical research has consistently revealed Ashwagandha's potential to decrease oxidative stress, suppress neuroinflammation, inhibit amyloid-beta aggregation, and enhance mitochondrial function. Furthermore, it promotes neuronal regeneration and synaptic plasticity, both of which are essential for cognitive resilience. These mechanistic findings are supported by clinical trials, which show that Ashwagandha standardized extracts effectively improve memory, attention, and thinking skills in MCI and early dementia with a strong safety profile. Its adaptogenic action also supports stress reduction and hippocampal neurogenesis that indirectly stop neurodegenerative advancement. Despite these encouraging findings, larger, well-performed placebo-controlled studies are still required to establish standard dosing regimens, confirm long-term safety, and establish efficacy in a variety of dementia populations. Standardization of extract preparations with equivalent bioactive content, in particular withanolides, will be necessary for reproducibility and clinical practice. Future research should focus on closing the current translational gap, facilitating the evidence-based application of the traditional medicinal plant in contemporary neurotherapeutics.

REFERENCES

Afewerky, H.K., Ayodeji, A.E., Tiamiyu, B.B., Orege, J.I., Okeke, E.S., Oyejobi, A.O., & Adeyemi, S.B. (2021). Critical review of the *Withania somnifera* (L.) Dunal: Ethnobotany, pharmacological efficacy, and commercialization significance in Africa. *Bulletin of the National Research Centre*, *45*, 176.

Asuku, A.O., Ayinla, M.T., Ajibare, A.J., & Krishnamurthy, R. (2024a). Modulatory and Neuroprotective Roles of Marine Green Spirulina on the Brain and Neurodegeneration. In: *Marine Greens: Environmental, Agricultural, Industrial, and Biomedical Applications*. CRC Press, Taylor and Francis group, Boca Raton, 17: 89. https://doi.org/10.1201/9781003369738

Asuku, A.O., Ayinla, M.T., Gbonjubola, O.O., Babatunde, S.S., Olajide, T.S., & Oyerinde, T.O. (2024b). Neurocognitive Aspects of Dementia. In: *Nutrition in Brain Aging and Dementia. Nutritional Neurosciences*. Springer, Singapore.

Asuku, A.O., Ogungbangbe, G.O., & Olajide, T.S. (2025). Effect of Happiness and Positive Emotions on Dementia. In *Neuroscience of Feelings. Progress in Brain Research*, Elsevier, https://doi.org/10.1016/bs.pbr.2025.03.001

Atluri, V.S.R., Tiwari, S., Rodriguez, M., Kaushik, A., Yndart, A., Kolishetti, N., Yatham, M., & Nair, M. (2020). Inhibition of amyloid-beta production, associated neuroinflammation, and histone deacetylase 2-mediated epigenetic modifications prevent neuropathology in Alzheimer's disease *in vitro* model. *Frontiers in Aging Neuroscience*, *11*, 342.

Ayinla, M.T., & Asuku, A.O. (2025). The modulatory role of 6-gingerol-rich portion of ginger on brain oxidative stress, inflammation and neurotransmission in chronic unpredicted mild stress-induced depressed male Wistar rats. *Avicenna Journal of Phytomedicine*. Epub ahead of print.

Aziz, M.A., Khan, A.H., Adnan, M., & Ullah, H. (2018). Traditional uses of medicinal plants used by Indigenous communities for veterinary practices at Bajaur Agency, Pakistan. *Journal of Ethnobiology and Ethnomedicine*, *14*(1), 1–18.

Bai, R., Guo, J., Ye, X.Y., Xie, Y., & Xie, T. (2022). Oxidative stress: The core pathogenesis and mechanism of Alzheimer's disease. *Ageing Research Reviews*, *77*, 101619.

Barua, N., Ibn Aziz, M.A., Tareq, A.M., Sayeed, M.A., Alam, N., & Alam, N.U. (2020). *In vivo* and *in vitro* evaluation of pharmacological activities of *Adenia trilobata* (Roxb.). *Biochemistry and Biophysics Reports*, *23*, 100772.

Bashir, A., Nabi, M., Tabassum, N., Afzal, S., & Ayoub, M. (2023). An updated review on phytochemistry and molecular targets of *Withania somnifera* (L.) Dunal (Ashwagandha). *Frontiers in Pharmacology*, *14*, 1049334.

Bhat, J.A., Akther, T., Najar, R.A., Rasool, F., & Hamid, A. (2022). *Withania somnifera* (L.) Dunal (Ashwagandha): Current understanding and future prospect as a potential drug candidate. *Frontiers in Pharmacology*, *13*, 1029123.

Bhattacharya, S., & Muruganandam, A. (2023). Ashwagandha root extract reverses stress-induced hippocampal neurodegeneration via BDNF/CREB activation. *Neuropharmacology*, *225*, 109398.

Brar, G.K., & Malhotra, M. (2021). Ashwagandha (*Withania somnifera*)—A herb with versatile medicinal properties empowering human physical and mental health. *Journal of Pre-Clinic and Clinical Research*, *15*(3), 129–133.

Dar, N.J., & Ahmad, M. (2020). Neurodegenerative diseases and *Withania somnifera* (L.): An update. *Journal of Ethnopharmacology*, *256*, 112769.

Das, R., Rauf, A., Akhter, S., Islam, M.N., Emran, T.B., Mitra, S., Khan, I.N., & Mubarak, M.S. (2021). Role of withaferin A and its derivatives in the management of Alzheimer's disease: Recent trends and future perspectives. *Molecules*, *26*(12), 3696.

Dutta, R., Khalil, R., Green, R., Mohapatra, S.S., & Mohapatra, S. (2019). *Withania somnifera* (Ashwagandha) and Withaferin A: Potential in integrative oncology. *International Journal of Molecular Sciences*, *20*(21), 5310.

Er, B., Ozmen, B., Sahin, E., Orhan, C., Sahin, N., Morde, A.A., Padigaru, M., & Sahin, K. (2025). A novel Ashwagandha (*Withania somnifera*) formulation mitigates sleep deprivation-induced cognitive impairment and oxidative stress in a rat model. *Biomolecules*, *15*(5), 710.

Farooqui, A.A., Farooqui, T., Madan, A., Ong, J.H., & Ong, W.Y. (2018). Ayurvedic medicine for the treatment of dementia: Mechanistic aspects. *Evidence-Based Complementary and Alternative Medicine*, *2018*, 2481076.

Gladen-Kolarsky, N., Monestime, O., Bollen, M., Choi, J., Yang, L., Magaña, A.A., & Gray, N.E. (2024). *Withania somnifera* (Ashwagandha) improves spatial memory, anxiety, and depressive-like behavior in the 5xFAD mouse model of Alzheimer's disease. *Antioxidants*, *13*(10), 1164.

Gregory, J., Vengalasetti, Y.V., Bredesen, D.E., & Rao, R.V. (2021). Neuroprotective herbs for the management of Alzheimer's disease. *Biomolecules*, *11*(4), 543.

Joshi, M.K. (2024). *Ashwagandha-An Herbal Product Used as a Neuroprotective Agent* (Master's thesis, Adelphi University).

Kamal, H.I., Patel, K., Brdak, A., Heffernan, J., & Ahmad, N. (2022). Ashwagandha as a unique cause of thyrotoxicosis presenting with supraventricular tachycardia. *Cureus*, *14*(3), e23494.

Khalil, H.M.A., Eliwa, H.A., El-Shiekh, R.A., Al-Mokaddem, A.K., Hassan, M., Tawfek, A.M., & El-Maadawy, W.H. (2021). Ashwagandha (*Withania somnifera*) root extract attenuates hepatic and cognitive deficits in thioacetamide-induced rat model of hepatic encephalopathy via induction of Nrf2/HO-1 and mitigation of NF-kappaB/MAPK signaling pathways. *Journal of Ethnopharmacology*, *277*, 114141.

Kuboyama, T., Tohda, C., & Komatsu, K. (2023). *Withania somnifera* suppresses neuroinflammation and oxidative stress to promote hippocampal neurogenesis in chronic stress models. *Journal of Ethnopharmacology*, *311*, 116452.

Kumar, A., Venugopal, S., Jnanesha, A.C., & Lal, R.K. (2023). Agricultural-based challenges, genetic enhancement, and obstacles to an industrially important medicinal plant, ashwagandha (*Withania somnifera* (L.) Dunal): A review. *Ecological Genetics and Genomics*, *28*, 100183.

Leonard, M., Dickerson, B., Estes, L., Gonzalez, D.E., Jenkins, V., Johnson, S., …, & Kreider, R.B. (2024). Acute and repeated ashwagandha supplementation improves markers of cognitive function and mood. *Nutrients*, *16*(12), 1813.

Lerose, V., Ponticelli, M., Benedetto, N., Carlucci, V., Lela, L., Tzvetkov, N.T., & Milella, L. (2024). *Withania somnifera* (L.) Dunal, a potential source of phytochemicals for treating neurodegenerative diseases: A systematic review. *Plants*, *13*(6), 771.

Lopresti, A.L., Smith, S.J., & Drummond, P.D. (2019). An investigation into the stress-relieving and pharmacological actions of an ashwagandha (*Withania somnifera*) extract: A randomized, double-blind, placebo-controlled study. *Medicine*, *98*(37), e17186.

Lopresti, A.L., Smith, S.J., & Drummond, P.D. (2023). *Withania somnifera* improves hippocampal neurogenesis and memory in stressed adults: A randomized, double-blind, placebo-controlled trial. *Journal of Psychopharmacology*, *37*(8), 789–801.

Mikulska, P., Malinowska, M., Ignacyk, M., Szustowski, P., Nowak, J., Pesta, K., …, & Cielecka-Piontek, J. (2023). Ashwagandha (*Withania somnifera*)—Current research on the health-promoting activities: A narrative review. *Pharmaceutics*, *15*(4), 1057.

Naik, K.S., Muralidhar, G., Rao, B.S., & Wankhade, K. (2024). An 8-week oral therapy with Ashwagandha (*Withania somnifera*) root extract (600 mg/day) improves frailty and quality of life in elderly: A prospective, randomized, placebo-controlled, efficacy and safety study. *medRxiv*. https://doi.org/10.1101/2024.10.15.24315515

National Center for Complementary and Integrative Health (NCCIH). (2023 March). Ashwagandha. U.S. Department of Health & Human Services. National Institutes of Health. https://www.nccih.nih.gov/health/ashwagandha

Ng, Q.X., Loke, W., Foo, N.X., Tan, W.J., Chan, H.W., Lim, D.Y., & Yeo, W.S. (2020). A systematic review of the clinical use of *Withania somnifera* (Ashwagandha) to ameliorate cognitive dysfunction. *Phytotherapy Research*, *34*(3), 583–590.

Paul, S., Chakraborty, S., Anand, U., Dey, S., Nandy, S., Ghorai, M., …, & Dey, A. (2021). *Withania somnifera* (L.) Dunal (Ashwagandha): A comprehensive review on ethnopharmacology, pharmacotherapeutics, biomedicinal and toxicological aspects. *Biomedicine & Pharmacotherapy*, *143*, 112175.

Polumackanycz, M., Petropoulos, S.A., Śledziński, T., Goyke, E., Konopacka, A., Plenis, A., & Viapiana, A. (2023). *Withania somnifera* L.: Phenolic compounds composition and biological activity of commercial samples and its aqueous and hydromethanolic extracts. *Antioxidants*, *12*(3), 550.

Połumackanycz, M., Forencewicz, A., Wesołowski, M., & Viapiana, A. (2020). Ashwagandha (*Withania somnifera* L.)—The plant with proven health-promoting properties. *Farmacja Polska*, *76*(8), 442–447.

Ratnaparkhi, H.H., Gajbhiye, P.R., Kohale, N.B., Rathod, S.B., Kadam, J.S., & Katare, K.D. (2023). Historical, chemical and cardiovascular perspective on Ashwagandha. *International Journal of Humanities Social Science and Management*, *3*(2), 455–462.

Raziya Banu, M., Ibrahim, M., Prabhu, K., & Rajasankar, S. (2020). Withaferin-A protects the nigral dopamine neuron and recovers motor activity in aged rats. *Cells Tissues Organs*, *208*(1-2), 59–65.

Saha, P., Ajgaonkar, S., Maniar, D., Sahare, S., Mehta, D., & Nair, S. (2024). Current insights into transcriptional role (s) for the nutraceutical *Withania somnifera* in inflammation and aging. *Frontiers in Nutrition*, *11*, 1370951.

Salve, J., Pate, S., Debnath, K., & Langade, D. (2019). Adaptogenic and anxiolytic effects of Ashwagandha root extract in healthy adults: A double-blind, randomized, placebo-controlled clinical study. *Cureus*, *11*(12), e6466.

Santhanu, K., & Senthil, K. (2021). Therapeutic potential of *Withania somnifera* (Linn) Dunal (Ashwagandha) in historical perspective and pharmacological evidence. *Annals of Ayurvedic Medicine*, *10*(2), 135–147.

Sharma, A.K., Basu, I., & Singh, S. (2018). Efficacy and safety of Ashwagandha root extract in subclinical hypthyroid patients: A double-blind, randomized placebo-controlled trial. *Journal of Alternative and Complementary Medicine* (New York, N.Y), *24*(3), 243–248.

Singh, N., Bhalla, M., de Jager, P., & Gilca, M. (2023). *Withania somnifera* (Ashwagandha): Current research on neuroprotective and neurodegenerative effects. *Phytomedicine*, *109*, 154582.

Speers, A.B., Cabey, K.A., Soumyanath, A., & Wright, K.M. (2021). Effects of *Withania somnifera* (Ashwagandha) on stress and the stress-related neuropsychiatric disorders anxiety, depression, and insomnia. *Current Neuropharmacology*, *19*, 1468–1495.

Syed, A.A., Reza, M.I., Singh, P., Thombre, G.K., & Gayen, J.R. (2021). *Withania somnifera* in neurological disorders: Ethnopharmacological evidence, mechanism of action and its progress in delivery systems. *Current Drug Metabolism*, *22*(7), 561–571.

Tancreda, G., Ravera, S., & Panfoli, I. (2025). Preclinical evidence of *Withania somnifera* and *Cordyceps* spp.: Neuroprotective properties for the management of Alzheimer's disease. *International Journal of Molecular Sciences*, *26*(11), 5403.

Tiwari, S., Atluri, V., Kaushik, A., Yndart, A., & Nair, M. (2019). Alzheimer's disease: Pathogenesis, diagnostics, and therapeutics. *International Journal of Nanomedicine*, *14*, 5541–5554.

Wiciński, M., Fajkiel-Madajczyk, A., Kurant, Z., Liss, S., Szyperski, P., Szambelan, M., …, & Sadowska-Krawczenko, I. (2024). Ashwagandha's multifaceted effects on human health: Impact on vascular endothelium, inflammation, lipid metabolism, and cardiovascular outcomes—A review. *Nutrients*, *16*(15), 2481.

Zhang, L., Wang, Y., & Kumar, V. (2024). Neuroprotective mechanisms of Ashwagandha root extract in a mouse model of chronic stress. *Journal of Ethnopharmacology*, *328*.

9 *Ocimum gratissimum* for Managing and Treating Dementia

Grace Ayobami Fajemidagba, Niyi Ebenezer Omojadegbe, Abraham Olufemi Asuku, and Maryam Tayo Ayinla

9.1 INTRODUCTION

Dementia is a progressive disorder affecting multiple cognitive domains, leading to decline in cognitive function, memory loss, and behavioral impairment of sufficient severity to interfere with normal daily functional abilities of an affected individual as the disease progresses (Hafiz et al., 2023; Asuku et al., 2025; Choudry et al., 2025). Almost 55 million people suffer from dementia, with about 10 million new cases being added almost every year (Hafiz et al., 2023). Due to its rapid growth of prevalence and impact on patients' quality of life, one of the major health challenges globally is dementia.

Despite the advancement in the diagnosis and management of dementia, there are no drugs with dementia-modifying properties available yet, nor is there any curative treatment; likewise, currently available treatment options focus only on symptomatic management and present with undesirable side effects, limiting existing therapies (Perneczky, 2019; Ayinla et al., 2020; Hafiz et al., 2023). These limitations have consequently motivated research efforts to find other management therapies against dementia. In light of this, compounds in medicinal plants claimed to promote mental and cognitive functions are being studied for their potential in the treatment of dementia.

Eighty percent of people worldwide, according to WHO, rely on traditional therapy for their health care, with most of these medicines primarily plant products because of their easy availability, low cost, safety, and people's faith in them (Amita & Shalini, 2014; Ugbogu et al., 2021; Olaniyi et al., 2025).

Ocimum gratissimum, commonly known as African basil, scent leaf, sweet basil, clove basil, tea-bush, or fever plant, is a perennial plant and medicinal herb that has been utilized for centuries for treating and managing various diseases (Udi et al., 2025). It contains various phytochemicals and essential oils that have been reported to possess pharmacological properties, including scavenging free radicals, stimulating detoxifying enzymes to render harmful chemicals harmless, stimulating chemicals to maintain a body state of well-being, and anti-inflammatory, hepatoprotective, anticancer, antihypertensive, antidiabetic, antidiarrheal, and antimicrobial properties, which promote its therapeutic use (Obho, 2006; Ugbogu et al., 2021; Edo et al., 2023).

It has been used traditionally for treating a diverse range of diseases such as diarrhea, microbial infections, respiratory tract infections, rheumatism, hemorrhoids, anemia, dermatological conditions (Ogundoju et al., 2023; Bangar et al., 2024), epilepsy, high fever, headache, pneumonia, cough, fever, conjunctivitis, inflammation, pain, diabetes, mental illness, cancer (Ugbogu et al., 2021), treatment of memory disorders, anxiety, and depression (Tesi et al., 2022; Udi et al., 2025).

O. gratissimum's conventional use in the management of brain and mental function-related disorders has led to various scientific investigations into its potential for the management of dementia, of which numerous studies have reported its neuroprotective properties, ability to enhance memory, reduce anxiety, and protect against cognitive impairment (Udi et al., 2025), therefore highlighting

DOI: 10.1201/9781003657385-9

its significance as a potential natural source of a complementary or alternative agent for the treatment and management of dementia.

9.2 *OCIMUM GRATISSIMUM*

O. gratissimum is a perennial herbaceous shrub commonly known as African basil, clove basil, African teabush, and scent leaf or fever plant (Figure 9.1). It is indigenous to Africa's tropical and subtropical areas, including Nigeria, Kenya, Ghana, Ivory Coast, Cameroon, Liberia, Benin, Togo, Sierra Leone, Congo, Uganda, Angola, and Tanzania, and Ethiopia, and has also been introduced and naturalized in other countries like South America, Southeast Asia, and the Caribbean (Udi et al., 2025).

O. gratissimum is cultivated for various purposes, mainly its medicinal and nutritional purposes. It is added to food as a spice and natural flavoring agent (Mann, 2012; Ugbogu et al., 2021), in the preparation of teas and infusions, and in traditional medicines for treating different disease conditions (Ugbogu et al., 2021).

The plant possesses various pharmacological properties, including anti-inflammatory, antioxidative, antidiarrheal, anesthetic, anti-stress, antinociceptive, antidiabetic, chemotherapeutic, gastroprotective, antipyretic, hepatoprotective, anti-mutagenic, anti-anemic, insecticidal, and antiseptic, antimicrobial properties ranging from antifungal, antibacterial, antiviral, anthelminthic, and antiparasitic properties, as well as other pharmacological activities, validating its extensive medicinal use (Ugbogu et al., 2021; Udi et al., 2025).

9.2.1 Botanical Description of *Ocimum gratissimum*

O. gratissimum is an herbaceous plant with distinctive slightly camphoraceous scent. It is about one to three meters in height with branched erect square-shaped stems and green-colored ovate to oblong-shaped leaves of about 3—4 cm long and 1–2 cm broad (Jahanger et al., 2023; Udi et al., 2025). The plant stomata are mostly present on the lower surface of the leaf and consist of verticillaster inflorescence made up of purple to pinkish color flowers arranged along the upper part of the stems, fruits consisting of, one-seeded dry nutlets enclosed in the calyx and reddish black in color small seeds (Zareiyan & Khajehsharifi, 2022) (Figure 9.2).

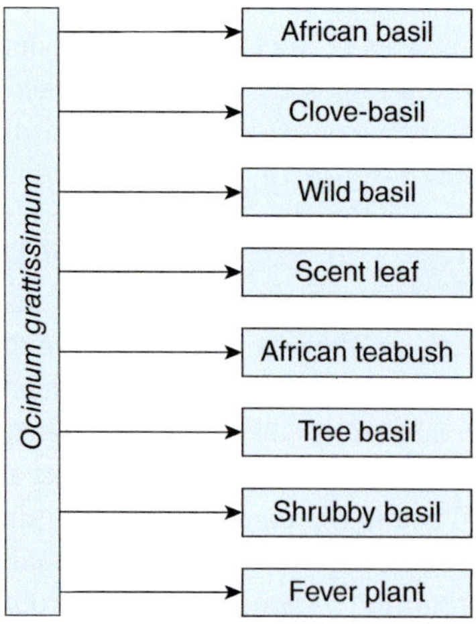

FIGURE 9.1 Some common names of *Ocimum gratissimum*.

FIGURE 9.2 *Ocimum gratissimum* plant.

9.2.1.1 *Ocimum gratissimum* Taxonomical Classification

O. gratissimum is a member of the genus *Ocimum*, named from 'ozo,' a Greek word meaning 'to smell,' because of the strong odor of the species that are in the genus (Monga al., 2017; Udi et al., 2025). The genus contains about 30–60 species, with *O. gratissimum* and a number of other species in the genus that have a similar clove-like scent categorized under the section Basilicum (Jahanger et al., 2023; Udi et al., 2025). The taxonomic hierarchy of *O. gratissimum* is given by ITIS (Integrated Taxonomic Information System), as shown in Table 9.1.

9.2.2 Phytochemical Composition of *Ocimum gratissimum*

The O. gratissimum phytochemical screening report has revealed that the plant possesses many bioactive compounds, including essential oils and phytochemicals with several beneficial pharmacological properties.

Phytochemicals present include the phenolic compounds like gallic acid, sinapic acid, ellagic acid, rosmarinic acid, caffeic acid, L-chicoric acid, myricetin, salvigenin, methyleugenol, L-caftaric, trans-ferulic acid, triterpenes such as ursolic acid, pomolic acid, oleanolic and tormentic acid, and flavonoids, including rutin, quercetin, quercitrin, quercetin 3-O-glucoside, cirsimaritin, luteolin 7-O-glucoside, vicenin-2, epicatechin, xanthomicrol, apigenin, kaempferol, 7,4,0-dimethyl ether,

TABLE 9.1
Taxonomical Classification of *Ocimum gratissimum*

Kingdom	Plantae
Sub-kingdom	Viridiplantae
Infra-kingdom	Streptophyta (land plants)
Super-division	Embryophyta
Division	Tracheophyta (vascular plants)
Sub-division	Spermatophytina (seed plants)
Class	Magnoliopsida (Dicotyledons)
Sub-class	Asteridae
Super-order	Asteranae
Order	Lamiales
Family	Lamiaceae (Mint family)
Genus	*Ocimum* (basil)
Species	*Ocimum gratissimum*

luteolin 5-O-glucoside, isovitexin, vitexin, nepetoidin A, nevadensin, hymenoxin, basilimoside, isothymusin, and morin (Ugbogu et al., 2021; Edo et al., 2023).

O. gratissimum essential oil contains compounds such as oxygenated monoterpenes, including carvacrol, thymol, thymol methyl ether, estragole, cis- and trans-sabinene hydrate, 1.8-cineole, trans-thujone, linalool, borneol, umbellulone, citronellal, p-cymene-7-ol, terpinen-4-ol, and α-terpineol; hydrocarbonated monoterpenes, including α-terpinene, α-pinene, α-thujene, β-phellandrene, p-menthane-1,3,8-triene, β-pinene, p-cymene, γ-terpinene, p-cymenene, terpinolene, trans-β-ocimene, β-myrcene, camphene, limonene, sabinene α- and δ-3-carene; oxygenated sesquiterpenes, including 1,2-epoxydehumulene, 3,7-(11)-eudesmadiene, spathulenol and caryophyllene oxide; hydrocarbonated sesquiterpenes like α-trans-bergamotene, α-copaene, α-humulene, α-guaiene, α-selinene, β-selinene, β-caryophyllene, β-elemene, β-bourbonene, δ-cadinene, germacrene D, γ-selinene, γ-elemene, and 7-epi-α-selinene (Ugbogu et al., 2021) (Figure 9.3).

Many of these phytochemicals and essential oil compounds exhibit various biological activities, such as antioxidative activities, anti-inflammatory effects, neuroprotective effects, the ability to modulate neurotransmitter levels, the inhibition of amyloid-beta aggregation, the reduction of neuronal apoptosis, and the improvement of cognitive functions, which are all properties relevant to dementia treatment.

9.2.3 Medicinal Uses of *Ocimum gratissimum*

O. gratissimum has been widely utilized in treating various diseases across different regions of the world. In Nigeria, the plant is used for treating respiratory ailment, diarrhea, fever, epilepsy mental illness, and as anthelminthic (Effraim et al, 2003; Akinmoladun et al., 2007; Ugbogu et al., 2021). It is also used for treating headache, skin, and ophthalmic problem, pneumonitis, component of malaria treatment, preparation of mouthwash, prevention of postnatal sepsis in addition to other therapeutic use (Ugbogu et al., 2021; Edo et al., 2023).

In Cameroon, *O. gratissimum* infusions are used as tonic and treatment of chest or respiratory issues (Ugbogu et al., 2021), its root decoction is used as a sedative for children in Brazil and several formulation of the plant in Cote d'Ivoire are used for the treatment of ear infections, eyes problems, and skin diseases (Ugbogu et al., 2021). In Benin republic, it is utilized in the

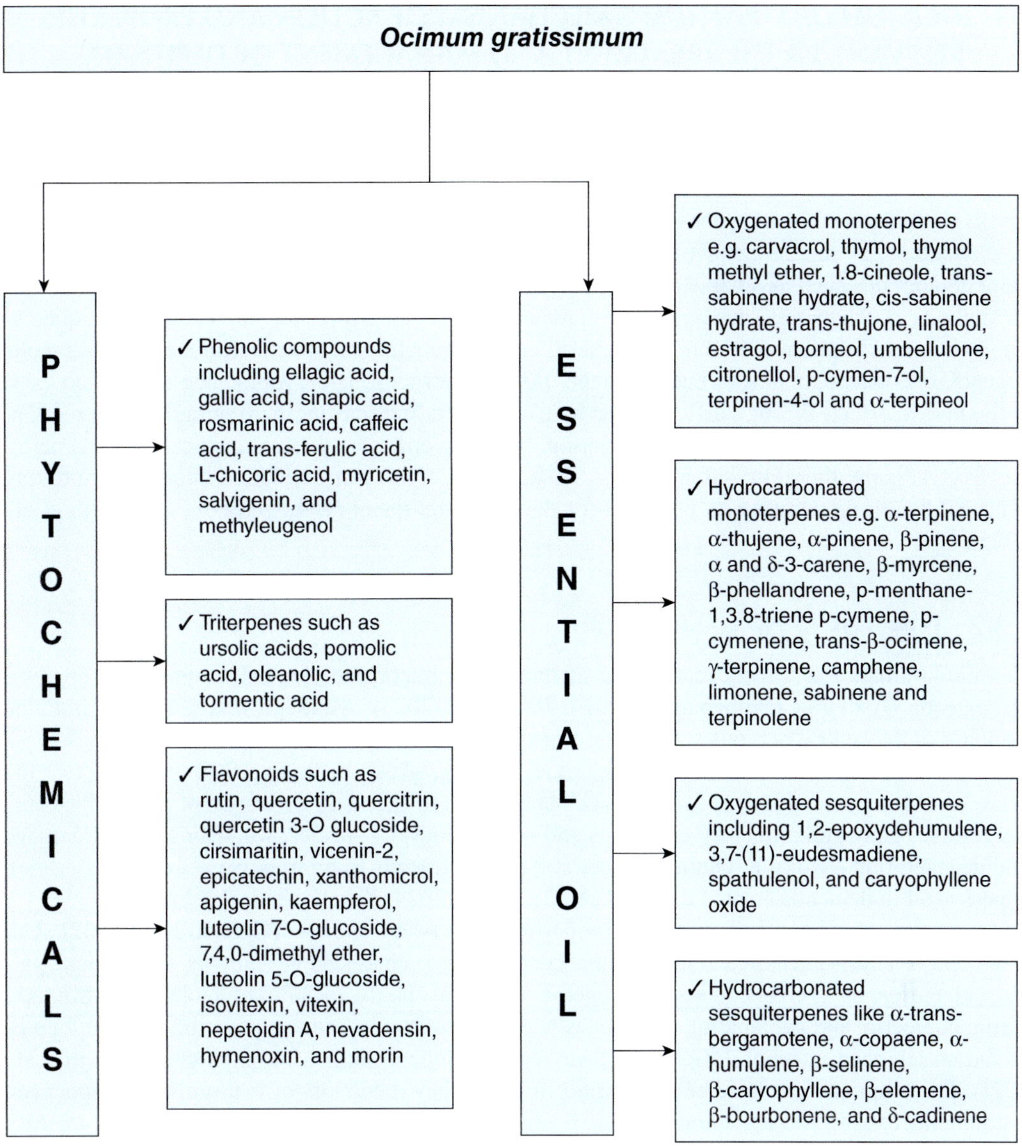

FIGURE 9.3 Phytochemicals and essential oil of *Ocimum gratissimum*.

treatment of pelvic ache, colic, emesis, dysmenorrhea, fungal infection, wound infections, dystopias, diarrhea and hemorrhoids, anginas, cough, asthma, fever, malnutrition, headaches, and hepatitis (Ugbogu et al., 2021); in Kenya and Sub-Saharan Africa, *O. gratissimum* is used to treat coughs, convulsions, sore eyes, infections, abdominal pains, barrenness, fever, rectum prolapse, and menstruation regulation (Matasyoh et al., 2007); in Togo, the plants are used for the treatment of cough, hematuria, and urethritis (Ugbogu et al., 2021); and in India, it is used as diaphoretic and for treating sunstroke, headache, and influenza (Sagar et al., 2025). The plant has also been traditionally employed to treat ulcers (Edo et al., 2023), for wound healing (Orafidiya et al., 2006), wound-dressing, prevention of postnatal sepsis, preparation of mouthwash, treatment of sore throat, tonsillitis, component of malaria treatment, etc. (Agholor et al., 2018; Ugbogu et al., 2021; Edo et al., 2023).

9.3 *OCIMUM GRATISSIMUM'S* MECHANISM OF ACTION AND PROPERTIES RELEVANT TO THE TREATMENT AND MANAGEMENT OF DEMENTIA

9.3.1 Antioxidant Activity

Oxidative stress plays a major role in dementia pathogenesis. It causes an imbalance between antioxidant defenses and production of reactive oxygen species (Akintoye et al., 2021), which is a major key factor in dementia pathogenesis (Ebrahim et al., 2024).

Phytochemical screening have shown that *O. gratissimum* possesses bountiful bioactive antioxidant compounds such as sinapic acids, gallic acid, caffeic acid, chlorogenic acid, triterpenes like oleanolic acid, flavonoids such as apigenin, nepetoidin, nevadensin, catechin, epicatechin, quercetin, rutin, and kaempferol, and its essential oil components like camphene, limonene, 1,8-cineole, eugenol, α-copaene, α-pinene and β-pinene, thymol carvacrol, β-caryophyllene and which exert antioxidative effects through different mechanisms, including scavenging free radicals, upregulating antioxidant enzymes, metal ions chelation, blocking enzyme systems responsible for generating free radicals, therefore safeguarding the neuronal cells from oxidative damage, improving neuronal health pinpointing its potential for dementia treatment (Ugbogu et al., 2021; Udi et al., 2025).

9.3.2 Inhibition of Neuroinflammation

Neuroinflammation is widely recognized as one of the major pathological mechanisms involved in dementia (Asuku & Ogungbangbe, 2024; Pan et al., 2024). Although it is a cellular immune response of the central nervous system that functions initially as a natural, innate, adaptive defense response to events such as infections, ischemia, injuries, stress, or trauma, helping to protect the brain, its chronic, sustained, or dysregulated state results in excessive activation of glial cells, causing release of pro-inflammatory cytokines and reactive oxygen species, worsening neural damage and degeneration (Asuku & Ogungbangbe, 2024), which leads to progressive neural deterioration experienced in dementia-related diseases (Adamu et al., 2024; Kamila et al., 2025).

O. gratissimum's anti-inflammatory properties have been reported (Ugbogu et al., 2021; Udi et al., 2025). This is attributed to its bioactive compounds, such as sinapic acid, rosmarinic acid, gallic acid, caffeic acid, ellagic acids, chlorogenic acids, flavonoids, including luteolin, catechin, apigenin, quercetin, and kaempferol, triterpenes (e.g., oleanolic acid), and its essential oil compounds, including sabinene, eugenol, 1,8-cineole, β-myrcene, limonene, and α- and β-pinene (Ugbogu et al., 2021). These compounds have been reported to inhibit key mediators of neuroinflammation, such as tumor necrosis factor-alpha, interleukin-1β, and interleukin-6 production, reducing inflammatory response and therefore protecting against neuronal damage and apoptosis (Ugbogu et al., 2021; Udi et al., 2025).

9.3.3 Neuroprotective Effects

Various studies have explored and reported the neuroprotective properties of *O. gratissimum*. This is primarily attributed to the diverse phytochemicals present in the plants, which act via various neuroprotective mechanisms, which include modulating levels of neurotransmitters, protecting neurons from oxidative damage and neuroinflammation, and consequently promoting neuronal health (Udi et al., 2025).

9.3.4 Acetylcholinesterase Inhibitory Activities

In disorders associated with dementia, such as Alzheimer's disease, Parkinson's disease, Lewy body dementia, and vascular dementia, cholinergic dysfunction has been linked to the impaired cognitive function and some behavioral symptoms, including agitation, psychosis, apathy, experienced in these disorders (Asuku et al., 2025).

O. gratissimum's ability to inhibit acetylcholinesterase has been reported by various studies. Different extracts of its leaves have been reported to possess cholinesterase inhibitory properties in vivo (Ayinla et al., 2019a, b, c), in vitro (Owokotomo et al., 2015), and in silico studies (Njan et al., 2023). Similarly, essential oils extracted from *O. gratissimum* seeds were reported to demonstrate acetylcholine esterase inhibition activities similar to galantamine, attributed to the presence of monoterpenes highly present in the oil, such as α-pinene and γ-terpinene, which have the ability to cross the blood-brain barrier, where they interact with the active site of cholinesterase enzymes (Owokotomo et al., 2015).

By inhibiting acetylcholine breakdown through its cholinesterase inhibitory activity, *O. gratissimum* can help maintain optimal levels of acetylcholine, therefore supporting cholinergic function and leading to improvement in cognitive performance and alleviation of behavioral symptoms of dementia associated with cholinergic dysfunction, such as agitation, apathy, sleep disturbance, psychosis, and confusion, offering potential treatment for people living with dementia.

9.3.5 Enhancing the Expression of Neurotrophic Factor

O. gratissimum has been reported to enhance the expression of neurotrophic factors such as brain-derived neurotrophic factor (BDNF) and nerve growth factor (NGF), which are essential proteins that play important roles in protecting neurons and promoting neural growth, development, and survival, in addition to enhancing neuronal plasticity, a major process involved in cognitive functions, emotions, and behavior (Rehman et al., 2019; Nagappan et al., 2020; Abdelhalim & Hanrahan, 2021; Udi et al., 2025).

9.3.6 Enhancement of Synaptic Plasticity

O. gratissimum's ability to enhance synaptic plasticity has been reported. It is reported to promote various synaptic plasticity enhancing mechanism such as promoting neurotransmitters release, improving neurotransmitter-receptor interactions, and various event that improve synaptic strength (Aigbomian et al., 2022; Ajiboye et al., 2024).

9.3.7 Alleviation of Behavioral Deficit and Cognitive Function

Studies evaluating the cognitive function of *O. gratissimum* have yielded positive results, with *O. gratissimum* found to alleviate behavioral deficits and improve learning processes, memory, and other cognitive functions in different experimental models (Freire et al., 2006; Thalla et al., 2012; Ajibola et al., 2015; Singh et al., 2016; Ajayi et al., 2017; Ayinla et al., 2019a; Njan et al., 2023).

9.3.8 Anxiolytic and Antidepressant Effects

Anxiety and depression are common symptoms of people with dementia (Kwak et al., 2017; Goodarzi et al., 2019; Yu et al., 2020); similarly, people with depression or anxiety are believed to have a higher risk of developing dementia (Santabárbara et al., 2020; Elser et al., 2023).

O. gratissimum anxiolytic and antidepressant properties have been reported, with these properties primarily attributed to its bioactive compounds such as eugenol, thymol, and linalool (Zareiyan & Khajehsharifi, 2022; Chinko & Orlu, 2023), which have been found to ameliorate anxiety and depression-like behaviors in various studies.

9.3.9 Antiapoptotic Potential

The ability of *O. gratissimum* to inhibit apoptosis has been reported (Figure 9.4). It was found to activate anti-apoptotic signaling pathways, such as the PI3K/Akt pathway, and also hinder the release of proapoptotic proteins, consequently inhibiting the production of caspase, a protease that plays an important role in cellular breakdown during apoptosis activity (Salemcity et al., 2023; Olorundare et al., 2024). Although this antiapoptosis mechanism of *O. gratissimum* was reported in other disorders apart from dementia (Chen et al., 2011; Venuprasad et al., 2014; Chao et al., 2017; Curtis et al., 2024; Udi et al., 2025), this mechanism might also be relevant to dementia, which is a disorder in which apoptosis also plays a role.

9.3.10 Enhancement of Blood–brain Barrier Integrity

Bioactive compounds in *O. gratissimum with the* ability to enhance blood–brain barrier integrity have been suggested. In addition to the ability of these compounds to modulate tight junctions, their ability to suppress the expression of metalloproteinase, a class of enzymes involved in the breakdown of extracellular matrix that could compromise the blood–brain barrier integrity, has been reported (Nangia-Makker et al., 2013; Perna et al., 2022; Njan et al., 2023; Udi et al., 2025).

9.3.11 Regulation of Mitochondrial Function

Oxidative stress and inflammation are processes that can cause mitochondrial dysfunction and neuronal death. *O. gratissimum's anti*oxidative and anti-inflammatory properties have been documented by numerous studies showing its ability to protect mitochondrial membrane from damage helping to maintain mitochondria function (Oyem et al., 2021; Ajiboye et al., 2024; Udi et al., 2025)

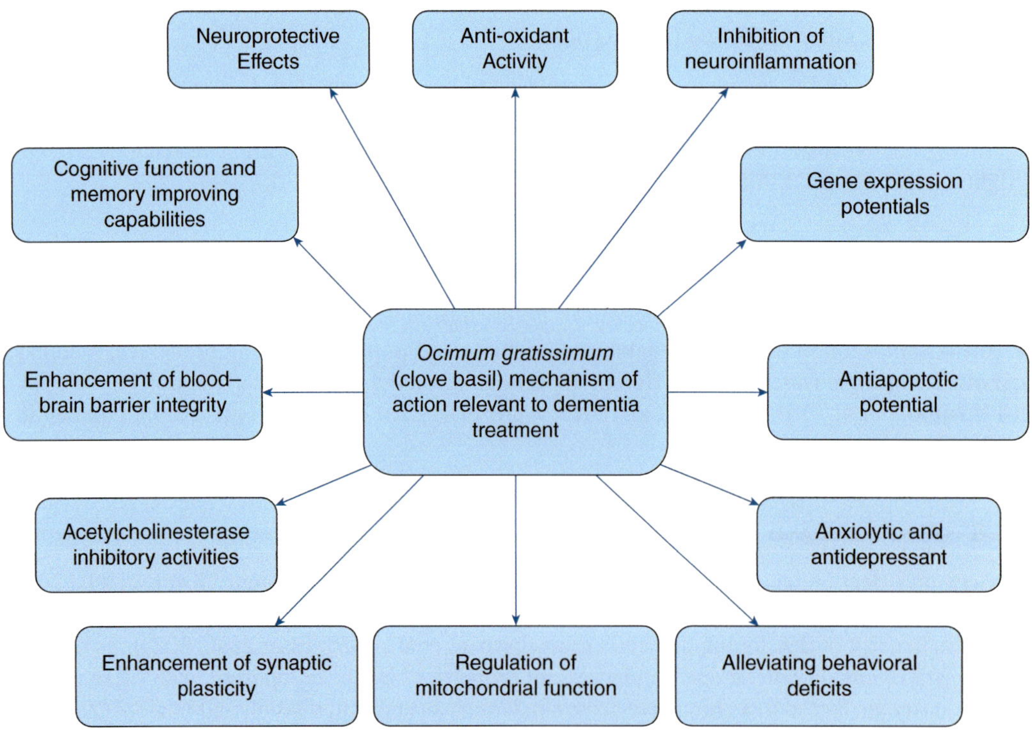

FIGURE 9.4 The mechanisms of actions of *Ocimum gratissimum* on dementia.

as shown in Figure 9.4. It has also been reported to promote mitochondrial respiration and ATP production which are important for energy supply needed for all body function, including neural function (Seyed et al., 2018, 2021; Salemcity et al., 2023).

9.3.12 Modulation of Gene Expression

O. gratissimum's ability to modulate the signaling of nuclear factor erythroid 2-related factor 2 (NRF2), a transcription factor that controls the expression of cytoprotective and antioxidant genes and is essential for protecting neurons from oxidative damage, has been reported (Iizuka et al., 2021), showing that *O. gratissimum* could play a role in the maintenance of neuronal development, survival, and plasticity.

9.4 FUTURE DIRECTION FOR RESEARCH IN RELATION TO THE USE OF *OCIMUM GRATISSIMUM* FOR THE MANAGEMENT AND TREATMENT OF DEMENTIA

Although the various mechanisms of action of *O. gratissimum* suggest that the plant and its active compounds could be useful in managing and treating dementia, longitudinal and mechanistic studies should be carried out to further investigate these mechanisms through which *O. gratissimum* exerts its effects and treatment potentials against dementia, determine the optimal dosing for its therapeutic use in dementia management, validate its efficacy for dementia treatment in humans, and investigate the optimal safe dosing for dementia management in humans.

9.5 CONCLUSION

Existing pharmaceutical treatments for dementia present with limitations considering their inability to cure the disorder and the associated side effects of their long-term use. *O. gratissimum* is a medicinal plant that contains many beneficial bioactive compounds, which act through multifaceted mechanisms, many of which are relevant to managing and treating dementia. Use of *O. gratissimum* alone or its synergy use with conventional dementia drugs or other natural compounds with also good anti-dementia properties could help provide a novel approach to dementia treatment and management.

REFERENCES

Abdelhalim, A., & Hanrahan, J. (2021). Biologically active compounds from Lamiaceae family: Central nervous system effects. *Studies in Natural Product Chemistry*, 68, 255–315.

Adamu, A., Li, S., Gao, F., & Xue, G. (2024). The role of neuroinflammation in neurodegenerative diseases: Current understanding and future therapeutic targets. *Frontiers in Aging Neuroscience*, 16, 1347987.

Agholor, K., Lucy, M. Y., Idris, A., Lucy, F. O., & Rakiya, Z. (2018). Antibacterial activity of Ocimum gratissimum (scent leaf) on some pathogenic gastrointestinal bacteria. *African Journal of Microbiology Research*, 12(40), 923–929.

Aigbomian, O.E., Senjobi, C.T., Onasanwo, S.A., Olatunde, M.O., & Fasola, T.R. (2022). Neurobehavioural and analgesic effect of *Ocimum gratissimum* Linn leaves essential oil in Wistar albino mice. *Biotechnology Acta*, 15(5), 47–57.

Ajayi, A.M., Ologe, M.O., Ben-Azu, B., Okhale, S.E., Adzu, B., & Ademowo, O.G. (2017). *Ocimum gratissimum* linn leaf extract inhibits free radical generation and suppressed inflammation in carrageenan-induced inflammation models in rats. *Journal of Basic and Clinical Physiology and Pharmacology*, 28(6), 531–541.

Ajibola, M.I., Ibrahim, R.B., Imam, A., Masud, M.A., Safiriyu, A., & Etibor, A.T. (2015). Neurodegenerative potential of the aqueous leaf extract of *Ocimum gratissimum:* A histological and biochemical study. *Anatomy Journal of Africa*, 4(2), 563–570.

Ajiboye, B.O., Famusiwa, C.D., Amuda, M.O., Afolabi, S.O., Ayotunde, B.T., & Adejumo, A.A. (2024). Attenuation of PI3K/AKT signaling pathway by *Ocimum gratissimum* leaf flavonoid-rich extracts in streptozotocin-induced diabetic male rats. *Biochemistry and Biophysics Reports*, 38, 101735.

Akinmoladun, A.C., Ibukun, E.O., Afor, E., Obuotor, E.M., & Farombi, E.O. (2007). Phytochemical constituent and antioxidant activity of extract from the leaves of *Ocimum gratissimum*. *Scientific Research and Essay*, 2, 163–166.

Akintoye, O.O., Ajibare, A.J., Fabunmi, O.A., Asuku, A.O., & Tajudeen, R.A. (2021). Effect of honey on oxidative stress in brain tissue of sleep deprived rats. *Nigerian Journal of Basic and Applied Science*, 29(1), 83–89. http://dx.doi.org/10.4314/njbas.v29i1.10

Amita, P., & Shalini, T. (2014). Concept of standardization, extraction and pre phytochemical screening strategies for herbal drug. *Journal of Pharmacognosy and Phytochemistry*, 2(5), 115–119.

Asuku, A.O., & Ogungbangbe, G.O. (2024). Neuroinflammation and Binge Eating. In: Neuroscientific Insights and Therapeutic Approaches to Eating Disorders. IGI Publisher, 10, 147–170. https://doi.org/10.4018/979-8-3693-3230-6.ch010

Asuku, A.O., Ogungbangbe, G.O., & Olajide, T.S. (2025). Effect of Happiness and Positive Emotions on Dementia. In: Neuroscience of Feelings. Progress in Brain Research, Elsevier, https://doi.org/10.1016/bs.pbr.2025.03.001

Ayinla, M.T., Asuku, A.O., Bayo-Olugbami, A., Ayeni, A.O., Abiola, A.A., & Owoyele, B.V. (2020). Melatonin and vitamin C modulate cholinergic neurotransmission and oxidative stress in scopolamine–induced rat model of memory impairment. *Journal of African Association of Physiological Sciences*, 8(1), 20.

Ayinla, M.T., Fajemidagba, G.A., Owoyele, B.V., & Oyewole, L.A. (2019a). Effect of n-hexane extracts of *Ocimum gratissimum* and *Momordica charantia* leaves on learning and memory in scopolamine-induced rat model of dementia. *LASU Journal of Medical Sciences*, 4(1).

Ayinla, M.T., Owoyele, B.V., Abdulmajeed, W., & Imam, A. (2019b). Effects of aqueous extract of *Ocimum gratissimum* and *Momordica charantia* on learning and memory in scopolamine-induced dementia rats model. *IBRO Reports*, *7*, S2–S51. https://doi.org/10.1016/j.ibror.2019.09.026

Ayinla, M.T., Uthman, Y., Bayo-Olugbami, A., Oyewole, A.L., & Owoyele, B.V. (2019c). Effects of ethyl-acetate leaf extracts of *Ocimum gratissimum* and *Momordica charantia* on memory in scopolamine-induced dementia rats. *Nigerian Journal of Biochemistry and Molecular Biology*, 34(1 &2), 28–37.

Bangar, S.B., Bankar, S.S., Vibhute, A.S., Yadav, A.R., Pednekar, H.R., Dsilva, L.S., & Patil, S. (2024). Review on therapeutic approach of anti-acne treatment by *Azadirachta indica* and *Ocimum gratissimum*. *World Journal of Advanced Research and Reviews*, 22(2), 165–174.

Chao, P.Y., Lin, J.A., Ye, J.C., Hwang, J.M., Ting, W.J., Huang, C.Y., & Liu, J.Y. (2017). Attenuation of oxidative stress-induced cell apoptosis in Schwann RSC96 cells by *Ocimum gratissimum* aqueous extract. *International Journal of Medical Sciences*, 14(8), 764.

Chen, H.M., Lee, M.J., Kuo, C.Y., Tsai, P.L., Liu, J.Y., & Kao, S.H. (2011). *Ocimum gratissimum* aqueous extract induces apoptotic signalling in lung adenocarcinoma cell A549. *Evidence-Based Complementary and Alternative Medicine*, 2011, 739093.

Chinko, B.C., & Orlu, C.N. (2023). Antipyretic and anti-inflammatory effects of *Ocimum gratissimum* in male Wistar rats. *South Asian Research Journal of Natural Products*, 6(1), 28–38.

Choudry, M., Gould, M., & Ganti, L. (2025). Treatment of agitation in dementia: A systematic review. *International Journal of Emergency Medicine*, 18, 101. https://doi.org/10.1186/s12245-025-00902-7

Curtis, R.M., Wang, H.S., Luo, X., Dugo, E.B., Stevens, J.J., & Tchounwou, P.B. (2024). Fractionated leaf extracts of *Ocimum gratissimum* inhibit the proliferation and induce apoptosis of A549 lung adenocarcinoma cells. *Nutrients*, 16(16), 2737.

Ebrahim, L.K., Adel, S., & Hamza, S.A. (2024). Investigating the association between blood oxidative stress markers and dementia in Egyptian elderly women. *Egyptian Journal of Neurology, Psychiatry and Neurosurgery*, 60, 73.

Edo, G.I., Samuel, P.O., Ossai, S., Nwachukwu, S.C., Okolie, M.C., Oghenegueke, O., Asaah, E.U., Akpoghelie, P.O., Ugbune, U., Owheruo, J.O., Ezekiel, G.O., Onoharigho, F.O., & Agbo, J.J. (2023). Phytochemistry and pharmacological compounds present in scent leaf: A review. *Food Chemistry Advances*, 3, 100300. https://doi.org/10.1016/j.focha.2023.100300

Effraim, K.D., Jacks, T.W., & Sodipo, O.A. (2003). Histopathological studies on the toxicity of Ocimum gratissimum leaf extract on some organs of rabbit. *African Journal of Biomedical Research*, 6(1), 21–25.

Elser, H., Horváth-Puhó, E., & Gradus, J.L. (2023). Association of early-, middle-, and late-life depression with incident dementia in a Danish cohort. *JAMA Neurology*, 80(9), 949–958. https://doi.org/10.1001/jamaneurol.2023.2309

Freire, C.M.M., Marques, M.O.M., & Costa, M. (2006). Effects of seasonal variation on the central nervous system activity of Ocimum gratissimum L. essential oil. *Journal of Ethnopharmacology*, 105(1–2), 161–166.

Goodarzi, Z., Samii, L., Azeem, F., Sekhon, R., Crites, S., Pringsheim, T., Smith, E. E., Ismail, Z., & Holroyd-Leduc, J. (2019). Detection of anxiety symptoms in persons with dementia: A systematic review. *Alzheimer's & Dementia: Diagnosis, Assessment & Disease Monitoring*, 11, 340–347.

Hafiz, R., Alajlani, L., Ali, A., Algarni, G.A., Aljurfi, H., Alammar, O.A.M., Ashqan, M.Y., & Alkhashan, A. (2023). The latest advances in the diagnosis and treatment of dementia. *Cureus*, 15(12), e50522.

Iizuka, Y., Inomata, K., Nohara, R., Miyawaki, A., & Watanabe, Y. (2021). Activation of intracellular antioxidant master gene Nrf2 by the extracts of *Garcinia subelliptica, Ocimum gratissimum* L., and *Plectranthus ornatus. International Journal of Herbal Medicine*, 9, 54–60.

Jahanger, M.A., Patra, K.K., Kumari, S., Singh, A., Manika, N., Srivastava, R.P., & Singh, L. (2023). A glance at the phytochemical and ethno-pharmacological understanding of four *Ocimum* species. *Current Pharmaceutical Biotechnology*, 24(9), 1094–1107.

Kamila, P. et al. (2025). Effect of neuroinflammation on the progression of Alzheimer's disease and its significant ramifications for novel anti-inflammatory treatments. *IBRO Neuroscience Reports*, 18, 771–782.

Kwak, Y.T., Yang, Y.S., & Koo, M.-S. (2017). Anxiety in dementia. *Dementia and Neurocognitive Disorders*, 16(2), 33–39. https://doi.org/10.12779/dnd.2017.16.2.33

Mann, A. (2012). Phytochemical constituents and antimicrobial and grain protectant activities of clove basil (*Ocimum gratissimum*) grown in Nigeria. *International Journal of Plant Research*, 2(1), 51–58. https://doi.org/10.5923/j.plant.20120201.08

Matasyoh, L.G., Matasyoh, J.C., Wachira, F.N., Kinyua, M.G., Muigai, A.W.T., & Mukiama, T.K. (2007). Chemical composition and antimicrobial activity of the essential oil of Ocimum gratissimum L. growing in Eastern Kenya. *African Journal of Biotechnology*, 6(6): 760–765.

Monga, S., Dhanwal, P., Kumar, R., Kumar, A., & Chhokar, V. (2017). Pharmacological and physico-chemical properties of Tulsi (Ocimum gratissimum L.): An updated review. *The Pharma Innovation Journal*, 6(4), 181–186.

Nagappan, P.G., Chen, H., & Wang, D.Y. (2020). Neuroregeneration and plasticity: A review of the physiological mechanisms for achieving functional recovery postinjury. *Military Medical Research*, 7, 1–16.

Nangia-Makker, P., Raz, T., Tait, L., Shekhar, M.P., Li, H., Balan, V., & Raz, A. (2013). *Ocimum gratissimum* retards breast cancer growth and progression and is a natural inhibitor of matrix metalloproteases. *Cancer Biology & Therapy*, 14(5), 417–427.

Njan, A.A., Olaleye, E.O., Afolabi, S.O., Anoka-Ayembe, I., Gyebi, G.A., Nyamngee, A., & Ibrahim, H.D. (2023). Identification of neurotherapeutic constituents in *Ocimum gratissimum* with cholinesterase and mono amine oxidase inhibitory activities, using GC-MS analysis, in vitro, and in silico approaches. *Informatics in Medicine Unlocked*, 39, 101261.

Obho, G. (2006). Antioxidant and antimicrobial properties of ethanolic extract of *Ocimum gratissimum* leaves. *Journal of Pharmacology and Toxicology*, 1, 47–53.

Ogundoju, R.I., Oladunmoye, M.K., & Babatunde, O.J. (2023). GC-MS profiling and antibacterial efficacy of *Ocimum gratissimum* (Linn) against bacteria associated with gastroenteritis. *Microbes and Infectious Diseases*, 10, 12.

Olaniyi, O.M., Gbayisomore, T.J., Oluwagbamila, O.B., Adesewa, F.R., Oke, J.O., Adedotun, O.A., & Iteire, K.A. (2025). A review on some medicinal plants with therapeutic potential for Alzheimer disease and other related dementia (ADRD). *Journal of Experimental and Clinical Anatomy*, *22*(2).

Olorundare, O.E., Adeneye, A.A., Akinsola, A.O., Ajayi, A.M., Atolani, O., Soyemi, S.S., & Albrecht, R.M. (2024). Anti-apoptotic and antioxidant mechanisms may underlie the abrogative potential of *Ocimum gratissimum* Linn. leaf extract and fractions against trastuzumab-induced cardiotoxicity in Wistar rats. *Toxicology Reports*, 12, 200–214.

Orafidiya, L. O., Fakoya, F. A., Agbani, E. O., & Iwalewa, E. O. (2006). Vascular permeability-increasing effect of the leaf essential oil of Ocimum gratissimum Linn. as a mechanism for its wound healing property. *African Journal of Traditional, Complementary and Alternative Medicines*, 2(3), 253–258.

Owokotomo, I.A., Ekundayo, O., Abayomi, T.G., & Chukwuka, A.V. (2015). In-vitro anti-cholinesterase activity of essential oil from four tropical medicinal plants. *Toxicology Reports*, 2, 850–857. https://doi.org/10.1016/j.toxrep.2015.05.003

Oyem, J.C., Chris-Ozoko, L.E., Enaohwo, M.T., Otabor, F.O., Okudayo, V.A., & Udi, O.A. (2021). Antioxidative properties of *Ocimum gratissimum* alters lead acetate–induced oxidative damage in lymphoid tissues and hematological parameters of adult Wistar rats. *Toxicology Reports*, 8, 215–222.

Pan, J., Hu, J., Meng, D., Chen, L., & Wei, X. (2024). Neuroinflammation in dementia: A meta-analysis of PET imaging studies. *Medicine*, 103(18), e38086.

Perna, S., Alawadhi, H., Riva, A., Allegrini, P., Petrangolini, G., Gasparri, C., & Rondanelli, M. (2022). In vitro and in vivo anticancer activity of basil (*Ocimum* spp.): Current insights and future prospects. *Cancers*, 14(10), 2375.

Perneczky, R. (2019). Dementia treatment versus prevention. *Dialogues in Clinical Neuroscience*, 21(1), 43–51.

Rehman, M.U., Wali, A.F., Ahmad, A., Shakeel, S., Rasool, S., Ali, R., & Khan, R. (2019). Neuroprotective strategies for neurological disorders by natural products: An update. *Current Neuropharmacology*, 17(3), 247–267.

Sagar, A., Yash, M., Shalini, T.D., Vinay Kumar, R., & Gnana Babu, J. (2025). The healing potential of Ocimum gratissimum: An integrative review of its ethnomedicinal uses, phytochemistry, and pharmacological relevance. *International Journal of Pharmaceutical Research and Development*, 7(2), 454–469.

Salemcity, M.T., Olanlokun, J.O., Olowofolahan, A.O., Olojo, F.O., Adegoke, A.M., & Olorunsogo, O.O. (2023). Reversal of mitochondrial permeability transition pore and pancreas degeneration by chloroform fraction of *Ocimum gratissimum* (L) leaf extract in type 2 diabetic rat model. *Frontiers in Pharmacology*, 14, 1231826. https://doi.org/10.3389/fphar.2023.1231826

Santabárbara, J., Lipnicki, D.M., Olaya, B., Villagrasa, B., Bueno-Notivol, J., Nuez, L., López-Antón, R., & Gracia-García, P. (2020). Does anxiety increase the risk of all-cause dementia? An updated meta-analysis of prospective cohort studies. *Journal of Clinical Medicine*, *9*(6), 1791. https://doi.org/10.3390/jcm9061791

Seyed, M.A., Ayesha, S., Azmi, N., Al-Rabae, F.M., Al-Alawy, A.I., Al-Zahrani, O.R., Priyanka, C., Shivika, S., & Vikas, S. (2018). *Ocimum gratissimum*: A review on ethnomedicinal properties, phytochemical constituents, and pharmacological profile. *Biotechnological Approaches to Medicinal and Aromatic Plants: Conservation, Genetics, Improvement, and Utilization*, 10, 251–270.

Seyed, M. A., Ayesha, S., Azmi, N., Al-Rabae, F. M., Al-Alawy, A. I., Al-Zahrani, O. R., & Hawsawi, Y. (2021). The neuroprotective attribution of Ocimum basilicum: A review on the prevention and management of neurodegenerative disorders. *Future Journal of Pharmaceutical Sciences*, 7(139).

Singh, V., Kahol, A., Singh, I. P., Saraf, I., & Shri, R. (2016). Evaluation of anti-amnesic effect of extracts of selected Ocimum species using in-vitro and in-vivo models. *Journal of Ethnopharmacology*, 193, 490–499. https://doi.org/10.1016/j.jep.2016.10.026

Tesi, E.P., Ben-Azu, B., Mega, O.O., Mordi, J., Knowledge, O.O., Awele, E.D., & Eneni, O.A.E. (2022). Kolaviron, a flavonoid-rich extract ameliorates busulfan induced chemo-brain and testicular damage in male rats through inhibition of oxidative stress, inflammatory, and apoptotic pathways. *Journal of Food Biochemistry*, 46(4), e14071.

Thalla, S., Tammu, J., & Thalla, S.R. (2012). Nootropic activity of *Ocimum gratissimum* in streptozotocin induced amnesia. *Asian Journal of Research in Chemistry*, 5(12), 1437–1439.

Udi, O.A., Mega, O.O., & Adetomiwa, E.A. (2025). Exploring the cognitive-enhancing effects of *Ocimum gratissimum*: An ethnopharmacological and phytochemical review. *Discover Plants*, 2, 58. https://doi.org/10.1007/s44372-025-00127-1

Ugbogu, O.C., Emmanuel, O., Agi, G.O., Ibe, C., Ekweogu, C.N., Ude, V.C., Uche, M.E., Nnanna, R.O., & Ugbogu, E.A. (2021). A review on the traditional uses, phytochemistry, and pharmacological activities of clove basil (*Ocimum gratissimum* L.). *Heliyon*, 7(11), e08404. https://doi.org/10.1016/j.heliyon.2021.e08404

Venuprasad, M.P., Kandikattu, H.K., Razack, S., & Khanum, F. (2014). Phytochemical analysis of *Ocimum gratissimum* by LC-ESI–MS/MS and its antioxidant and anxiolytic effects. *South African Journal of Botany*, 92, 151–158.

Yu, O.-C., Jung, B., Go, H., Park, M., & Ha, I.-H. (2020). Association between dementia and depression: A retrospective study using the Korean National Health Insurance Service-National Sample Cohort database. *BMJ Open*, 10(10), e034924. https://doi.org/10.1136/bmjopen-2019-034924

Zareiyan, F., & Khajehsharifi, H. (2022). In-vitro phytochemical analysis of essential oil and methanolic and hydromethanolic extracts of *Ocimum gratissimum*. *Journal of Plant Biochemistry and Biotechnology*, 31(4), 894–906.

10 Green Tea for Managing and Treating Dementia

Eniola Oluwatomisin Adesanya, Oyaname Miracle Adejumo, Olabode Oluwadare Akintoye, Fabunmi A. Oyesanmi, and Ayodeji Johnson Ajibare

10.1 INTRODUCTION

Dementia is a progressive neurodegenerative syndrome that interferes with one's cognitive functions and quality of life (Asuku et al., 2024a). The commonest form of dementia is Alzheimer's disease, and other forms are vascular, Lewy body, and frontotemporal dementia (Asuku et al., 2022). The existing pharmacological therapy does not have much symptomatic action, which does not affect the course of the disease. Despite the positive aspects of the new approvals, they are also associated with enormous costs and safety issues that increase the importance of alternative therapies.

Green tea (*Camellia sinensis*) is a promising natural product due to its rich phytochemical content and evidence of preclinical neuroprotection. It also has an epidemiological history to explore since it is utilized globally especially in the East Asian population which have a lower occurrence of dementia. Green tea is an inexpensive and readily implementable intervention. Nonetheless, scientific investigation is needed to bridge the gap between epidemiological observation and preclinical promise and clinically meaningful effects.

Green tea consumption is high in East Asian countries like Japan, China, and Korea. The epidemiological investigations of these areas have continued to show the inverse correlation between the common consumption of green tea and the possibility of dementia or cognitive impairment (Park et al., 2011). A case in point is the Ohsaki Cohort Study in Japan where prevalence of green tea consumption is very high per capita and incidence of dementia was much lower in high green tea consumers (Tomata et al., 2016). This geographical association implies that there exists a possibility of protective effect which is to be investigated in various populations.

It is however important to note that the overall lifestyle and genetic and environmental factors which also predispose such populations to dementia ought to be taken into account. It is conceivable that the traditional East Asian diet that is in some instances rich in vegetables, fish, and soy and low in saturated fats in general may synergistically impact cognitive health (Okubo et al., 2017). It is also difficult to determine the precise effect of green tea alone because of the genetic predispositions, cultural practices, and social interaction. However, such consistent findings in these areas give a good reason to investigate the potential of using green tea as a culturally acceptable and available intervention in other regions of the world.

Furthermore, dementia is also prevalent in Western countries like Europe, where green tea is not consumed and such protective effects are less consistent in the Western world (Zhou et al., 2025). This shows the importance of culturally sensitive interventions and research in future to understand how positive effects of green tea consumption can be maximized in various environments across the world. The general promotion of green tea drinking as a healthy lifestyle practice may be a cost-effective alternative to the population health promotion, especially in the areas where access to costly pharmacological treatment is difficult to gain, yet it should be carefully counterbalanced with the local food culture and preferences.

DOI: 10.1201/9781003657385-10

10.2 PHYTOCHEMICAL COMPOSITION AND NEUROPROTECTIVE

10.2.1 Mechanisms of Green Tea

Green tea contains diverse phytochemicals, primarily catechins like epigallocatechin-3-gallate (EGCG), along with L-theanine and caffeine (Khan & Mukhtar, 2018; Natarajan et al., 2019). EGCG demonstrates neuroprotective effects through multiple, context-dependent mechanisms. It functions as a potent antioxidant, scavenging reactive oxygen species and upregulating endogenous antioxidant defenses, thereby mitigating oxidative stress and protecting mitochondrial function (Forester & Lambert, 2011) (Figure 10.1).

EGCG modulates neuroinflammation by inhibiting inflammatory pathways (e.g., NF-κB) and reducing pro-inflammatory cytokines, crucial for preserving neuronal integrity (Asuku et al., 2024b). It demonstrates anti-amyloidogenic and anti-tau properties, preventing Aβ aggregation and influencing tau phosphorylation, addressing core Alzheimer's pathologies (Payne et al., 2022). The bidirectional nature of neuroinflammation and neurodegeneration is acknowledged, with EGCG modulating both processes.

Additional mechanisms include neurotransmitter modulation (e.g., acetylcholinesterase inhibition by EGCG, dopamine/serotonin influence by L-theanine), support for mitochondrial function and neurogenesis (enhancing biogenesis and promoting hippocampal neurogenesis), and epigenetic

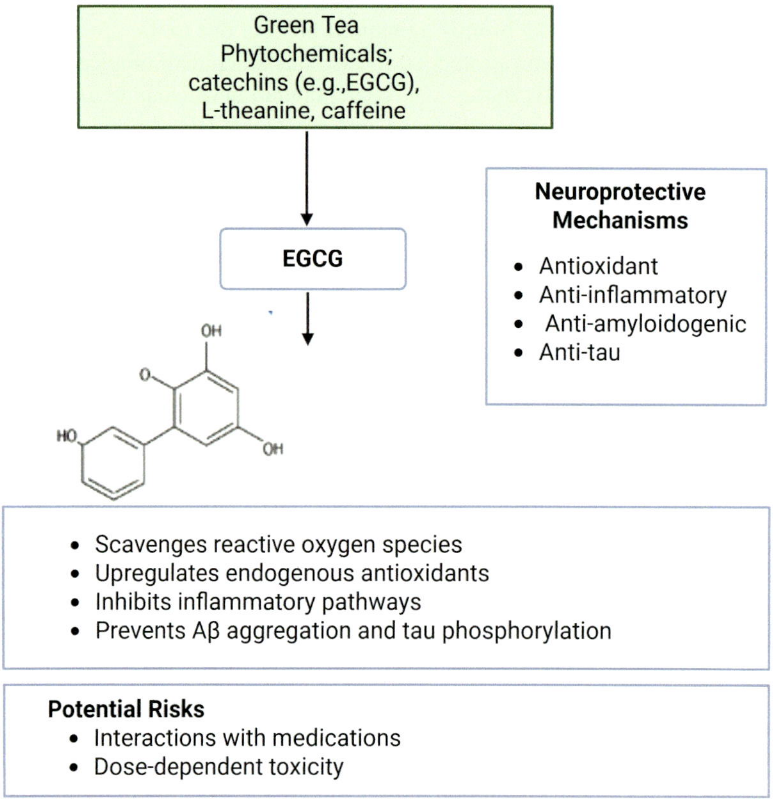

FIGURE 10.1 Phytochemical composition and neuroprotective mechanisms of green tea. (Green tea contains bioactive phytochemicals, including catechins such as epigallocatechin-3-gallate (EGCG), L-theanine, and caffeine. EGCG shows neuroprotective effects through antioxidant, anti-inflammatory, anti-amyloidogenic, and anti-tau mechanisms. These processes help in reduction of oxidative stress, modulate inflammation, and prevent aggregation of amyloid-beta and tau proteins, which are implicated in neurodegenerative diseases. It possesses potential risks such as drug interactions and dose-dependent toxicity.)

modulation (altering gene expression to upregulate neuroprotective genes like BDNF) (Mandel et al., 2008). The interaction with the gut–brain axis, where EGCG modulates gut microbiota to reduce dysbiosis and neuroinflammation, is gaining recognition.

Despite these promising mechanisms, the complexity and context-dependency of EGCG's actions require careful consideration. Potential negative interactions with medications (e.g., anticoagulants, statins) and dose-dependent adverse effects, including idiosyncratic hepatotoxicity, necessitate caution in clinical applications. These adverse events can occur across a wide dose range, with individual susceptibility and product formulation playing significant roles. While precise incidence rates are difficult to establish due to underreporting and variability in product use, reviews indicate over 100 documented cases of clinically apparent liver injury attributed to green tea extract (GTE) in the literature, with some cases linked to daily EGCG doses as low as 140 mg, though higher doses (e.g., >800 mg/day) are more frequently associated with risk (Oketch-Rabah et al., 2020). Liver monitoring is recommended in trials, especially for high-dose extracts, and the risk is often exacerbated when GTE is consumed on an empty stomach.

10.3 PRECLINICAL AND CLINICAL EVIDENCE

Green tea and EGCG's neuroprotective potential is supported by extensive preclinical and evolving clinical evidence.

10.3.1 Preclinical Evidence

In vitro studies demonstrate EGCG protects neurons from Aβ toxicity, inhibits Aβ/tau aggregation, and reduces oxidative stress. Animal models (e.g., transgenic AD mice) show that oral EGCG reduces cerebral Aβ plaques, improves spatial memory, enhances synaptic plasticity, and decreases oxidative stress/neuroinflammation (Islam et al., 2025). These consistent preclinical findings underscore green tea's neuroprotective capacity.

10.3.2 Clinical Evidence

Human observational studies consistently link regular green tea intake to reduced dementia risk, with reported risk reductions typically ranging from 10% to 40% in various cohorts. For instance, the Ohsaki Cohort Study, a large prospective cohort of elderly Japanese individuals, found that consuming ≥5 cups/day of green tea was associated with a hazard ratio of 0.73 (95% CI: 0.61–0.87) for incident dementia compared to consuming <1 cup/day, indicating a 27% risk reduction ($p < 0.001$) (Tomata et al., 2016). The original Japanese cohort study by Kuriyama et al. (2006) also found that higher green tea consumption was associated with a lower prevalence of cognitive impairment. A study on Korean population data showed that a combination of GTE and L-theanine improved memory and attention in subjects with mild cognitive impairment (Park et al., 2011). Another systematic review of Japanese cohorts synthesized similar findings, reinforcing the protective association (Kakutani et al., 2019).

Definitive Phase III trials for dementia typically require hundreds to thousands of participants (e.g., 500–2,000+) to achieve sufficient statistical power (e.g., 80%–90%) to detect clinically meaningful differences in cognitive outcomes or disease progression. Sample size calculations depend on expected effect size, variability of outcome measures, trial duration, and dropout rates (Ross et al., 2022). For instance, to detect a two-point difference in the ADAS-Cog score (a common cognitive outcome measure) with 80% power ($\beta = 0.20$) and a two-sided alpha of 0.05 ($\alpha = 0.05$), assuming a standard deviation of six points, approximately 284 participants per arm would be required for a randomized controlled trial. This calculation highlights the substantial number of participants needed to achieve statistical significance for even modest clinical effects in dementia trials.

Challenges in clinical translation include EGCG's low oral bioavailability, typically reported as 0.1%–0.3% in humans, and limited brain penetration (*in vitro* blood–brain barrier (BBB) permeability ~2.8%–4%, but low *in vivo* brain exposure). This low bioavailability is primarily due to extensive first-pass metabolism and active efflux by drug transporters. Achieving therapeutic concentrations often requires high doses, leading to adverse effects like idiosyncratic hepatotoxicity, which can occur across a wide dose range depending on individual susceptibility and product formulation. Liver monitoring is recommended in trials, especially for high-dose extracts.

Clinical trial inconsistencies stem from heterogeneous study populations, varied dosages, durations, and outcome measures. Publication bias, where positive results are favored, and small-study bias (e.g., funnel plot asymmetry) are observed in meta-analyses. Placebo effects, particularly with dietary interventions, also contribute. Failed trials often suffer from insufficient sample sizes, short durations misaligned with dementia progression, and insensitive outcome measures.

10.3.3 Failed Trials Analysis

Despite promising preclinical evidence, the translation of EGCG into clinically meaningful outcomes for Alzheimer's disease (AD) and dementia has been hampered by several challenges, leading to inconsistencies and negative results in human clinical trials. The primary reasons for these failures often revolve around issues of pharmacokinetics, trial design, and the inherent complexity of neurodegenerative diseases.

One significant hurdle is the low oral bioavailability and limited brain penetration of EGCG. Following oral administration, only a minimal amount of EGCG reaches systemic circulation and even less effectively crosses the BBB. This poor pharmacokinetic profile means that achieving therapeutically relevant concentrations of EGCG in the human brain, where it would exert its neuroprotective effects, often necessitates very high doses. However, such high doses are frequently associated with adverse effects, including gastrointestinal discomfort and idiosyncratic hepatotoxicity, which can occur across a wide dose range depending on individual susceptibility and product formulation. This creates a narrow therapeutic window, making it difficult to administer sufficient EGCG without inducing toxicity.

Clinical trial design limitations have also significantly contributed to the observed failures. Many studies investigating EGCG's efficacy in AD or dementia have suffered from several methodological shortcomings:

Insufficient Sample Sizes: Definitive Phase III trials for dementia typically require hundreds to thousands of participants to achieve sufficient statistical power to detect clinically meaningful differences. Many EGCG trials have been conducted with small cohorts, making it difficult to discern a true effect, especially if the effect size is modest.

Short Durations: Dementia is a progressive neurodegenerative condition that develops over many years. Clinical trials with short durations (e.g., less than 12–18 months) are often misaligned with the slow progression of the disease, making it challenging to observe significant changes in cognitive outcomes or disease progression within the study period.

Heterogeneous Study Populations: Variations in patient demographics, disease stage, genetic predispositions, and comorbidities can introduce significant variability, masking potential treatment effects. Standardizing study populations is crucial for obtaining clear results.

Insensitive Outcome Measures: The cognitive and functional assessment tools used in some trials may not be sensitive enough to detect subtle but clinically relevant improvements or slowing of decline, particularly in early stages of the disease or over shorter durations.

Furthermore, publication bias, where positive results are favored, and small-study bias are observed in meta-analyses of dietary interventions. The presence of placebo effects also complicates the interpretation of results in trials involving dietary supplements. The inherent complexity of AD, with its multifactorial etiology involving amyloid pathology, tauopathy, neuroinflammation, oxidative stress, and genetic factors, means that a single agent like EGCG may not be sufficient as a monotherapy to halt or reverse the disease process. Future research needs to address these limitations through improved delivery systems, combination therapies, and biomarker-guided trial designs to fully evaluate EGCG's therapeutic potential.

10.3.3.1 Economic Analysis

The economic implications of green tea as a dementia intervention warrant consideration. Current dementia treatments impose substantial costs: aducanumab costs approximately $28,000 annually per patient (Biogen, 2021), while cholinesterase inhibitors range from 1,000 to 3,000 annually (GoodRx, 2015). In contrast, green tea consumption costs approximately 50–200 annually, representing a potential 100-fold cost reduction. However, this comparison assumes equivalent efficacy, which remains unproven.

Cost-effectiveness analyses must consider implementation costs, including standardization, quality control, and monitoring requirements. Healthcare system integration would require training, infrastructure development, and regulatory compliance. Population-level accessibility varies significantly, with rural and low-income populations potentially benefiting most from affordable interventions. Economic modeling suggests that even modest cognitive benefits (e.g., six-month delay in nursing home placement) could yield substantial healthcare savings, potentially offsetting implementation costs within two to three years (Brettschneider et al., 2025). For instance, a cost-effectiveness analysis of a multi-domain dementia prevention program found an incremental cost-effectiveness ratio (ICER) of €17,149.23 per quality-adjusted life year (QALY), indicating that such interventions can be cost-effective within commonly accepted willingness-to-pay thresholds for healthcare interventions (McRae et al., 2021). Table 10.1 shows the clinical research modalities for green tea/EGCG in cognitive health.

TABLE 10.1
Summary of Clinical Research Modalities for Green Tea/EGCG in Cognitive Health

Research Modality	Primary Contribution	Illustrative Findings	Typical Sample Size/ Study Quality	Limitations/Considerations
In vitro studies	Elucidation of molecular mechanisms	EGCG inhibits Aβ aggregation; protects neurons from oxidative stress	N/A (cellular/molecular level)	High concentrations may not be physiologically relevant; *in vivo* translation requires further study
Animal models	Validation of systemic neuroprotection	Improved cognition; reduced Aβ/tau pathology; decreased neuroinflammation	Small N (dozens); short-term; high internal validity	Species differences; direct translation to human efficacy requires caution
Observational cohorts	Real-world epidemiological associations	Risk reductions typically 10%–40% in various cohorts	Large N (thousands to tens of thousands); long-term follow-up; moderate risk of bias	Association, not causation; potential confounding by lifestyle factors
Randomized controlled trials	Direct interventional evidence	Small, variable benefits; inconsistent across trials and frequently below clinical significance thresholds	Small N (often <100 participants); short durations (<12 months); high risk of bias	Small sample sizes; heterogeneity in protocols; potential for publication bias
Meta-analyses	Synthesis across multiple studies	Consistent beneficial association with cognitive health	Variable N; systematic review methodology; variable risk of bias	Aggregation of heterogeneous data may mask specific effects

10.4 TRANSLATIONAL CHALLENGES AND LIMITATIONS

Despite promising preclinical and emerging clinical evidence, integrating green tea constituents into mainstream dementia therapies faces significant hurdles.

10.4.1 Bioavailability and Pharmacokinetics

EGCG's low oral bioavailability and limited BBB penetration represent primary impediments. Following oral administration, minimal EGCG reaches systemic circulation and even less crosses the BBB. This is attributed to rapid gastrointestinal degradation, extensive first-pass metabolism, and active efflux by drug transporters. Consequently, achieving therapeutically relevant brain concentrations often requires high doses, introducing complications.

10.4.2 Compositional Variability and Standardization

Green tea's phytochemical composition varies considerably based on cultivar, geographical origin, cultivation practices, harvesting season, and processing methods. This variability challenges standardization and reproducible therapeutic effects. Clinical trials commonly use standardized extracts reporting EGCG percentage (e.g., 90% EGCG), and researchers should report batch assays (e.g., HPLC data) to ensure consistency.

10.4.3 Dosage, Safety, and Drug Interactions

Optimal therapeutic dosing for EGCG remains undefined. While moderate green tea consumption is generally safe, high-dose EGCG extracts have been associated with adverse effects, including gastrointestinal discomfort and idiosyncratic hepatotoxicity occurring across wide dose ranges. EGCG potentially interacts with anticoagulants and statins, necessitating careful clinical monitoring.

10.4.4 Clinical Study Design Limitations

Many studies suffer from small sample sizes, short durations, heterogeneous populations, and varied outcome measures. These limitations impact generalizability and interpretability. Confounding factors in observational studies make it difficult to draw definitive conclusions.

10.5 MECHANISTIC-CLINICAL TRANSLATION GAP

Despite strong in vitro and animal data supporting the claim that green tea catechins, and especially EGCG, are neuroprotective, at the moment it is hard to apply the information to a practical approach to the treatment of dementia. Even the concentrations of EGCG typically tested in laboratory models are significantly better than those which may be tested in humans, and the physiology of animals varies widely in genetic makeup, metabolism, and immune response. Besides this, decades-long multifactorial interactions of genetic, environmental, and comorbid factors drive dementia in a human being and cannot be captured in simplified models. Clinical benefit prediction is also made difficult by genetic, microbiome, medications, and lifestyle population variability. The short-term preclinical studies also fail to indicate the long-term natural history of the human disease, and therefore, the optimum dose, time, and long-term safety remain unknown.

10.6 COMBINATION THERAPIES

Dementia is a complex pathology, and therefore, EGCG monotherapy is unlikely to be sufficient. One can also add EGCG to other natural neuroprotectants, i.e., curcumin, resveratrol, or omega-3 fatty acids, which can act synergistically, i.e., in a synergistic manner via antioxidant, anti-inflammatory, and neurotrophic effects (Molino et al., 2016). It can be used together with other medicines, such as donepezil or memantine, in preclinical experiments; however, the drug–herb interactions should be applied with caution (Kumar et al., 2017). Wider interventions may consist of combining EGCG and lifestyle interventions, Mediterranean diet, exercise, and cognitive training, which would provide a multidimensional prevention and control strategy (Forcano et al., 2025). Such combinations must be clinically tested in order to show their safety and efficacy.

10.7 FUTURE DIRECTIONS

Overcoming the existing translational challenges and maximizing the therapeutic potential of these natural compounds will require innovative approaches in drug development, clinical trial design, and personalized medicine.

1. *Enhancing Bioavailability and Brain Penetration*: A major focus for future research must be on developing novel delivery systems to improve EGCG's oral bioavailability and its ability to cross the BBB. Strategies such as nanotechnology-based formulations (e.g., nanoparticles, liposomes, micelles), phytosomes, and prodrug approaches could significantly enhance EGCG's systemic absorption and brain exposure, allowing for lower, safer, and more effective doses (Smith et al., 2010). Research into specific transporters and efflux pumps at the BBB could also lead to methods for bypassing these barriers or inhibiting their activity, thereby increasing EGCG concentrations in the brain.
2. *Precision Dosing and Personalized Medicine*: The observed idiosyncratic hepatotoxicity and variable responses to EGCG highlight the need for personalized dosing strategies. Future research should aim to identify biomarkers (e.g., genetic polymorphisms, gut microbiome composition) that predict individual susceptibility to adverse effects or responsiveness to EGCG (James et al., 2015). This would enable a precision medicine approach, tailoring EGCG interventions to individual patient profiles, thereby maximizing efficacy and minimizing risks. Pharmacogenomic studies could elucidate genetic variations influencing EGCG metabolism and transport, guiding personalized therapeutic regimens.
3. *Well-Designed Clinical Trials*: The inconsistencies in past clinical trials underscore the necessity for more rigorous and standardized study designs. Future trials should feature larger sample sizes, longer durations (aligned with the slow progression of dementia), and homogeneous patient populations, perhaps stratified by disease stage or genetic risk factors (Cummings et al., 2020). The use of sensitive and objective outcome measures, including validated cognitive assessments, neuroimaging biomarkers (e.g., amyloid PET, tau PET, fMRI), and fluid biomarkers (e.g., CSF Aβ, tau, neurofilament light chain), will be crucial for detecting subtle but clinically meaningful effects (Blennow & Zetterberg, 2018). Adaptive trial designs and pragmatic clinical trials could also facilitate more efficient and real-world relevant evaluations.
4. *Exploring Combination Therapies and Synergistic Effects*: As discussed, the multifactorial nature of dementia suggests that combination therapies may be more effective than monotherapy. Future research should systematically investigate the synergistic effects of EGCG with other neuroprotective compounds (both natural and synthetic) and conventional anti-dementia drugs (Shabbir et al., 2020). This includes preclinical studies to identify optimal combinations and mechanisms of synergy, followed by well-designed clinical trials to assess their safety and efficacy. The potential for EGCG to enhance the effects of existing treatments or mitigate their side effects warrants further exploration.
5. *Standardized Extracts and Quality Control*: Ensuring the consistency and quality of GTEs used in research and clinical practice is paramount. Future efforts should focus on developing and adhering to stringent standardization protocols, including comprehensive phytochemical profiling (e.g., HPLC–MS) of all batches used in studies (Napolitano et al., 2014). This will reduce variability in results and enhance the reproducibility and translatability of findings.

By addressing these future directions, researchers can move closer to fully understanding and harnessing the therapeutic potential of green tea and its components in the fight against dementia, ultimately leading to more effective prevention and treatment strategies.

10.8 CONCLUSION

Green tea, particularly its key catechin EGCG, emerges as a mechanistically supported natural neuroprotectant with potential to delay or mitigate dementia progression. Preclinical and observational study has consistently demonstrated benefits in reducing oxidative stress, neuroinflammation, and pathological protein aggregation, central processes in neurodegeneration. Despite this promise, clinical translation has been limited by EGCG's low oral bioavailability, inconsistent trial methodologies, and safety considerations, including idiosyncratic hepatotoxicity.

Overcoming these barriers requires development of enhanced delivery systems, rigorous standardization, and large-scale, biomarker-informed clinical trials with appropriate cognitive and functional endpoints. Additionally, combination strategies involving other neuroprotective agents and lifestyle interventions should be explored to address dementia's multifactorial nature. Given its accessibility and affordability, green tea holds significant promise as a complementary intervention for dementia, particularly in populations lacking access to costly pharmacotherapies. Future research must focus on translating green tea's neuroprotective mechanisms into clinically meaningful outcomes to fully harness its therapeutic potential.

REFERENCES

Asuku, A.O., Adeyemo, M.B., Adeyemo, R.O., Agoro, S.A., Odugbesan, O.A., & Ayoola, S.A. (2022). Applications of stem cell therapy in neurodegenerative diseases. *European Journal of Biotechnology and Bioscience*, *10*(2): 17–24, 2321–9122.

Asuku, A.O., Ajibare, A.J., Ajagun, E.J., Babatunde, S.S., Olajide, T.S., Oyerinde, T.O., Ogungbangbe, G.O., & Odugbesan, O.A. (2024b). Herbal Extracts in Anti-Aging Products. In: Eco-Friendly Skin Solutions for Natural Cosmeceuticals. IGI Publisher. http://doi.org/10.4018/979-8-3693-5473-5.ch012

Asuku, A.O., Ayinla, M.T., Gbonjubola, O.O., Babatunde, S.S., Olajide, T.S., & Oyerinde, T.O. (2024a). Neurocognitive Aspects of Dementia. In: Nutrition in Brain Aging and Dementia: Nutritional Neurosciences. Springer, Singapore. https://doi.org/10.1007/978-981-97-4117-5

Biogen. (2021, December 20). Biogen announces reduced price for ADUHELM® to improve access for patients. https://investors.biogen.com/news-releases/news-release-details/biogen-announces-reduced-price-aduhelmr-improve-access-patients

Blennow, K., & Zetterberg, H. (2018). Biomarkers for Alzheimer's disease: Current status and future prospects. *Journal of Internal Medicine*, *284*(4): 379–92. https://doi.org/10.1111/joim.12816

Brettschneider, C., Buczak-Stec, E., Witte, V., et al. (2025). Cost-effectiveness of a multicomponent intervention against cognitive decline. *Alzheimer's & Dementia (New York, N.Y.)*, *11*(1): e12579. https://alz-journals.onlinelibrary.wiley.com/doi/abs/10.1002/trc2.70028

Cummings, J., Lee, G., Ritter, A., Sabbagh, M., & Zhong, K. (2020). Alzheimer's disease drug development pipeline: 2020. *Alzheimer's & Dementia (New York, N.Y.)*, *6*(1): e12050. https://pubmed.ncbi.nlm.nih.gov/32637699/

Forcano, L., Soldevila-Domenech, N., Boronat, A., Sánchez-Benavides, G., Puig-Pijoan, A., Lorenzo, T., & de la Torre, R (2025). A multimodal lifestyle intervention complemented with epigallocatechin gallate to prevent cognitive decline in APOE-ε4 carriers with subjective cognitive decline: A randomized, double-blinded clinical trial (PENSA study). *The Journal of Prevention of Alzheimer's Disease*. Advance online publication. https://doi.org/10.1016/j.tjpad.2025.100271

Forester, S.C., & Lambert, J.D. (2011). The role of antioxidant versus pro-oxidant effects of green tea polyphenols in cancer prevention. *Molecular Nutrition & Food Research*, *55*(6): 844–54. https://doi.org/10.1002/mnfr.201000641

GoodRx. (2015). Popular cholinesterase inhibitors list, drug prices and coupons. https://www.goodrx.com/classes/cholinesterase-inhibitors

Islam, M.R., Rauf, A., Akter, S., Akter, H., Al-Imran, M.I.K., Islam, S., et al. (2025). Epigallocatechin 3-gallate-induced neuroprotection in neurodegenerative diseases: Molecular mechanisms and clinical insights. *Molecular and Cellular Biochemistry*. Advance online publication. https://doi.org/10.1007/s11010-025-05211-4

James, K.D., Forester, S.C., & Lambert, J.D. (2015). Dietary pretreatment with green tea polyphenol, (-)-epigallocatechin-3-gallate reduces the bioavailability and hepatotoxicity of subsequent oral bolus doses of (-)-epigallocatechin-3-gallate. *Food and Chemical Toxicology*, *76*: 103–8. https://doi.org/10.1016/j.fct.2014.12.009

Kakutani, S., Watanabe, H., & Murayama, N. (2019). Green tea intake and risks for dementia, Alzheimer's disease, mild cognitive impairment, and cognitive impairment: A systematic review. *Nutrients*, *11*(5): 1165. https://doi.org/10.3390/nu11051165

Khan, N., & Mukhtar, H. (2018). Tea polyphenols in promotion of human health. *Nutrients*, *10*(1): 39. https://doi.org/10.3390/nu11010039

Kumar, A., Singh, A., & Aggarwal, A. (2017). Therapeutic potentials of herbal drugs for Alzheimer's disease—An overview. *Indian Journal of Experimental Biology*, *55*(2): 63–71.

Kuriyama, S., Hozawa, A., Ohmori, K., Shimazu, T., Matsui, T., Nakaya, N., et al. (2006). Green tea consumption and cognitive function: A cross-sectional study from the Tsurugaya Project. *American Journal of Clinical Nutrition*, *83*(2): 355–63. https://www.sciencedirect.com/science/article/pii/S0002916523292916

Mandel, S., Amit, T., Kalfon, L., Reznichenko, L., & Youdim, M. (2008). Targeting multiple neurodegenerative diseases etiologies with multimodal-acting green tea catechins. *Journal of Nutrition*, *138*(8): 1578S–83S. https://doi.org/10.1093/jn/138.8.1578S

McRae, I., Zheng, L., Bourke, S., Cherbuin, N., & Anstey, K.J. (2021). Cost-effectiveness of dementia prevention interventions. *Journal of Prevention of Alzheimer's Disease*, *8*(4): 511–8. https://www.sciencedirect.com/science/article/pii/S2274580724003923

Molino, S., Dossena, M., Buonocore, D., Ferrari, F., Venturini, L., Ricevuti, G., & Verri, M. (2016). Polyphenols in dementia: From molecular basis to clinical trials. *Life Sciences*, *161*: 69–77. https://doi.org/10.1016/j.lfs.2016.07.021

Napolitano, J., Gödecke, T., Lankin, D., Jaki, B., McAlpine, J., Chen, S.-N., & Pauli, G. (2014). Orthogonal analytical methods for botanical standardization: Determination of green tea catechins by qNMR and LC-MS/MS. *Journal of Pharmaceutical and Biomedical Analysis*, *88*: 23–31. https://doi.org/10.1016/j.jpba.2013.06.017

Natarajan, S., Chandran, S.P., Khan, S., Natarajan, P., & Rengarajan, K. (2019). Versatile health benefits of catechin from green tea (*Camellia sinensis*). *Current Nutrition & Food Science*, *15*(3): 210–8. https://doi.org/10.2174/1573401313666171003150503

Oketch-Rabah, H., Roe, A., Rider, C., Bonkovsky, H., Giancaspro, G., Navarro, V., & Ko, R (2020). United States Pharmacopeia (USP) comprehensive review of the hepatotoxicity of green tea extracts. *Toxicology Reports*, *7*: 796–806. https://doi.org/10.1016/j.toxrep.2020.02.008

Okubo, H., Inagaki, H., Gondo, Y., Kamide, K., Ikebe, K., Masui, Y., et al. (2017). Association between dietary patterns and cognitive function among 70-year old Japanese elderly: A cross-sectional analysis of the SONIC study. *Nutrition Journal*, *16*: 50. https://doi.org/10.1186/s12937-017-0273-2

Park, S.K., Jung, I.C., Lee, W.K., Lee, Y.S., Park, H.K., Go, H.J., et al. (2011). A combination of green tea extract and L-theanine improves memory and attention in subjects with mild cognitive impairment: A double-blind placebo-controlled study. *Journal of Medicinal Food*, *14*(4): 334–43. https://www.liebertpub.com/doi/abs/10.1089/jmf.2009.1374

Payne, A., Nahashon, S., Taka, E., Adinew, G.M., & Soliman, K (2022). Epigallocatechin-3-gallate (EGCG): New therapeutic perspectives for neuroprotection, aging, and neuroinflammation for the modern age. *Biomolecules*, *12*(3): 371. https://doi.org/10.3390/biom12030371

Ross, E.L., Weinberg, M.S., & Arnold, S.E. (2022). Cost-effectiveness of aducanumab and donanemab for early Alzheimer disease in the US. *JAMA Neurology*, *79*(5): 493–501. https://jamanetwork.com/journals/jamaneurology/fullarticle/2790531

Shabbir, U., Rubab, M., Tyagi, A., & Oh, D.-H. (2020). Curcumin and its derivatives as theranostic agents in Alzheimer's disease: The implication of nanotechnology. *International Journal of Molecular Sciences*, *22*(1): 196. https://doi.org/10.3390/ijms22010196

Smith, A.J., Giunta, B., Bickford, P.C., Fountain, M.D., Tan, J., & Shytle, R (2010). Nanolipidic particles improve the bioavailability and alpha-secretase inducing ability of epigallocatechin-3-gallate (EGCG) for the treatment of Alzheimer's disease. *International Journal of Pharmaceutics*, *389*(1–2): 207–12. https://doi.org/10.1016/j.ijpharm.2010.01.012

Tomata, Y., Sugiyama, K., Kaiho, Y., Honkura, K., Watanabe, T., Zhang, S., et al. (2016). Green tea consumption and the risk of incident dementia in elderly Japanese: The Ohsaki Cohort 2006 Study. *American Journal of Clinical Nutrition*, *104*(5): 1415–22. https://www.sciencedirect.com/science/article/pii/S1064748116301774

Zhou, S., Zhu, Y., Ren, N., Wu, M., & Liu, Y. (2025). The association between green tea consumption and cognitive function: A meta-analysis of current evidence. *Neuroepidemiology*. Advance online publication. https://doi.org/10.1159/000543784

11 6-Gingerol for Managing and Treating Dementia

Adedamola Aminat Bayo-Olugbami, Tolulope Timothy Arogundade, Emmanuel Taiwo George, Abraham Olufemi Asuku, Olawale Olaleye Obembe, Philip Adeyemi Adeniyi, and Bamidele Victor Owoyele

11.1 INTRODUCTION

Dementia is an increasingly prevalent age-related neurodegenerative syndrome that currently affects tens of millions worldwide and is projected to rise substantially by mid-century. Alzheimer's disease (AD) is the leading cause of dementia, accounting for approximately 70–80% of cases, and presents a significant unmet medical and socioeconomic challenge. Existing pharmacotherapies provide symptomatic benefit for some patients but have limited impact on the progressive neurodegeneration that underlies cognitive decline, highlighting the need for new multi-target strategies (Kim et al., 2024).

Pathologically, AD is characterised by extracellular accumulation of amyloid-β (Aβ) peptides and intraneuronal neurofibrillary tangles composed of hyperphosphorylated tau (Asuku et al., 2025). These core lesions trigger a cascade of maladaptive processes, including oxidative stress, chronic neuroinflammation, synaptic dysfunction, impaired trophic signalling, and cholinergic deficits, that collectively drive neuronal loss and cognitive impairment (Kim et al., 2024). Because AD is a multifactorial disorder, interventions that modulate several pathogenic nodes simultaneously are of particular interest.

Natural products have attracted renewed attention as sources of multi-modal agents with favourable safety profiles. *Zingiber officinale* (ginger) has a long history of dietary and medicinal use, and phytochemical analyses identify a family of phenolic compounds (gingerols, shogaols, etc.) as principal bioactive constituents (Ahmed et al., 2021; Yahyazadeh et al., 2021). Among these, 6-gingerol is the most abundant and has demonstrated diverse pharmacological activities in preclinical studies, including antioxidant, anti-inflammatory, anti-amyloidogenic, and neurotrophic effects (Roudsari et al., 2021; Asuku et al., 2024a; Ayinla & Asuku, 2025). Its lipophilic character and reported gastrointestinal absorption support reasonable oral bioavailability, and several studies indicate that 6-gingerol can access the central nervous system (CNS), a critical consideration for any candidate targeting dementia (Durojaye et al., 2019).

This chapter synthesises the mechanistic and preclinical evidence for 6-gingerol in models relevant to AD and related dementias. We summarise cellular and animal data on antioxidant, anti-inflammatory, anti-amyloidogenic, and cholinergic/neurotrophic mechanisms, and emphasise pharmacokinetic and translational gaps that must be addressed before clinical evaluation. The aim is to provide an evidence-based appraisal of 6-gingerol's potential as a multi-target neuroprotective agent and to identify priorities for future translational research (Durojaye et al., 2019; Ahmed et al., 2021; Kim et al., 2024).

11.2 MECHANISMS OF ACTIONS OF 6-GINGEROL IN DEMENTIA MANAGEMENT

Several beneficial properties and the possible mechanistic implications of 6-gingerol have been documented. However, the focus of this chapter is narrowed only to those mechanisms that have the potential of preventing or reducing the impacts of the major players implicated in the development and progression of AD.

DOI: 10.1201/9781003657385-11

11.2.1 Antioxidant Properties

Oxidative stress is a central mediator of neuronal injury in AD, amplifying Aβ toxicity, impairing synaptic function, and promoting mitochondrial dysfunction. Preclinical studies consistently report that 6-gingerol exerts antioxidant actions across chemical, cellular, and animal models, acting both as a direct free-radical scavenger and as an inducer of endogenous antioxidant defences (Zeng et al., 2013; Lin et al., 2014; Ahmed et al., 2021).

In chemical assays and simple cell-free systems, 6-gingerol demonstrates dose-dependent radical-scavenging activity against common reactive species, supporting a primary antioxidant capacity that complements its cellular effects (Lin et al., 2014). In neuronal and glial cell models exposed to Aβ or other oxidative insults, pre-treatment with 6-gingerol reduces reactive oxygen species (ROS) accumulation, lowers markers of lipid peroxidation (e.g., MDA), and decreases lactate dehydrogenase (LDH) release, measures that correlate with improved cell viability and reduced apoptosis (Lee et al., 2011; Zeng et al., 2015; Ighodaro & Akinloye, 2018). At the mitochondrial level, 6-gingerol preserves membrane potential, limits cytochrome-c release, and sustains ATP production in challenged neurons, indicating protection of a key organelle implicated in AD pathogenesis (Lee et al., 2011).

Notably, several studies report that 6-gingerol activates endogenous cytoprotective pathways rather than acting solely as a stoichiometric antioxidant. Activation of the Nrf2 signalling axis and upregulation of downstream targets (HO-1, γ-glutamylcysteine ligase) have been observed following 6-gingerol treatment, accompanied by restored intracellular glutathione and increased activities of superoxide dismutase and catalase in brain tissue (Durojaye et al., 2019; Ahmed et al., 2021). These effects provide a durable, gene-regulated antioxidant response that is more likely to confer in vivo neuroprotection than short-lived direct scavenging alone.

Animal studies reinforce the translational potential: rodents treated with ginger extracts or purified 6-gingerol show reduced hippocampal lipid peroxidation, improved antioxidant enzyme activities, and preservation of synaptic markers after Aβ or toxin exposure (Zeng et al., 2013). However, important caveats temper enthusiasm. Many *in vitro* experiments use relatively high micromolar concentrations that may not be achieved in the human brain following oral dosing, and whole-extract studies complicate attribution of effects specifically to 6-gingerol. Moreover, while Nrf2 activation is a plausible mechanism, direct causal evidence (for example, using Nrf2 knockout models or selective Nrf2 antagonists) is limited.

To strengthen the antioxidant hypothesis and aid translation, future studies should prioritise: (i) rigorous Pharmacokinetics/Pharmacodynamics (PK/PD) characterisation to determine brain exposure after oral administration; (ii) dose-response studies that align *in vitro* concentrations with achievable in vivo levels; (iii) use of genetic or pharmacological loss-of-function models (e.g., Nrf2–/–) to test the mechanism; and (iv) standardised biochemical endpoints (Reduced/Oxidized glutathione (GSH/GSSG) ratio, enzyme activities, lipid peroxidation) reported alongside behavioural outcomes. Together, these steps will clarify whether 6-gingerol's antioxidant actions can meaningfully mitigate oxidative drivers of neurodegeneration in humans (Zeng et al., 2015; Durojaye et al., 2019; Ahmed et al., 2021) (Figure 11.1).

11.2.2 Anti-Inflammatory Effects

Neuroinflammation is primarily driven by activated microglia and astrocytes and sustained by pro-inflammatory cytokines, and is a core contributor to neurodegeneration in AD. Preclinical evidence indicates that 6-gingerol attenuates inflammatory signalling in both peripheral and central immune models, via multiple convergent mechanisms that include inhibition of NF-κB activation, suppression of MAPK pathways, modulation of inflammasome activity, and upstream effects on cellular energy sensors such as AMPK (Durojaye et al., 2019; Ahmed et al., 2021).

In vitro studies using microglial (e.g., BV2) and astrocytic models report that pre-treatment with 6-gingerol reduces lipopolysaccharide (LPS)- or Aβ-induced release of classical pro-inflammatory mediators (IL-1β, TNF-α, IL-6, iNOS-derived NO, and COX-2-dependent prostaglandins) and

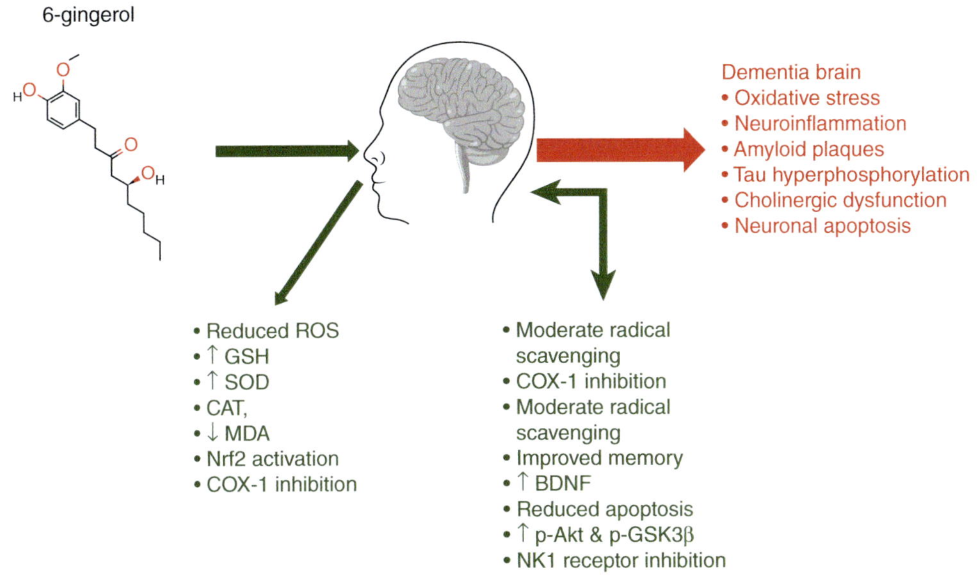

FIGURE 11.1 Schematic overview of proposed neuroprotective mechanisms of 6-gingerol in Alzheimer's disease models. 6-Gingerol mitigates oxidative stress via Nrf2 activation, reduces neuroinflammation via NF-κB inhibition and AMPK modulation, interferes with Aβ aggregation and APP processing, and enhances cholinergic and neurotrophic signalling (BDNF/CREB/Akt). Arrows indicate presumed direction of effect; dashed arrows indicate putative/indirect pathways supported mainly by preclinical evidence.

lowers expression of inducible inflammatory enzymes. Mechanistically, these anti-inflammatory effects are commonly associated with decreased nuclear translocation of NF-κB p65 and reduced phosphorylation of p38 and JNK MAPKs, changes that are consistent with a shift away from an M1/pro-inflammatory microglial phenotype towards a less reactive state (Ahmed et al., 2021; Pan et al., 2024). Several reports also note restoration of anti-inflammatory or reparative mediators (e.g., increased IL-10 or Arg1 expression) following 6-gingerol exposure, although these findings are less consistent across models.

In vivo, 6-gingerol or ginger extracts decrease systemic and central inflammatory markers in rodent models of neuroinflammation and Aβ toxicity. Treated animals show lower hippocampal and cortical levels of pro-inflammatory cytokines, reduced microglial and astrocyte activation by immunohistochemistry (Iba1, GFAP), and attenuation of sickness-like behaviour and cognitive deficits in tasks sensitive to neuroinflammatory challenge (Lee et al., 2011; Zeng et al., 2013; Maduako & Farombi, 2025). Some of these studies implicate modulation of the NLRP3 inflammasome, an important mediator of IL-1β maturation in AD, as part of 6-gingerol's repertoire, although this area requires further biochemical validation.

Potential upstream regulators have also been proposed. Activation of AMPK by 6-gingerol has been observed in metabolic and inflammatory models and could contribute to suppressed mTORC1 signalling and reduced inflammatory cytokine production; however, the causal chain (AMPK → mTORC1 → inflammasome/NF-κB) has not been established in CNS-specific models and should be presented cautiously (Durojaye et al., 2019). Similarly, claims that 6-gingerol directly inhibits pattern-recognition receptor signalling (e.g., TLR4) are suggestive but currently supported mainly by *in vitro* correlational data rather than loss-of-function experiments.

Important caveats temper the translational interpretation. Many mechanistic studies use supra-physiological concentrations or complex whole-extract preparations, making it difficult to ascribe effects specifically to 6-gingerol at doses achievable in the human brain. Distinguishing central anti-inflammatory actions from indirect peripheral immunomodulation (e.g., reduced systemic cytokines that secondarily affect the brain) is rarely addressed. Moreover, the heterogeneity of microglial states in vivo and the dynamic time course of neuroinflammatory responses mean that timing, dose, and the chosen disease model strongly influence outcomes.

To advance translational confidence, future work should prioritise: (i) PK/PD studies that define brain exposure relative to effective *in vitro* concentrations; (ii) experiments using genetic or pharmacological loss-of-function approaches (for example, NLRP3–/– mice, microglia-specific NF-κB inhibition) to test mechanism causally; (iii) standardised in vivo neuroinflammation paradigms (acute LPS, chronic Aβ models) with concurrent behavioural, histological (Iba1/GFAP), and molecular (cytokine panels, inflammasome readouts) endpoints; and (iv) higher-resolution phenotyping of microglia (flow cytometry, single-cell transcriptomics) to determine whether 6-gingerol truly reprogrammes microglial states towards neuroprotective phenotypes. Complementary translational measures, such as peripheral cytokine profiling and neuroinflammation PET (e.g., TSPO) in larger animal models, would also help bridge to human studies.

In summary, preclinical data support a multi-modal anti-inflammatory action for 6-gingerol that plausibly contributes to its neuroprotective profile. Nevertheless, clear demonstration of central target engagement at clinically achievable exposures, and causal linkage between specific anti-inflammatory mechanisms and preserved cognition, remain essential next steps before clinical testing in dementia populations (Durojaye et al., 2019; Ahmed et al., 2021; Pan et al., 2024).

11.2.3 Anti-Amyloidogenic Effects and Cholinergic System Modulation

The accumulation and aggregation of Aβ peptides are an early and driving event in AD pathogenesis, triggering synaptic dysfunction, neuroinflammation, and downstream tau pathology that together promote neuronal loss and cognitive decline (Asuku et al., 2024b). Evidence for 6-gingerol's actions against amyloid-driven toxicity and its modulation of cholinergic signalling comes from complementary in-silico, cellular, and animal studies, though human data are currently lacking.

In-silico and target-screening studies suggest that 6-gingerol can interact with multiple AD-relevant proteins, including acetylcholinesterase (AChE), MMP-2, PTGS-2, and CASP-3, forming stable hydrogen-bonding and hydrophobic contacts that imply potential modulation of enzyme activity and cell-death pathways (Pan et al., 2024). Such docking results are hypothesis-generating and require biochemical validation, but they provide a mechanistic rationale for the compound's pleiotropic effects.

Robust *in vitro* evidence supports a direct neuroprotective effect of 6-gingerol against Aβ toxicity. In PC12 neuronal cells, pre-treatment with 6-gingerol (effective concentrations reported at 80–200 μM) attenuated Aβ1-42-induced apoptosis, reduced markers of oxidative stress (ROS, MDA, NO, LDH), and increased antioxidant enzyme activity (e.g., SOD). Mechanistically, these studies report preservation of mitochondrial integrity and activation of pro-survival signalling, notably increased phosphorylation of Akt and inhibitory phosphorylation of GSK-3β, which together counteract apoptotic cascades and processes linked to tau hyperphosphorylation and APP misprocessing (Lee et al., 2011; Zeng et al., 2015).

In vivo and *ex vivo* studies provide convergent support. Ginger root extract or purified 6-gingerol reduces Aβ-related behavioural deficits and biochemical markers in rodent models, and high-dose extract has been associated with restored antioxidant enzyme levels and reduced lipid peroxidation in hippocampal tissue (Zeng et al., 2013). These effects are accompanied by improved synaptic markers and reduced neuronal loss in treated animals, consistent with an anti-amyloidogenic and neuroprotective profile.

Beyond anti-amyloid actions, 6-gingerol favourably modulates cholinergic neurotransmission, a critical therapeutic axis in AD. Preclinical work shows that 6-gingerol can inhibit AChE activity and increase acetylcholine levels, while upregulating choline acetyltransferase (ChAT) and choline transporter expression in neuronal models, effects that correlate with improvements in learning and memory tasks in rodents (Lim et al., 2014; Maralla & Reddy, 2014). These findings indicate that

TABLE 11.1
Mechanisms by Which 6-Gingerol Is Proposed to Act in Dementia-Related Models

Mechanism	Pathway/Target	Key Effect	Evidence Type	Implications for Dementia	Key References
Antioxidant	ROS scavenging; Nrf2 activation	↑GSH, ↑SOD/CAT; ↓MDA	*In vitro*/In vivo	Reduces oxidative stress-induced neuronal injury, prevents apoptosis	Zeng et al. (2013)
Anti-inflammatory	NF-κB, AMPK, COX-2	↓IL-1β, ↓TNF-α, ↓IL-6	*In vitro*/In vivo	Mitigates chronic neuroinflammation	Lee et al. (2011), Ali et al. (2013), Han et al. (2022)
Anti-amyloidogenic	Aβ aggregation; APP processing	↓Aβ aggregation; improved synaptic markers	*In vitro*/In vivo	May reduce plaque formation and synaptic dysfunction	Pan et al. (2024)
Cholinergic and neurotrophic modulation	AChE inhibition; ↑BDNF; CREB/Akt	Improved learning and memory, ↑synaptic plasticity	In vivo (rodent)	Cognitive enhancement via neurotransmission and trophic support	Kim et al. (2018), Pan et al. (2024)
Receptor modulation	NK-1, dopamine, serotonin receptors	↓Excitotoxicity; ↓oxidative stress and behavioural regulation	Clinical/ in vivo	Addresses oxido-inflammation and excitotoxicity	Marx et al. (2017), Cheng et al. (2020)

6-gingerol may protect neurons from amyloid-induced injury and ameliorate the cholinergic deficits contributing to cognitive symptoms.

While the preclinical data are encouraging, several caveats should be emphasised. Many mechanistic studies use relatively high compound concentrations or whole-extract preparations; docking predictions require biochemical confirmation; and rodent doses do not directly translate to safe, efficacious human regimens. Crucially, there are no randomised clinical trials testing 6-gingerol for cognitive outcomes in AD, so translational claims must remain tentative. Future work should prioritise biochemical validation of predicted targets, standardised PK/PD and blood–brain barrier penetration studies, and early-phase clinical trials using cognitive and biomarker endpoints to determine whether the compound's anti-amyloidogenic and pro-cholinergic effects translate to humans (Table 11.1).

11.3 CLINICAL TRIALS INVOLVING THE USE OF 6-GINGEROL

According to the clinical trials reported by Marx et al. (2017), Konmun et al. (2017), and the preclinical study of Ullah et al. (2015), 6-gingerol showed anti-emetic propensity by decreasing substance P and serotonin levels, as well as inhibiting the expression of neurokinin-1 (NK-1) and 5-HT_3 receptors (Cheng et al., 2020). While there is no published clinical evidence that tests the neuroprotective impact of 6-gingerol on AD in humans explicitly, clinical reports of its anti-emetic effect and the implicated mechanism above have provided a theoretical basis for plausible positive clinical outcomes in AD patients. The theoretical basis for this argument follows thus:

Overstimulation of NK-1 receptors, which primarily bind the neuropeptide substance P, has been implicated in the pathogenesis of dementia through its role in promoting neuroinflammation, oxidative stress, and excitotoxicity; all of which cause neurodegeneration (Cinzia, 2025). Substance P is widely distributed in brain regions associated with cognition, such as the hippocampus and cortex. Excessive release during chronic stress, injury, or inflammation activates

NK-1 receptors (Muñoz & Coveñas, 2014). This overstimulation leads to excessive release of pro-inflammatory cytokines, activation of microglia and astrocytes, and increased production of ROS (Johnson et al., 2017), consequently leading to neuronal damage and synaptic dysfunction. Furthermore, NK-1 receptor activation is linked to enhanced glutamatergic transmission (Garcia-Recio & Gascón, 2015), which, when dysregulated, can cause calcium overload and excitotoxicity – a process known to damage neurons in AD and other dementias. These cumulative effects of sustained NK-1 receptor activation may accelerate amyloid-beta deposition, tau hyperphosphorylation, and neuronal loss, thereby contributing significantly to cognitive decline and dementia progression. The ability of 6-gingerol to inhibit NK-1 receptors presents a promising avenue for managing dementia, as it may counteract the neuroinflammatory and excitotoxic processes associated with NK-1 overstimulation. Also, by attenuating the activity of substance P at NK-1 receptors, 6-gingerol could help reduce the cascade of pro-inflammatory cytokine release, oxidative stress, and neuronal damage contributing to the cognitive decline associated with dementia.

While these pharmacological interactions are promising, especially from the anti-emetic perspectives, further studies and clinical trials are required to examine the therapeutic effect of 6-gingerol specifically on dementia patients.

11.4 CONCLUSION

In summary, this chapter highlights the therapeutic potential of 6-gingerol in managing dementia (with emphasis on AD) through its antioxidant, anti-inflammatory, anti-amyloidogenic, and procholinergic neuroprotective properties. Its multi-targeted mechanisms offer promising avenues for future research and development of safer and natural interventions for the prevention and treatment of dementia. However, despite encouraging preclinical findings, translation to humans is hindered by limited pharmacokinetic and toxicity data at doses relevant to CNS effects, variability across animal models, and the absence of dementia-focused clinical trials. Future work should prioritise rigorous PK/PD studies, standardised behavioural protocols, and early-phase clinical trials focused on safety, tolerability, and cognitive biomarkers.

BIBLIOGRAPHY

Abolaji, A.O., Ojo, M., Afolabi, T.T., Arowoogun, M.D., Nwawolor, D. & Farombi, E.O. (2017). Protective properties of 6-gingerol-rich fraction from *Zingiber officinale* (ginger) on chlorpyrifos-induced oxidative damage and inflammation in the brain, ovary and uterus of rats (ginger) on chlorpyrifos-induced oxidative damage and inflammation in the brain, ovary and uterus of rats *Chemico-Biological Interactions*, *270*:15–23.

Adetuyi, B.O. & Farombi, E. (2021). 6-gingerol, an active constituent of ginger, attenuates lipopolysaccharide-induced oxidation, inflammation, cognitive deficits, neuroplasticity, and amyloidogenesis in rat. *Journal of Food Biochemistry*, *45*(4):e13660.

Ahmed, S.H.H., Gonda, T. & Hunyadi, A. (2021). Medicinal chemistry inspired by ginger: Exploring the chemical space around 6-gingerol. *RSC Advances*, *11*(42):26687–26699.

Alharbi, K.S., Nadeem, M.S., Afzal, O., Alzarea, S.I., Altamimi, A.S.A., Almalki, W.H., Mubeen, B., Iftikhar, S., Shah, L. & Kazmi, I. (2022). Gingerol, a natural antioxidant, attenuates hyperglycemia and downstream complications. *Metabolites*, *12*(12):1274.

Ali, S.K., Hamed, A.R., Soltan, M.M., Hegazy, U.M., Elgorashi, E.E., El-Garf, I.A. & Hussein, A.A. (2013). In-vitro evaluation of selected Egyptian traditional herbal medicines for treatment of Alzheimer disease. *BMC Complementary and Alternative Medicine*, *13*(1):121.

Alsahli, M.A., Almatroodi, S.A., Almatroudi, A., et al. (2021). 6-Gingerol, a major ingredient of ginger attenuates diethylnitrosamine-induced liver injury in rats through the modulation of oxidative stress and anti-inflammatory activity. *Mediators of Inflammation*, *1*:6661937.

Asuku, A.O., Ayinla, M.T., Ajibare, A.J. & Olajide, T.S. (2024a). Mercury chloride causes cognitive impairment, oxidative stress and neuroinflammation in male Wistar rats: The potential protective effect of 6-gingerol-rich fraction of *Zingiber officinale* via regulation of antioxidant defense system and reversal of pro-inflammatory markers increase. *Brain Research*, *1826*:148741. ISSN 0006-8993. https://doi.org/10.1016/j.brainres.2023.148741

Asuku, A.O., Ayinla, M.T., Gbonjubola, O.O., Babatunde, S.S., Olajide, T.S. & Oyerinde, T.O. (2024b). Neurocognitive Aspects of Dementia. In: *Nutrition in Brain Aging and Dementia. Nutritional Neurosciences*. Springer, Singapore. https://doi.org/10.1007/978-981-97-4117-5_5

Asuku, A.O., Ogungbangbe, G.O. & Olajide, T.S. (2025). Effect of Happiness and Positive Emotions on Dementia. In: *Neuroscience of Feelings, Progress in Brain Research*. Elsevier. https://doi.org/10.1016/bs.pbr.2025.03.001

Ayinla, M.T. & Asuku, A.O. (2025). The neurotoxic effects of lead acetate and the abrogating actions of 6-gingerol-rich extract of ginger via modulation of antioxidant defence system, pro-inflammatory markers, and apoptotic cascade. *Naunyn-Schmiedeberg's Archives of Pharmacology*, *398*, 9341–9356. https://doi.org/10.1007/s00210-025-03873-x

Bolisetty, S. & Jaimes, E.A. (2013). Mitochondria and reactive oxygen species: Physiology and pathophysiology. *International Journal of Molecular Sciences*, *14*(3):6306–6344.

Cheng, Q., Feng, X., Meng, Q., et al. (2020). 6-gingerol ameliorates cisplatin-induced Pica by regulating the TPH/MAO-A/SERT/5-HT/5-HT3 receptor system in rats. *Drug Design, Development and Therapy*, *14*:4085–4099.

Cinzia, S. (2025). Substance P in Alzheimer's and Parkinson's disease. *Substance P*, pp. 467–485.

Cork, G.K., Thompson, J. & Slawson, C. (2018). Real talk: The interplay between the mTOR, AMPK, and hexosamine biosynthetic pathways in cell signaling. *Frontiers in Endocrinology*, *9*:522.

Denzer, I., Münch, G., Pischetsrieder, K. & Friedland, K. (2016). S-allyl-L-cysteine and isoliquiritigenin improve mitochondrial function in cellular models of oxidative and nitrosative stress. *Food Chemistry*, *194*:843–848.

Dugasani, S., Pichika, M.R., Nadarajah, V.D., Balijepalli, M.K., Tandra, S. & Korlakunta, J.N. (2010). Comparative antioxidant and anti-inflammatory effects of 6-gingerol, [8]-gingerol, [10]-gingerol and 6-shogaol. *Journal of Ethnopharmacology*, *127*(2):515–520.

Durojaye, O.A., Nwanguma, B.C., Joshua, P.E., Ogidigo, J.O., Cosmas, S., Asomadu, R.O. & Obeta, J.N. (2019). Evaluation of 6-gingerol and its modified analogues as therapeutic candidates against *Schistosoma mansoni* phosphofructokinase. *VacciMonitor*, *28*(1):38–47.

Fiedorowicz, M. & Grieb, P. (2012). Chapter 6: Nitrooxidative Stress and Neurodegeneration. In: A. Gonzalez-Quevedo, ed. *Brain Damage Bridging between Basic Research and Clinics*. InTech, Rijeka.

Garcia-Recio, S. & Gascón, P. (2015). Biological and pharmacological aspects of the NK1-receptor. *BioMed Research International*, *1*:495704.

Haass, C. & Selkoe, D.J. (2007). Soluble protein oligomers in neurodegeneration: Lessons from the Alzheimer's amyloid β-peptide. *Nature Reviews Molecular Cell Biology*, *8*(2):101–112.

Han, X., Liu, P., Zheng, B., Zhang, M., et al. (2022). 6-gingerol exerts a protective effect against hypoxic injury through the p38/Nrf2/HO-1 and p38/NF-κB pathway in H9c2 cells. *The Journal of Nutritional Biochemistry*, *104*:08975.

Hayati, R.F., Better, C.D., Denis, D., Komarudin, A.G., et al. (2021). 6-gingerol inhibits chikungunya virus infection by suppressing viral replication. *BioMed Research International*, 6623400.

Ho, L., Qin, W., Stetka, B.S. & Pasinetti, G.M. (2006). Is there a future for cyclo-oxygenase inhibitors in Alzheimer's disease? *CNS Drugs*, *20*(2):85–98.

Ighodaro, O.M. & Akinloye, O.A. (2018). First line defence antioxidants-superoxide dismutase (SOD), catalase (CAT) and glutathione peroxidase (GPX): Their fundamental role in the entire antioxidant defence grid. *Alexandria Journal of Medicine*, *54*(4):287–293.

Johnson, M.B., Young, A.D. & Marriott, I. (2017). The therapeutic potential of targeting substance P/NK-1R interactions in inflammatory CNS disorders. *Frontiers in Cellular Neuroscience*, *10*:296.

Kim, A.Y., Al Jerdi, S., MacDonald, R. & Triggle, C.R. (2024). Alzheimer's disease and its treatment–yesterday, today, and tomorrow. *Frontiers in Pharmacology*, *15*:1399121.

Kim, C.Y., Seo, Y., Lee, C., Park, G.H. & Jang, J.H. (2018). Neuroprotective effect and molecular mechanism of 6-Gingerol against scopolamine-induced amnesia in C57BL/6 mice. *Evidence-Based Complementary and Alternative Medicine*, *2018*(1):8941564.

Konmun, J., Danwilai, K., Ngamphaiboon, N., Sripanidkulchai, B., Sookprasert, A. & Subongkot, S. (2017). A phase II randomized double-blind placebo-controlled study of 6-gingerol as an anti-emetic in solid tumor patients receiving moderately to highly emetogenic chemotherapy. *Medical Oncology*, *34*(4):69.

Lee, C., Park, G.H., Kim, C.Y. & Jang, J.H. (2011). 6-gingerol attenuates β-amyloid-induced oxidative cell death via fortifying cellular antioxidant defense system. *Food and Chemical Toxicology*, *49*(6):1261–1269.

Lim, S., Moon, M., Oh, H., et al (2014). Ginger improves cognitive function via NGF-induced ERK/CREB activation in the hippocampus of the mouse. *Journal of Nutritional Biochemistry*, *25*(10):1058–1065.

Lin, J., Li, X., Chen, L., Lu, W., Chen, X., Han, L. & Chen, D. (2014). Protective effect against hydroxyl radical-induced DNA damage and antioxidant mechanism of 6-gingerol: A Chemical Study. *Bulletin of the Korean Chemical Society*, *35*(6):1633–1638.

Maduako, I.C. & Farombi, E.O. (2025). 6-gingerol inhibition of NF-κB expression in diethylnitrosamine-initiated, 2-acetylaminofluorene-promoted liver dysfunction and inflammation involves induction of antioxidant enzymes and suppression of pro-inflammatory cytokines. *International Journal of Gastroenterology Practice*, *1*(4): 1–10.

Maralla, S. & Reddy, K.S. (2014). Effect of ginger on acetylcholine activity in brain and its possible role in learning and memory during ethanol withdrawal. *Asian Journal of Pharmaceutical and Clinical Research*, *7*(2):192–197.

Marx, W., Ried, K., McCarthy, A.L., et al. (2017). Ginger-mechanism of action in chemotherapy-induced nausea and vomiting: A review. *Critical Reviews in Food Science and Nutrition*, *57*(1):141–146.

Mohamed, A.R., Suganthy, N., Aarifa, M.A., et al. (2021). Alzheimer's disease: A molecular view of β-amyloid induced morbific events. *Biomedicines*, *9*(9):1126.

Muñoz, M. & Coveñas, R. (2014). Involvement of substance P and the NK-1 receptor in human pathology. *Amino Acids*, *46*(7):1727–1750.

Ogino, K. & Wang, D.H. (2007). Biomarkers of oxidative/nitrosative stress: An approach to disease prevention. *Acta Medica Okayama*, *61*(4):181–189.

Pan, Y., Li, Z., Zhao, X., Du, Y., et al. (2024). Screening of active substances regulating Alzheimer's disease in ginger and visualization of the effectiveness on 6-gingerol pathway targets. *Foods*, *13*(4):612.

Park, J.C., Han, S.H. & Mook-Jung, I. (2020). Peripheral inflammatory biomarkers in Alzheimer's disease: A brief review. *BMB Reports*, *53*(1):10–17.

Pázmándi, K., Szöllősi, A.G. & Fekete, T. (2024). The "root" causes behind the anti-inflammatory actions of ginger compounds in immune cells. *Frontiers in Immunology*, *15*:1400956.

Roudsari, N.M., Lashgari, N., Momtaz, S., et al. (2021). Ginger: A complementary approach for management of cardiovascular diseases. *Biofactors*, *47*(5):933–951.

Suto, T. & Karonitsch, T. (2020). The immunobiology of mTOR in autoimmunity. *Journal of Autoimmunity*, *110*:102373.

Tripathi, S., Maier, K.G., Bruch, D. & Kittur, D.S. (2007). Effect of 6-gingerol on pro-inflammatory cytokine production and costimulatory molecule expression in murine peritoneal macrophages. *Journal of Surgical Research*, *138*(2):209–213

Ullah, I., Subhan, F., Ayaz, M., Shah, R., et al. (2015). Anti-emetic mechanisms of *Zingiber officinale* against cisplatin induced emesis in the pigeon; behavioral and neurochemical correlates. *BMC Complementary and Alternative Medicine*, *15*(1):34.

Yahyazadeh, R., Rahimi, V.B., Yahyazadeh, A., et al. (2021). Promising effects of gingerol against toxins: A review article. *Biofactors*, *47*(5):885–913.

Yang, G., Gao, X., Jiang, L. & Sun, X. (2017). 6-gingerol prevents MEHP-induced DNA damage in human umbilical vein endothelia cells. *Human & Experimental Toxicology*, *36*(11):1177–1185.

Zeng, G.F., Zhang, Z.Y., Lu, L., et al. (2013). Protective effects of ginger root extract on Alzheimer disease-induced behavioral dysfunction in rats. *Rejuvenation Research*, *16*(2):124–133.

Zeng, G.F., Zong, S.H., Zhang, Z.Y., et al. (2015). The role of 6-gingerol on inhibiting amyloid β protein-induced apoptosis in PC12 cells. *Rejuvenation Research*, *18*(5):413–421.

Zhang, F., Zhang, J.G., Yang, W., et al. (2018). 6-gingerol attenuates LPS-induced neuroinflammation and cognitive impairment partially via suppressing astrocyte overactivation. *Biomedicine & Pharmacotherapy*, *107*:1523–1529.

12 Quercetin for Managing and Treating Dementia

Olugbemi Tope Olaniyan and Duyilemi Chris Ajonijebu

12.1 INTRODUCTION

Quercetin is described as a plant bioflavonoid with very potent memory-enhancing and neuroprotective properties in animal and human models for neurodegeneration studies. The brain effects of quercetin are more pronounced on the hippocampus and cortex through the suppression of reactive oxygen species, inflammatory cytokines production, lipid peroxidation and restoring normal cholinergic transmission, brain-derived neurotrophic factor (BDNF) levels, upregulation of antioxidants, modulation of numerous apoptotic/survival signaling pathways like phosphatase and tensin homolog, cyclic adenosine monophosphate (cAMP) response element protein, mitogen-activated protein kinase (MAPK), and mammalian target of rapamycin (Khan et al., 2018). In addition to the mechanisms, quercetin also expresses sirtuin-1 in many organs with several beneficial health effects, such as cardiovascular protection, cataract prevention, anticancer activity, anti-inflammatory effects, and antiviral activity. The advantage offered by quercetin is the ability to pass through blood–brain barrier, which again shows a positive effect for the treatment of neurodegenerative diseases. Quercetin can pass through the rigid phospholipid bilayer of membranes, decrease the interaction of the membrane with the amyloid beta (Aβ) peptide, and prevent neurotoxicity (Odubela et al., 2025). The chemical structure of quercetin is reported to contain 15 carbon atoms in its structure, with a heterocyclic pyran ring and two benzene rings. The development of quercetin derivatives involves oxidation or substitution of the carbonyl group on position C-4 of ring C with 4-oxo-flavonoid or heterocyclic pyran or benzene ring substitution with different subclasses of flavonoids. General quercetin is a flavone and aglycone with an OH group on positions 3, 5, 7, 3′, and 4. To form a quercetin glycoside derivative, there must be an attachment of glucose, rutinose, or rhamnose (glycosyl group) on position 3, replacing the OH group. Quercetin and its derivative quercetin glycoside are very soluble in lipids and alcohol due to the presence of the glucosyl group. Generally, quercetin is named from the Latin word quercetum, which can be isolated from various plant sources like leaves, fruits, stems, seeds, and bud extracts. Quercetin is from the family Compositae, Rhamnaceae, Solanaceae, and Passiflorae with several pharmaceutical activities. In the intestine, quercetin undergoes sulfuric acid and glucuronic acid conjugation. Quercetin undergoes intracellular deconjugation by y-glucuronidase at the site of inflammation in the central nervous system to increase penetration in neurodegenerative disorders (Khan et al., 2020). Based on the physiological applications of quercetin and its derivatives, the aim of this chapter is to give an overview of the management of dementia using quercetin.

12.2 QUERCETIN FOR MANAGING AND TREATING DEMENTIA

Alshammari et al. (2021) reported the role of quercetin in hippocampal neurotoxicity using cadmium chloride in male Wistar rats. In their study, the authors reported that there was activation of sirtuin-1, improvement in the dental gyrus and the spatial structures, and memories. Furthermore, they also confirmed the suppression of inflammation, apoptosis, acetylcholinesterase, and oxidative stress by quercetin. Quercetin as an antioxidant was able to elevate the tissue endogenous antioxidant levels, such as glutathione and manganese superoxide dismutase, BDNF, and acetylcholine. The authors concluded by revealing that quercetin was able to increase the expression of

DOI: 10.1201/9781003657385-12

hippocampus, decrease acetylation of Nrf2, p65, p53, inhibit of acetylcholinesterase and NF-κB, and activities of sirtuin-1.

Dong-Mei et al. (2014) reported that the use of quercetin for chronic treatment of mitochondrial and cognitive functions was able to reduce the plaque burden in neurodegenerative diseases. In their study, the authors revealed that mitochondrial dysfunction is induced by Aβ in neurodegenerative diseases. Quercetin, a bioflavonoid with a potent neuroprotective role, was investigated by the authors using 20 and 40 mg kg^{-1} day^{-1} doses of the bioflavonoid for about 16 days in a mouse model. From the results, they revealed that quercetin improved memory and learning deficits, ameliorated the mitochondrial dysfunction in the brain, and reduced scattered senile plaques in the hippocampus. Furthermore, the bioflavonoid was able to increase the mitochondrial membrane potential, suppress reactive oxygen species, increase activation of AMP-activated protein kinase activity, and increase the generation of mitochondrial adenosine triphosphate (ATP) levels, particularly in the 40 mg kg group. The authors suggested that the role of quercetin in improving mitochondrial dysfunctions and reducing amyloid plaque is through the activation of the AMP-activated protein kinase pathway.

Yu et al. (2020) revealed the role of quercetin in neurodegenerative diseases in a cell model using $A\beta_{25-35}$ in PC12 cells. The authors utilized different concentrations of quercetin to investigate the biological activity of quercetin, such as 0, 10, 20, 40, and 80 μmol L^{-1} doses. From the cell survival rate and cell morphological results using 3(4,5 – dimethylthiazol-2-yl-)-2,5-diphenyltetrazolium bromide (MTT) assay, it was discovered that there was cell damage, a reduced number, swollen, protruding, rounded, and retracting cells, with suppressed adherent function with $A\beta_{25-35}$ treatment. The authors revealed that quercetin enhanced cell survival and proliferation rate, increased the antioxidant capacity, and reduced lipid peroxidation damage. The authors concluded that quercetin offers a neuroprotective property in neurodegenerative conditions, thus an important biomolecule in the management of neurodegenerative diseases. Hoxha et al. (2022) reported that neurodegenerative diseases affect the memory, cognitive function, and daily lifestyle of patients. In recent years, attention has been shifted to the use of plant phenolic phytochemicals such as quercetin in the management and treatment of dementia. These plants containing these phytochemicals have been found to be beneficial with promising results. Quercetin is a flavonoid and a polyphenol with the ability to reduce cell damage, inactivate glyceraldehyde-3-phosphate dehydrogenase and free radicals, inhibit AChE, suppress β amyloid aggregation, produce anti-inflammatory effects, and suppress tau aggregation.

Quercetin, also known as 3,3′,4′,5,7-pentahydroxyflavone, is found in many fruits and vegetables with many pharmacological roles, particularly as a neuroprotective agent. Recently, quercetin has been found to display a notable positive effect on learning, memory, and cognitive behavior in neurodegenerative diseases. Quercetin has been found to accumulate in the brain and brain tissue (Ishisaka et al., 2011), improving recognition memory, learning, attenuating mitochondrial dysfunction, and reducing scattered senile plaques. Ishisaka et al. (2011) reported that quercetin suppresses reactive oxygen species generation and increases AMP-activated protein kinase activity in the mouse model of neurodegeneration. Furthermore, quercetin was reported to reduce the presence of presenilin, Aβ secretion, suppress eukaryotic translation initiation factor 2a phosphorylation, and reduce the activation of transcription factor 4 expression, thus enhancing growth arrest and DNA-damaged inducible gene expression in the hippocampus. These roles result in a delay in the deterioration of learning and memory. In addition, quercetin was reported to inhibit protein phosphatase 2C alpha activated by cholesterol, NF-κB p65 nuclear translocation, decreased tauopathy, β-amyloidosis, microgliosis and astrogliosis, and MDA levels in the hippocampus dentate gyrus cell.

Recently, some herbal molecules in the management of neurodegenerative diseases were screened using in vitro investigation and molecular docking. Among the multiple herbal molecules screened for acetylcholinesterase, β-secretase, and butyrylcholinesterase inhibitors, quercetin was found to be the best molecule in the modulation of cholinergic pathways in neurodegenerative diseases. Karthika et al. (2022) showed that neurodegenerative diseases can be managed using a combination of bioflavonoids. The authors described the complexity of the blood–brain barrier as the reason for the multiple biomarkers combating the management of neurodegenerative diseases. The nontoxicity of quercetin makes it an important candidate for the management of neurogenerative diseases. The authors utilized quercetin with other phytochemicals to formulate a nano-based compound as a natural-based product with a high level of antioxidant for neurodegenerative diseases.

Sengupta and Begum (2025) revealed that current pharmacological management of dementia as a neurodegenerative disease involves the use of quercetin. Quercetin is a natural flavonoid with physiological properties such as antioxidant, anti-amyloid, and anti-inflammatory effects for the management of dementia. The authors revealed that due to the property of the blood–brain barrier, the application of quercetin becomes difficult, but the advancement in the development of nanotechnology offers an innovative approach in the treatment of neurodegenerative diseases. The development of polymeric nanoparticles, liposomes, and solid lipid nanoparticles using quercetin can be explored for improving cognitive performance, Aβ inhibition, and oxidative stress in neurodegeneration. Nishimura et al. (2017) reported that quercetin exhibits antioxidant, anti-dysglycemic, and anti-dyslipidemic activities in enhancing cognitive functions in dementia. The role of quercetin can be attributed to the ability to activate SIRT-1, resulting in neuroprotective properties. Ademosun et al. (2016) reported the antioxidant and anticholinesterase properties of quercetin. The glycosylated conjugate and quercetin inhibit acetylcholinesterase, butyrylcholinesterase, and lipid peroxidation, thus making quercetin a good candidate for the management of oxidative-stress-induced neurodegenerative diseases. Lu et al. (2018) revealed the protective role of quercetin in neurodegenerative diseases. Quercetin is a flavonoid that can ameliorate the pathology of dementia through reduction in astrogliosis and increased Aβ clearance.

Alexander et al. (2023) revealed that neurodegenerative diseases can impair memory and cognitive functions, particularly among the elderly. These dysfunctions can result from senile plaque, neuroinflammation, neurofibrillary tangle formation, damaged neuroplasticity, and oxidative stress. The authors noted that the current approach in the management of dementia is poor with many limitations, thus the search for alternative solutions. In their study, it was revealed that herbal medicine using quercetin could possibly be the best approach in the management of dementia. Quercetin is a medicinal phytochemical present in plants as a polyphenol that possesses the ability to facilitate many physiological mechanisms in ameliorating dementia, such as inhibition of Aβ plaque formation, suppression of oxidative stress, reduction in tau hyperphosphorylation, protein kinase B/phosphoinositide3-kinases (AKT/PI3K) pathways, and modulation of BDNF. The authors concluded by suggesting that quercetin suppresses neuroinflammation, restores proneuroplastic signaling pathways, and enhances proper cellular functions. Khan et al. (2020) showed that quercetin, as a flavonoid, produces beneficial pharmacological effects such as neuroprotection. In their study, the authors noted that quercetin is found in fruits and vegetables and is useful in many physiological mechanisms like protecting the neurons, reducing lipid peroxidation, antioxidant properties, inhibition of fibril formation of amyloid-proteins, anti-inflammatory cascade pathways, and counteracting cell lysis.

Yao et al. (2020) reported that quercetin, as a bioflavonoid, can inhibit tau protein and enhance neuroprotection in neurodegenerative diseases. The authors utilized plasma exosomes-loaded quercetin to increase the bioavailability of the drug in managing cognitive dysfunctions in okadaic-acid-induced dementia in a mouse model. The quercetin inhibits cyclin-dependent kinase-5-mediated phosphorylation of tau protein and suppresses the formation of insoluble neurofibrillary tangles. Sriraksa et al. (2012) reported that quercetin can be utilized for cognitive enhancement in rats. The authors revealed that cognitive impairment is induced by oxidative stress, which can be ameliorated using quercetin due to its antioxidant, anti-inflammatory, and increased neuron density. Odubela et al. (2025) revealed that dementia in neurodegenerative diseases is caused by activation of glycogen synthase kinase 3 beta, resulting in extracellular Aβ plaques generation and intracellular neurofibrillary tangles formation, as shown in Table 12.1. These processes lead to mitochondrial dysfunction and oxidative stress in the neuron; thus, the authors suggested that quercetin bioflavonoid phytochemical possesses anti-inflammatory and antioxidant roles in the management of dementia, thereby improving survival rate and locomotor activity.

TABLE 12.1
Quercetin for the Management of Dementia

SN	Mechanism	Authors
1	Suppression of reactive oxygen species, inflammatory cytokines production, lipid peroxidation and restoring normal cholinergic transmission, BDNF levels, upregulation of antioxidants, modulation of numerous apoptotic/survival signaling pathways like phosphatase and tensin homolog, cAMP response element protein, mitogen-activated protein kinase	Khan et al. (2018) Alexander et al. (2023) Alsemeh et al. (2019) Zaplatic et al. (2019) Su et al. (2024)
2	Decrease the interaction of the membrane with the Aβ peptide, and prevent neurotoxicity	Odubela et al. (2025) Wróbel-Biedrawa et al. (2022), Hala et al. (2023)
3	Intracellular deconjugation by y-glucuronidase of inflammation	Khan et al. (2020) Chiang et al. (2023)
4	Suppression of inflammation, apoptosis, acetylcholinesterase, and oxidative stress	Alshammari et al. (2021) Nagu et al. (2023) Salehi et al. (2020)
5	Ameliorate the mitochondrial dysfunction	Dong-Mei et al. (2014)
6	Cell survival and proliferation rate, increased the antioxidant capacity, and reduced the lipid peroxidation damage	Yu et al. (2020), Adnan et al. (2025) Zu et al. (2021)
7	Reduce cell damage, inactivate glyceraldehyde-3-phosphate dehydrogenase and free radicals, inhibit AChE, suppress β amyloid aggregation, produce anti-inflammatory effects, and suppress tau aggregation	Hoxha et al. (2022),
8	Attenuate mitochondrial dysfunction and reduce scattered senile plaques	Ishisaka et al. (2011)
9	Antioxidant, anti-amyloid, and anti-inflammatory	Karthika et al. (2022) Sengupta and Begum (2025)
10	Activates SIRT-1	Nishimura et al. (2017)
11	Inhibit acetylcholinesterase, butyrylcholinesterase, and lipid peroxidation	Ademosun et al. (2016)
12	Reduction in astrogliosis and increased Aβ clearance	Lu et al. (2018)
13	Suppressing the formation of insoluble neurofibrillary tangles	Yao et al. (2020)

Nagu et al. (2023) reported that the current lack of therapeutic intervention for the treatment of dementia is due to the unclear and multidimensional pathogenesis of the disease. In their study, it was revealed that developing sustainable and innovative therapeutic agents in the management of dementia will create opportunities for the management of the disease (Adetunji et al., 2023a; Olaniyan et al., 2023a, 2023b). The use of herbal medicine seems to be an innovative approach utilizing phytochemical-rich compounds such as quercetin with an inhibitory role as anti-inflammation, antioxidants, anticholinesterase, blocking β-Secretase, and butyrylcholinesterase. Through molecular docking, it was revealed that quercetin possesses strong receptor affinity and cholinergic modulatory pathways, thus suggesting the neuroprotective role in neurodegenerative diseases. Adnan et al.

(2025) and Adetunji et al. (2023b) demonstrated that the progressive nature of neurodegenerative diseases can be suppressed using phytochemicals like quercetin and a nutrigenomics approach with the ability to suppress Aβ peptide accumulation, mitochondrial dysfunction, cholinergic deficits, and oxidative stress. From their results, it was demonstrated that quercetin enhanced cell viability and mitochondrial integrity, conferring neuroprotective properties on quercetin in dementia.

Khan et al. (2018) demonstrated that multiple neurodegenerative diseases are due to chronic neuroinflammatory response, causing cognitive dysfunctions. Quercetin, as a natural bioflavonoid, is abundant in vegetables and fruits with many biological activities. The authors revealed that the mitochondrial apoptotic pathway and neuronal degeneration, which are regulated by Bax/Bcl2, activation of cytochrome c, cleaving PARP-1 and caspases-3, 7, and 9 activities across the hippocampal and cortical regions of the mouse brain, were inhibited by quercetin. Quercetin was able to reverse neurodegeneration in the brains of mice. Nishihira et al. (2021) reported that flavonoids such as quercetin, could suppress age-related cognitive and learning dysfunctions. In their study, the potential functions of quercetin on cognitive function were investigated among 70 healthy Japanese individuals. From the results, it was discovered that the flavonoid improved cognitive functions, mood, and depression. Babaei et al. (2018) revealed that quercetin, with the chemical formula 3,3,4,5,7-pentahydroxyflavone, is a rich bioflavonoid with several pharmacological effects in the body. The authors investigated the role of quercetin in neurodegenerative diseases. The effect of the flavonoid on memory and learning involves several pathways, thus it can serve as a potential nutraceutical neurotherapeutic agent in the management of dementia.

Wróbel-Biedrawa et al. (2022) revealed that the most common flavonoid is quercetin, with several physiological roles in the body, such as anti-inflammatory, anti-anxiolytic, and antioxidant. The authors also revealed that quercetin possesses the ability to inhibit hyaluronidase, lipoxygenases, and cyclooxygenases. Alsemeh et al. (2019) revealed that organophosphates are compounds that can damage the brain due to their neurotoxicity effect, resulting in neurodegenerative diseases like dementia. The authors reported that the toxicity effect of organophosphates can be ameliorated using quercetin, a bioflavonoid found in vegetables and fruits. The therapeutic role of quercetin has been attributed to the antioxidant, anti-inflammatory, and cell survival-promoting properties, as shown in Figure 12.1. Organophosphate exerts the neurotoxicity effect by causing functional and

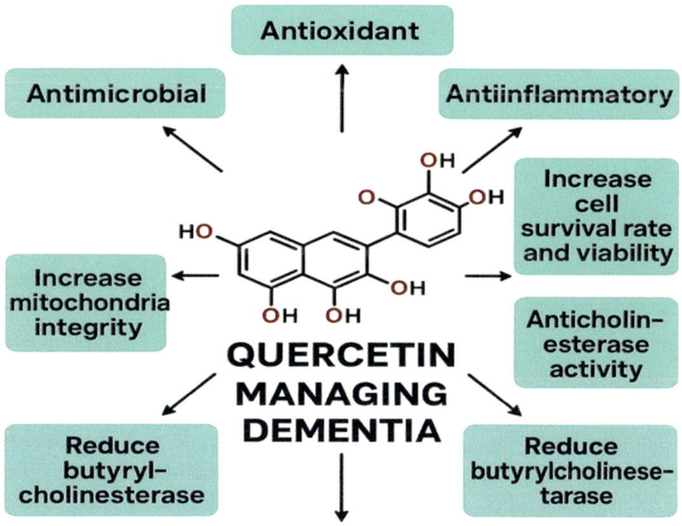

FIGURE 12.1 Mechanism of action of quercetin in dementia management.

structural damage to the cell membrane through MDA-activating apoptotic marker (caspase-3, 7, and 9), downregulation of synaptophysin, and astrogliosis marker glial fibrillary acidic protein. Zaplatic et al. (2019) reported that quercetin has antioxidant properties for the management of dementia. The phytochemical possesses the ability to scavenge hydroxyl radical (OH–), superoxide anion (O_2–), and hydrogen peroxide (H_2O_2) in neurodegenerative diseases. Neuro-anti-inflammatory response involves downregulation of cytokines through nuclear factor (erythroid-derived 2)-like 2 (Nrf2), protein kinase C, paraoxonase-2, c-Jun N-terminal kinase (JNK), PI3K/Akt, and MAPK-signaling cascade pathways.

Hala et al. (2023) demonstrated that neurodegeneration involves a decline in cognitive functions, and the use of quercetin can significantly ameliorate the detrimental effects of the aluminum-chloride-induced dementia. They concluded that quercetin offers neuroprotective effects, significantly reducing amyloid precursor protein levels, presenilin I, β-amyloid-converting enzyme 1 (BACE1), and elevating the expression of ADAM17 in the brain. Karthika et al. (2021) demonstrated that neurodegenerative diseases have increased globally over the last few years. Therefore, there is a need to search for an alternative solution for management due to the limitations of orthodox drugs with several side effects, as shown in Table 12.1. The authors noted that polyphenolic flavonoid quercetin nanoparticles offer neuroprotective effects against dementia by inhibition of fibril formation of Aβ plaque and inflammatory response. Hayashib et al. (2023) showed that dementia is linked to depression, causing changes in behavior and mood. The dysfunctional cognition is associated with an emotional condition. The authors revealed that quercetin is an important phytochemical with recent applications in neurodegenerative treatment. They noted that the brain of a mouse treated with quercetin showed improved cognitive functions.

The antioxidant capacity of quercetin makes it very powerful in the management of dementia. Quercetin possesses the ability to scavenge free radicals and regulate many biochemical pathways for neuronal survival and neurogenesis. Quercetin can suppress the formation of misfolded protein aggregates involved in inflammation and neuronal death along the Nrf2-ARE pathway (Olaniyan et al., 2020). Joshi et al. (2022) showed that memory loss in neurodegenerative diseases is linked to the accumulation of β-amyloid and tau plaques with detrimental effects on the neuron. The authors reported that quercetin inhibits beta-secretase enzyme activity and promotes neuroprotective effects. Behrasi et al. (2024) demonstrated that depression-linked conditions, such as dementia, are becoming prevalent globally. That management has been unclear and limited, with several side effects using neuropsychiatric drugs. Recently, quercetin was identified by the authors as a potent phytochemical with strong anti-inflammatory and antioxidant activity for the management of dementia. Quercetin suppressed tumor necrosis factor and galectin-3, thereby reducing inflammatory responses.

Su et al. (2024) reported that the maturation and proportion of NK cells binding to the MYH9 protein in mice for the enhancement of learning and cognitive function are promoted by quercetin administration. In their study, it was revealed that quercetin, as a bioflavonoid, possesses the ability to act as an antioxidant, anticancer, anti-inflammatory, senescent cell-clearing, and neuroprotective agent in neurodegenerative diseases like dementia, as shown in Figure 12.1. The authors concluded that quercetin is able to stimulate cognitive functions through interaction with the immune system in the hippocampal tissue in mice. Zu et al. (2021) revealed that the physiological properties of quercetin have attracted the attention of scientists for the therapeutic management of neurodegenerative diseases. Quercetin was reported by the authors to possess antioxidant, hypoglycemic, hypolipidemic, and anti-inflammatory activity as shown in Figure 12.1. Through molecular docking and dynamic simulation, it was discovered that quercetin stimulates the KEGG and MAPK signaling pathways. Sabarathinam et al. (2024) revealed that bioflavonoids like quercetin, and its derivative quercetin 3-O-glucuronide, are promising therapeutic agents for neuroinflammation-induced dementia. In their study, using molecular docking and dynamics, the bioactive phytochemical quercetin was seen to have multiple target sites for therapeutic potential.

Salehi et al. (2020) revealed that quercetin and its derivatives are phytochemicals found naturally in plants known for their promising biological effects in the management of neurodegeneration. This phytochemical acts as anti-inflammatory, antimicrobial, and antioxidant in dementia for increased bioavailability-associated drug-delivery nano-systems as shown in Figure 12.1. Chiang et al. (2023) showed that neuroinflammation is associated with neurodegenerative diseases, neurotoxicity, and permanent brain damage. Quercetin was revealed to have potential health benefits in modulating AMP-activated protein kinase, NLRP3, and NF-κB inflammasome pathways, modulating oxidative stress and inflammation in neurodegeneration. Zhu et al. (2022) demonstrated that the treatment of dementia can involve the use of multi-target drugs using flavonoid or combination techniques, but this has not yielded the best results due to low bioavailability and complex metabolic processes. The authors revealed that nanoparticles using quercetin can enhance bioavailability and reduce Aβ aggregation.

12.3 CONCLUSION

Quercetin, as a bioflavonoid, has numerous physiological applications due to its physiochemical properties, such as anti-inflammatory, antioxidant, anticholinesterase, antidiabetic, antiviral, antiproliferative, and anti-anxiolytic. The role of quercetin in the management of dementia was highlighted in this book chapter. The physiological mechanisms of quercetin in the management of dementia include reduction in Aβ aggregation, suppression of oxidative stress and inflammation, senescent cell-clearing, immune system modulation, blockage of β-secretase, butyrylcholinesterase, and anticholinesterase.

REFERENCES

Ademosun, A.O., Oboh, G., Bello, F., & Ayeni, P.O. (2016). Antioxidative properties and effect of quercetin and its glycosylated form (rutin) on acetylcholinesterase and butyrylcholinesterase activities. *Journal of Evidence-Based Complementary & Alternative Medicine*, 21(4), NP11–NP17. doi: 10.1177/2156587215610032.

Adetunji, C.O., Olugbemi, T.O., Maksim, R., Mohammad, A.S., Oluwatosin, A.I., Olulope, O.A., Abel, I. et al. (2023b). Roles of nutrigenomics in drug discovery and development. Role of Nutrigenomics in Modern-day Healthcare and Drug Discovery. Elsevier, 277–299.

Adetunji, C.O., Olugbemi, T.O., Olorunsola, A., Ayobami, D., Mayowa, J.A., & Alex, E. (2023a). Classification of neurodegenerative disorders using machine learning techniques. Artificial Intelligence for Neurological Disorders. Academic Press, 261–273.

Adnan, M., Siddiqui, A.J., Bardakci, F., Surti, M., Badraoui, R., & Patel, M. (2025). Neuroprotective potential of quercetin in Alzheimer's disease: Targeting oxidative stress, mitochondrial dysfunction, and amyloid-β aggregation. *Frontiers in Pharmacology*, 16, 1593264. doi: 10.3389/fphar.2025.1593264.

Alexander, C., Parsaee, A., & Vasefi, M. (2023). Polyherbal and multimodal treatments: Kaempferol and quercetin-rich herbs alleviate symptoms of Alzheimer's disease. *Biology*, 12, 1453. doi: 10.3390/biology12111453.

Alsemeh, A.E., Eman, M.A., & Abdelghany, N.H.H. (2019). Beneficial effect of quercetin against dimethoate induced cerebellar cortex injury in adult male albino rat: Histological and immunohistochemical study. *The Medical Journal of Cairo University*, 87(8), 5121–5134.

Alshammari, G.M., Wahidah, H., Al-Qahtani, M.A., Alshuniaber, Abu, E.A.Y., Abdullrahman, S.A., Laila, N.H., Maha, H.A. et al. (2021). Quercetin improves the impairment in memory function and attenuates hippocampal damage in cadmium chloride-intoxicated male rats by suppressing acetylcholinesterase and concomitant activation of SIRT1 signaling. *Journal of Functional Foods*, 86(2021), 104675.

Babaei, F., Mohammadreza, M., & Marjan, N. (2018). Quercetin in food: Possible mechanisms of its effect on memory. *Journal of Food Science*, 83(9), 2280–2287. doi: 10.1111/1750-3841.14317.

Behrasi, F., Mirshekar, M.A., Montazerifar, F., Tarbiat nobari, N., Gholipour, D., Arabmoazzen, S., & Veisi, A. (2024). Quercetin attenuates depressive-like behavior in a rat model of Alzheimer's disease: Involvement of galectin-3 signaling pathway. *Research Journal of Pharmacognosy*, 11(3), 67–75.

Chiang, M.-C., Tsai, T.-Y., & Wang, C.-J. (2023). The potential benefits of quercetin for brain health: A review of anti-inflammatory and neuroprotective mechanisms. *International Journal of Molecular Sciences*, 24, 6328. doi: 10.3390/ijms24076328.

Dong-Mei, W., San-Qiang, L., Wen-Lan, W., Xiao-Ying, Z., Yong, W., & Hong-Ying, Y. (2014). Effects of long-term treatment with quercetin on cognition and mitochondrial function in a mouse model of Alzheimer's disease. *Neurochemical Research*. doi: 10.1007/s11064-014-1343-x.

Hala, A.E., Asmaa, M., Elfiky, A.A.M., Khadiga, S., Ibrahim, & Mohamed, A.G. (2023). Neuroprotective effect of quercetin through targeting key genes involved in aluminum chloride induced Alzheimer's disease in rats. *Egyptian Journal of Basic and Applied Sciences*, 10(1), 174–184. doi: 10.1080/2314808X.2022.2164136.

Hayashib, Y., Fuminori, H.T., Kiyomi, N., Takuma, I., Masayuki, M., Takayoshi, S., Jun, N., Masuko, K., & Toshiyuki, N. (2023). Continuous intake of quercetin-rich onion powder may improve emotion but not regional cerebral blood flow in subjects with cognitive impairment. *Heliyon*, 9(2023), e18401.

Hoxha, D., Biljana, B., Gjoshe, S., & Gresa, H. (2022). Phenolic phytochemicals for treatment of Alzheimer's condition – A review. *Macedonian Pharmaceutical Bulletin*, 68(2), 3–15. doi: 10.33320/maced.pharm.bull.2022.68.02.001.

Ishisaka, A., Ichikawa, S., Sakakibara, H., Piskula, M.K., Nakamura, T., Kato, Y., & Terao, J. (2011). Accumulation of orally administered quercetin in brain tissue and its antioxidative effects in rats. *Free Radical Biology and Medicine*, 51(7), 1329–1336. doi: 10.1016/j.freeradbiomed.2011.06.017.

Joshi, A., Neha, V., Himanshu, S., & Vikas, K. (2022). Quercetin compound analysis to develop treatment for dementia associated with Alzheimer's disease in humans: In-silico study. *Journal of Drug and Alcohol Research*, 11(2022), 1–7.

Karthika, C., Rokeya, P.A., Habibur, A., Priti, R., Ghulam, T., Mohamed, A., Abdel-Daim, M., Syed, S.H., Areha, A., & Simona, B. (2022). Potential innovation against Alzheimer's disorder: A tricomponent combination of natural antioxidants (vitamin E, quercetin, and basil oil) and the development of its intranasal delivery. *Environmental Science and Pollution Research*, 29, 10950–10965. doi: 10.1007/s11356-021-17830-7.

Karthika, M.R.G., Mathan, S., Shaiju, S.D., Karthika, M., Mathan, R.G., & Shaiju, S. (2021). Neuroprotective role of quercetin in Alzheimer's disease. *International Journal of Pharmaceutical Research and Applications*, 6(2), 378–391.

Khan, A., Ali, T., Rehman, S.U., Khan, M.S., Alam, S.I., Ikram, M., Muhammad, T., Saeed, K., Badshah, H., & Kim, M.O. (2018). Neuroprotective effect of quercetin against the detrimental effects of LPS in the adult mouse brain. *Frontiers in Pharmacology*, 9, 1383. doi: 10.3389/fphar.2018.01383.

Khan, H., Hammad, U., Michael, A., Wai, S.C., & Esra, K.A. (2020). Neuroprotective effects of quercetin in Alzheimer's disease. *Biomolecules*, 10, 59. doi: 10.3390/biom10010059.

Lu, Y., Qi, L., & Qin, Y. (2018). Quercetin enrich diet during the early-middle not middle-late stage of Alzheimer's disease ameliorates cognitive dysfunction. *American Journal of Translational Research*, 10(4), 1237–1246. https://pmc.ncbi.nlm.nih.gov/articles/PMC5934583

Nagu, P., Amjad, K.A.P., & Vineet, M. (2023). Identification of herbal molecules for the treatment of Alzheimer's disease through a combination of molecular docking and in-vitro analysis. *Journal of Scientific & Industrial Research*, 82, 504–514. doi: 10.56042/jsir.v82i05.1076.

Nishihira, J., Mie, N., Masanori, K., Kagami-Katsuyama, H., Hiroki, H., Toshiyuki, N., Takato, M., & Masuko, K. (2021). The effect of 24-week continuous intake of quercetin-rich onion on age-related cognitive decline in healthy elderly people: A randomized, double-blind, placebo-controlled, parallel-group comparative clinical trial. *Journal of Clinical Biochemistry and Nutrition*, 69(2), 203–215. doi: 10.3164/jcbn.21-17.

Nishimura, M., Ohkawara, T., Nakagawa, T., Muro, T., Sato, Y., Satoh, H., Kobori, M., & Nishihira, J.A. (2017).) Randomized, double-blinded, placebo-controlled study evaluating the effects of quercetin-rich onions on cognitive function in elderly subjects. *Functional Foods in Health and Disease*, 7(6), 353–374.

Odubela, O.O., Olanrewaju, A.J., & Daniel, E.D. (2025). Synergistic therapeutic potential of curcumin and quercetin co-administration in Alzheimer's disease: Insights from transgenic drosophila melanogaster models. *IPS Journal of Drug Discovery Research and Reviews*, 3(1), 5–14. doi: 10.54117/ijddrr.v3i1.25.

Olaniyan, O.T., Adetunji, C.O., Dare, A., Adeyomoye, O., Adeniyi, M.J., & Alex, E. (2023a). Clinical applications of deep learning in neurology and its enhancements with future predictions. Artificial Intelligence for Neurological Disorders. Publisher Academic Press, 209–224.

Olaniyan, O.T., Adetunji, C.O., Dare, A., Adeyomoye, O., Adeniyi, M.J., & Alex, E. (2023b). New trends in deep learning for neuroimaging analysis and disease prediction. Artificial Intelligence for Neurological Disorders. Publisher Academic Press, 275–287.

Olaniyan, O.T., Okoli, B., Adetunji, C.O., Bretet, P., Adebayo, F., Dare, A., Okotie, G., Ibitoye, O., Eweoya, O., & Mali, P.C. (2020). Quercetin modulates granulosa cell mRNA androgen receptor gene expression in dehydroepiandrosterone-induced polycystic ovary in Wistar rats via metabolic and hormonal pathways. *Journal of Basic and Clinical Physiology and Pharmacology*, 31(4), 20190076.

Sabarathinam, S. (2024). Unraveling the therapeutic potential of quercetin and quercetin-3-O-glucuronide in Alzheimer's disease through network pharmacology, molecular docking, and dynamic simulations. *Scientific Reports*, 14, 14852. doi: 10.1038/s41598-024-61779-9.

Salehi, B., Laura, M., Lianet, M., Sharifi-Rad, J., Shahira, M.E., Mohamed, A.S., Rana, M.M., El Mahdy, N.M., Ceyda, S.K., Oksana, S., Sharifi-Rad, M., Farukh, S., Natália, M., Miquel, M., & William, C.C. (2020). Therapeutic potential of quercetin: New insights and perspectives for human health. *ACS Omega*, 5, 11849–11872.

Sengupta, S., & Begum, J. (2025). Quercetin as a neuroprotective agent in Alzheimer's disease: A comprehensive review. *International Journal of Engineering Research and Applications*, 15(5), 120–134.

Sriraksa, N., Jintanaporn, W., Supaporn, M., Somsak, T., Kamoltip, B., & Kowit, C. (2012). Cognitive-enhancing effect of quercetin in a rat model of Parkinson's disease induced by 6-hydroxydopamine. Hindawi Publishing Corporation. *Evidence-Based Complementary and Alternative Medicine*, Article ID 823206, 9 pages. doi:10.1155/2012/823206.

Su, T., Haitao, S., Mengyuan, H., Shanshan, Y., Xue, G., Ce, H., Liuling, G., Hao, W., Shengyu, F., Taotao, M., Meili, Z., Qing, L., Fengjiao, H., Jian-Kang, Z., Jianbo, Z., Hongbin, L., & Hailiang, L. (2024). Quercetin promotes the proportion and maturation of NK cells by binding to MYH9 and improves cognitive functions in aged mice. *Immunity & Ageing*, 21, 29. doi: 10.1186/s12979-024-00436-1.

Wróbel-Biedrawa, D., Grabowska, K., Galanty, A., Sobolewska, D., & Podolak, I. (2022). A flavonoid on the brain: Quercetin as a potential therapeutic agent in central nervous system disorders. *Life*, 12, 591. doi: 10.3390/life12040591.

Yao, Q., Lin, G., Yibing, J., Yijie, S., Haijuan, S., & Liang, Z. (2020). Brain delivery of quercetin-loaded exosomes improved cognitive function in AD mice by inhibiting phosphorylated tau-mediated neurofibrillary tangles. *Drug Delivery*, 27(1), 745–755. doi: 10.1080/10717544.2020.1762262.

Yu, X., Yicai, L., & Xiaohua, M. (2020). Effect of quercetin on PC12 Alzheimer's disease cell model induced by Aβ25-35 and its mechanism based on Sirtuin1/Nrf2/HO-1 pathway. *Hindawi BioMed Research International*, Article ID 8210578, 10. doi: 10.1155/2020/8210578.

Zaplatic, E., Bule, M., Shah, Z.A., Sahab, U., & Kamal, N. (2019). Molecular mechanisms underlying protective role of quercetin in attenuating Alzheimer's disease, *Life Sciences*. doi: 10.1016/j.lfs.2019.03.055.

Zhu, C., Yuheng, Y., Xiaowen, L., Xingyu, C., Xucong, L., & Xiaoping, W. (2022). Develop potential multi-target drugs by self-assembly of quercetin with amino acids and metal ion to achieve significant efficacy in anti-Alzheimer's disease. *Nano Research*, 15(6), 5173–5182. doi: 10.1007/s12274-021-4066-8.

Zu, G., Keyun, S., Ling, L., Xiuli, Z., Tao, H., & Hailiang, H., (2021). Mechanism of quercetin therapeutic targets for Alzheimer disease and type 2 diabetes mellitus. *Scientific Reports*, 11, 22959. doi: 10.1038/s41598-021-02248-5.

13 Lion's Mane for Managing and Treating Dementia

Abdulmajeed Opeyemi Agboola, Ryu Ki-Jin, Abdulwarith Ayomide Bamidele, Uthman Okikiola Adebayo, and Rukayat Omowumi Salahudeen

13.1 INTRODUCTION

13.1.1 Brief Overview of Dementia and Its Global Impact

The most prevalent subtype of dementia is Alzheimer's disease, which is an essential type of neurocognitive disease marked by a decrease in memory, functions, language, and the capacity to carry out daily tasks (Asuku et al., 2025a; World Health Organization, 2021). The number of persons with dementia is estimated to be in the tens of millions, and this figure is expected to elevate due to population aging and increasing life expectancies (Alzheimer's Association, 2023; World Health Organization, 2021). Dementia has a significant social and economic impact, including loss of productivity, demands for informal caregiving, and medical expenses that strain families and healthcare systems, especially in low- and middle-income countries with limited resources and official support (World Health Organization, 2021). Timely diagnosis and access to care are made more difficult by presenting variability, overlaying comorbidities, and stigma. As a result, public health solutions prioritize multisectoral planning, caregiver support, and early detection to lessen the increasing human and financial costs. Framework designs in prevention and policies can be utilized in changing demographics globally, and having a comprehensive understanding of dementia (Alzheimer's Association, 2023; World Health Organization, 2021).

13.1.2 Importance of Exploring Alternative and Natural Therapeutic Options

Dementia cases can be uniquely moderated by pharmaceutical therapies, which also cannot alter the disease's growth. This limitation has sparked interest in natural and complementary therapeutic techniques that can be utilized with traditional care. Studies and consensus guidelines have demonstrated progress in cognition and quality of life through non-pharmacological strategies (Picone et al., 2024). Due to their biological plausibility, relative accessibility, inflammation, and metabolic contributors to natural compounds and nutraceuticals (i.e., omega-3 fatty acids and B-vitamin strategies) are being subjected to ongoing research. Nevertheless, extensive large-scale trials are still required to confirm effectiveness and safety. In areas where access to specialized services is limited, it is sacrosanct to make findings on these options to target modifiable risk factors at the population level and provide culturally appropriate and cost-effective adjuncts (Picone et al., 2024; World Health Organization, 2019).

13.2 LION'S MANE MUSHROOM: BACKGROUND

13.2.1 Lion's Mane Mushroom (*Hericium erinaceus*)

Hericium erinaceus, which is also known as lion's mane mushroom, with scientific discoveries, providing sustainable health solutions, which serve as drivers that increase natural products with anti-inflammatory, antioxidant, and antibacterial qualities (Allegra, 2019; Rodríguez-Yoldi, 2021).

 DOI: 10.1201/9781003657385-13

Diseases, such as cancer, diabetes, neurological diseases, and cardiovascular disorders, are caused by chronic inflammation and oxidative stress (OS) (Contato et al., 2021; Rybczyńska-Tkaczyk et al., 2023). Synthetic medications, which are consistently utilized in conventional therapies, can have serious adverse effects and pose long-term health hazards (Leyane et al., 2022; Neganova et al., 2021). Furthermore, bioactive substances extracted from natural sources are considered by researchers as possible supplemental treatments (Blagov et al., 2024; Contato et al., 2020). These substances, which were obtained from fungi, plants, and marine life, have shown evidence in the prevention and treatment of diseases because they can neutralize reactive oxygen species (ROS) (Khare et al., 2021; Sun & Shahrajabian, 2023).

13.2.2 Characteristics, Habitat, and Chemical Composition of *H. erinaceus*

13.2.2.1 Taxonomy and Morphology

The Kingdom Fungi, Phylum Basidiomycota, Class Agaricomycetes, Order Russulales, and Family Hericiaceae are all home to *H. erinaceus* (Jahedi et al., 2024; Juszczyk et al., 2021; Lee et al., 2022) (Figure 13.1). Among the species in the Hericium genus are *Hericium coralloides* and *Hericium americanum*, but *H. erinaceus* is notable for its unique shape and therapeutic qualities (Almjalawi et al., 2024). Because of its unusual look, it is frequently referred to as "lion's mane," "monkey head," or "pom-pom mushroom" (Valu et al., 2021). As it matures, the white-to-cream-colored fruiting body of *H. erinaceus* yields to yellow or brown. *H. erinaceus* lacks a clear morphological distinction in contrast to more common edible mushrooms, which have a cap-and-stem structure (Almjalawi et al., 2024).

FIGURE 13.1 Scientific classification of lion mane mushroom (*Hericium erinaceus*).

13.2.3 Botanical Description and History of Use in Traditional Medicine

H. erinaceus is a weak parasitic wood-decay fungus which are saprophytic in nature and recognized by a cushion-like fruiting body that is covered in spines or pendent "teeth." Fresh basidiomes are white, turning cream to yellow as they age; in temperate woods, they form a brief lateral attachment to hardwood substrates (usually beech or oak) and have dangling spines that are usually more than 1 cm long and can reach several centimeters. Micromorphologically, *H. erinaceus* produces smooth, subglobose to short-ellipsoid amyloid spores that are approximately 5–7 × 4–5 μm in size and possess a monomitic hyphal system with clamp connections (Contato & Conte-Junior, 2025).

Based on habitat specialization, they were pressure on veteran hardwoods, the species, which is endemic to Europe, North America, and East Asia, is protected and rare in several areas of Europe (Contato & Conte-Junior, 2025). It is listed as a tonic for the stomach and spleen in traditional Chinese medicine (TCM)'s traditional pharmacopeias; these indications roughly correspond to contemporary ideas of convalescence and digestive assistance (Contato & Conte-Junior, 2025). Hot-water decoctions, powders, and later mycelial preparations entered contemporary functional-food markets as a result of cultivation techniques developed in China and Japan that allowed for regular culinary and therapeutic use (Surendran et al., 2025).

Lion's mane's traditional profile is still focused on advancing vigor and digestion while being a unique "toothed" fungus that is ecologically associated with old hardwoods. However, modern research in the plant combines this ethnomedical background with pharmaceutical discoveries (such as nerve growth factor (NGF)-modulating terpenoids) (Corana et al., 2019).

13.2.4 Key Bioactive Compounds (Hericenones, Erinacines, Polysaccharides)

Cyathane-type diterpenoid erinacines, β-glucan-rich polysaccharides, and aromatic/meroterpenoid hericenones are the three main types of compounds responsible for *H. erinaceus*'s bioactivity. Isolated from fruiting bodies, hericenones (e.g., C–H, K, L) are phenolic derivatives with oxidized geranyl side chains. They have been shown to have anti-inflammatory and antioxidant properties as well as to induce Nerve growth factor (NGF) in vitro (Al-Azzawi & Bandr, 2023; Contato & Conte-Junior, 2025).

The majorities of erinacines are made by mycelia and have a distinctive 5–6–7 tricyclic cyathane skeleton. A number of these members can pass through the blood–brain barrier and increase neurotrophic signaling (Spangenberg et al., 2025; Yao et al., 2021). Preclinical syntheses show that erinacine C and A are frequent exemplars of NGF/Brain-Derived Neurotrophic Factor (BDNF) regulation, Nrf2 antioxidant pathway activation, and neuroinflammation mitigation. *H. erinaceus* polysaccharides (HEPs) are made up of heterogeneous β-glucans and associated glycans (mostly glucose and galactose, with smaller amounts of mannose and fucose). They are usually extracted using hot water and then alcohol precipitation (Ge et al., 2025). The intestinal barrier is protected, gut-microbiota modification occurs, antioxidant and anti-tumor activities are documented, immunomodulatory mechanisms (e.g., macrophage/Th1–Th2 balance), and hypoglycemic and hypolipidemic effects are documented (Ge et al., 2025; Li et al., 2021).

In a nutshell, hericenones and erinacines work together to produce lipophilic, neurotrophin-oriented effects, while HEPs use metabolic and innate/adaptive immunological pathways to produce wide-ranging systemic effects. Production decisions are also influenced by this chemical division of labor: cultured mycelial products are preferred for erinacine enrichment (Contato & Conte-Junior, 2025; Doar et al., 2025; Zaitseva et al., 2024).

13.2.5 General Health Benefits of *H. erinaceus*

H. erinaceus has a significant effect on the gastrointestinal system, including experimentally demonstrated anti-ulcer and anti-*Helicobacter pylori* actions (Spangenberg et al., 2025). While hypersensitivity interactions (e.g., with anticoagulants or antidiabetic drugs) have been reported, safety profiles for culinary use. Prudence is advised in pregnancy, lactation, and transplant settings, pending robust data (Smith et al., 2025).

According to preclinical data, polysaccharide fractions have anti-tumor activity and microbiota modulation. These properties are frequent with conventional digestive indications and demonstrate antioxidant effects (Li et al., 2021). *Neurocognitive Endpoints*: Terpenoids (hericenones, erinacines) elevate NGF/BDNF signaling, reduce neuroinflammation/OS, and improve cognition, along with elevated serum BDNF (Contato & Conte-Junior, 2025; Szućko-Kociuba et al., 2023).

Lion's mane is a useful food with neurotrophic and gut-immune processes, but translation necessitates careful planning (Contato & Conte-Junior, 2025; Doar et al., 2025; Smith et al., 2025).

13.3 MECHANISMS OF ACTION IN DEMENTIA MANAGEMENT

13.3.1 Stimulation of Nerve Growth Factor (NGF) Synthesis

Neurotrophin is an integral factor that is essential for survival and phenotypic stability of basal forebrain cholinergic neurons (Asuku et al., 2025b), and it unconsciously disappeared in many dementias. Restoring cholinergic signaling and delayed cognitive decline are the aims of therapeutic approaches that either replicate the trophic activities of NGF. The reason for the interest in natural chemicals and fungal metabolites is that a number of small molecules increase the manufacture of NGF in glial and neuronal cells, which stimulate intracellular signaling (Szućko-Kociuba et al., 2023).

The bioactive terpenoids erinacines and hericenones, which were isolated from *H. erinaceus*, have been shown to improve NGF in vivo and stimulate NGF production in astroglial cultures. These effects have been connected to enhanced cholinergic indicators and behavior in animal models. NGF-based therapeutic paradigms, in general, include gene therapy, small-molecule stimulators, and targeted delivery of recombinant NGF; each strategy must strike a compromise between delivery difficulties and off-target effects due to NGF's pleiotropic properties (Gálvez et al., 2022).

Based on this mechanism, treatments that stimulate endogenous NGF synthesis function ahead of synaptic maintenance by promoting axonal and dendritic growth, supporting cholinergic neuron survival, and indirectly influencing tau and amyloid-related disorders through enhanced neuronal resilience. Despite the encouraging preclinical fact, translation calls for safety testing and methods to produce long-lasting, brain-region-specific NGF elevations free of adverse effects.

13.3.2 Neuroprotective Effects and Reduction of Oxidative Stress

The main cause of neuronal defect in dementia is OS, which is caused by an excess of ROS that destroys proteins, lipids, and mitochondrial DNA, disrupts synaptic function, and starts cell death pathways (Asuku et al., 2024). Thus, the goals of neuroprotective interventions are to decrease the generation of ROS, strengthen natural antioxidant defenses (such as Nrf2-driven enzymes), balance mitochondrial function, and prevent downstream apoptotic signaling (Galla et al., 2025). In cellular and animal models, natural polyphenols, flavonoids, and certain mushroom diterpenoids have shown multi-modal neuroprotection by regulating antioxidant response elements, scavenging free radicals, and ensuring balance in mitochondrial membrane potential (Spangenberg et al., 2025). Recent preclinical investigations have shown that several members of the erinacine family minimize neuronal death in toxin or amyloid models by activating Nrf2 signaling, increasing the generation of antioxidant enzymes, and reducing indicators of lipid peroxidation and protein carbonylation (Spangenberg et al., 2025). The ability to stabilize mitochondrial bioenergetics, restore redox homeostasis through gene regulation, and lessen oxidative damage to synaptic machinery are all essential components of neuroprotection. When paired with anti-amyloid or anti-tau, these methods can preserve cognition more effectively than single-target antioxidant supplementation (Galla et al., 2025).

13.3.3 Anti-Inflammatory Properties

The main pathology characteristics of dementia (amyloid, tau, and synaptic damage) are both responded to and amplified by chronic neuroinflammation, which is caused by peripheral immunological signals and prolonged microglial activation. Thus, anti-inflammatory methods in dementia care work by reprogramming innate immune responses, lowering the release of proinflammatory cytokines (such as Tumor necrotic factor-alpha (TNF-α) and IL-1β), limiting complement-mediated synapse pruning, and inhibiting the activation of inflammasomes (Corana et al., 2019). Certain Non-steroidal anti-inflammatory drugs (NSAIDs) in epidemiologic studies and experimental compounds that block some important signaling nodes (e.g., NLRP3 inflammasome) are examples of pharmacologic and nutritional interventions that reduce synaptic stripping and neuronal loss by changing microglia from a proinflammatory to a reparative phenotype (Surendran et al., 2025).

Eating certain diets, like the Mediterranean-style diets, and foods rich in omega-3 fatty acids and polyphenols, can help lower inflammation in the body and brain. This may help in protecting brain blood vessels and also slow down memory loss (Contato & Conte-Junior, 2025). The best way to treat harmful brain inflammation is to lower it while still allowing the brain to repair itself. Studies show that treatment works best if started early, that is, before inflammation causes serious damage and blocks the brain's ability to clear waste and rebuild.

13.3.4 Promotion of Neurogenesis and Synaptic Plasticity

Helping the brain grow new cells and strengthen its connections can repair damage caused by dementia. This involves building new neurons, strengthening existing links, and restoring memory networks. According to Contato and Conte-Junior (2025), neurotrophins (NGF, BDNF), activity-dependent signaling, and intracellular pathways (Cyclic AMP response element-binding proteins (CREB), mTOR, PI3K/Akt) that govern neuronal survival, dendritic spine formation, and long-term potentiation (LTP) all have an influence on neurogenesis in the adult hippocampus and the ability for synaptic remodeling. In preclinical models, interventions that increase BDNF/NGF, promote CREB-dependent transcription, or lessen inhibitory constraints on plasticity (such as excessive inflammation or oxidative damage) increase synaptic strength markers and also improve learning paradigm performance. The use of a lifestyle-based approach, such as exercise, an enriched environment, and cognitive training, together with trophic stimulation, seems to work in concert: in the early and prodromal stages of dementia, trophic chemicals provide a good molecular environment, and exercise promotes circuit-specific synaptic formation (Valu et al., 2021).

13.4 THERAPEUTIC POTENTIAL AND APPLICATIONS

13.4.1 Possible Role as Adjunct Therapy in Dementia Care

H. erinaceus's neuroprotective and neurotrophic qualities have drawn attention to its potential therapeutic use in the treatment of dementia. Since dementia is a complex illness that involves inflammation, OS, neurodegeneration, and synaptic dysfunction, treatments that can target several pathways may improve current care methods. Based on preclinical and clinical evidence, erinacines and hericenones, these two compounds found in lion's mane, help in the synthesis of NGF, support neurogenesis, and improve synaptic plasticity, all of which are essential for preserving cognitive function (Freund & Croughan, 2018; Mendoza et al., 2018).

Lion's mane supplementation has been shown in small-scale clinical trials to improve cognitive scores in older persons with moderate cognitive impairment. This suggests that it may be able to delay the advancement of the condition in its early stages (Saitsu et al., 2019). By improving neuroplasticity and offering anti-inflammatory and antioxidant properties that reduce neuronal damage, lion's mane may be used as a treatment given alongside the main one, known as pharmaceutical therapies (Menon et al., 2025). Additionally, its good safety record facilitates its incorporation into comprehensive dementia treatment plans in conjunction with medication, cognitive training, and lifestyle modifications. Lion's mane may be a useful supportive approach to enhance the quality of living and maintain function in dementia patients, even though it cannot replace traditional medications. However, before it can be implemented in clinical practice, further randomized trials are required to ensure consistent dose and long-term effectiveness.

13.4.2 Comparison with Conventional Treatments

The main ways that traditional dementia therapies, such as acetylcholinesterase inhibitors (donepezil, rivastigmine, galantamine) and N-methyl-D-aspartate (NMDA) receptor antagonists (memantine), relieve symptoms are by improving cholinergic transmission or controlling glutamatergic signaling. These medications have minor effects on brain-related and day-to-day functioning, but they don't change the underlying neurodegenerative processes. They also cause side effects such as fatigue, dizziness, and gastrointestinal problems (Alves, 2024). On the other hand, lion's mane seems to work via neurodegenerative processes, specifically promoting the production of NGF, neurite development, and the reduction of OS (Banerjee et al., 2024; Li et al., 2021).

This means that its effects might go beyond controlling symptoms and possibly even slowing the course of the illness. Furthermore, early human research shows enhanced cognitive test results with no side effects, establishing lion's mane as a safer substitute or addition to medication (Li et al., 2021). Conventional medications are recommended as first-line treatments because of established evidence from extensive trials; however, lion's mane is still in the trial phase and is primarily supported by preclinical research and tiny trials. Lion's mane targets neurotrophic signaling and neuroprotection, whereas conventional treatments restore neurotransmitter balance. This is a significant disparity in how they work. Combining the two approaches may significantly have synergistic effects, with lion's mane improving brain resilience and repair, and pharmaceutical medicines managing symptoms.

However, because there is currently inadequate evidence to substitute conventional drugs, integration requires standardized extracts, dose validation, and regulatory approval. Therefore, lion's mane has potential as a complementary therapy, but it is not yet on phase with traditional pharmaceutical therapies.

13.4.3 Use in Dietary Supplements and Functional Foods

In the nutraceutical sector, lion's mane has become well-known outside clinical settings as a dietary supplement promoted for neuroprotection and cognitive well-being. Its addition in fortified meals, teas, coffees, powders, and capsules shows the increased demand from consumers for natural ways to preserve brain function and postpone age-related brain decline (Contato & Conte-Junior, 2025). Long-term ingestion of lion's mane through dietary supplements may help maintain brain health because studies have shown that it contains bioactive chemicals that can support neuroplasticity and stimulate NGF (Banerjee et al., 2024; Li et al., 2021).

Lion's mane-enriched foods are especially desirable since they are easily included in regular meals and can be used as a preventative measure in populations at risk for brain impairment. Its status as a brain-boosting supplement is backed by pilot human studies that show an increase in subjective memory and cognitive performance (Corana et al., 2019). Concerns about uniformity and effectiveness among commercially accessible goods are raised by differences in product quality and extraction techniques. Clinical usage is limited by the absence of standardization of erinacine or hericenone concentrations in numerous supplements, in contrast to pharmaceutical formulations.

Regional differences in regulation and monitoring results in differences in labeling and safety procedures. However, lion's mane functional foods and supplements stand as a promising approach for supportive and preventive care for dementia management, as consumer demand continues to rise.

13.5 CHALLENGES AND LIMITATIONS

13.5.1 Gaps in Clinical Research and Standardization

H. erinaceus has shown encouraging results for the treatment of dementia, but there are still gaps in clinical research. Most of the evidence currently available is based on small-scale human trials and preclinical animal experiments, which limit the strength and generalizability of conclusions. For instance, the widely cited trial by Saitsu et al. (2019) showed improvements in patients with mild cognitive impairment, but its short duration and small sample size inhibit definitive claims about long-term effects. Other studies also show promising results but lack standard cognitive endpoints and diverse patient populations (Saitsu et al., 2019; Surendran et al., 2025). The lack of standard preparations is another point of setback. Numerous experiments use various lion's mane, fruiting body powders, and erinacine-enriched mycelium with varying amounts of bioactive substances such as erinacines and hericenones. Establishing repeatable clinical standards and cross-study comparisons is made more challenging by these differences (Banerjee et al., 2024; Li et al., 2021). The integration of lion's mane into clinical protocols is also hindered by the lack of information regarding its interactions with traditional dementia drugs. The use of standardized extracts and rigorous techniques in bigger, well-designed randomized controlled trials is necessary before lion's mane may be considered an evidence-based treatment option. Therefore, it is still important to fill in these research and standardization gaps before lion's mane may go from an experimental supplement to an approved adjunct therapy for dementia care.

13.5.2 Dosage, Formulation, and Bioavailability Issues

One of the significant challenges of *H. erinaceus* in treating dementia is the absence of evidence about the best dosage, formulation, and bioavailability. The supplements of lion's mane are purchased in a variety of forms, such as capsules, powders, teas, and functional foods, without consistent dose standards, different from alternative medications that have undergone pharmacokinetic and pharmacodynamic research. The safe and standard treatment window has not been established because clinical trials have utilized a wide range of doses (Saitsu et al., 2019).

In addition, there are investigations regarding the bioavailability of the bioactive compounds in lion's mane when taken orally due to their infrequent stability and absorption profiles (Li et al., 2021). In a recent study, these substances might be rapidly metabolized, which would restrict their capacity to enter the central nervous system at therapeutic doses (Banerjee et al., 2024). To increase absorption and stability, regular administration patterns such as standardized extract formulations and nanoparticles must go through tedious research strategies (Popoola et al., 2025).

Therefore, formulation decisions were complicated by the disparities between fruiting body and mycelial preparations, which had varying ratios of neuroactive substances. Clinicians and patients are left in the "State of Unknown" about how to use drugs effectively with adequate dosage recommendations. To initiate lion's mane as a recommendable therapeutic tool in dementia care, this formulation decision must be firm, undergoing deep pharmacological study.

13.6 CONCLUSION

Lion's mane mushroom (*H. erinaceus*) in treating dementia has been justified by evidence in maintaining brain function, serving as a natural remedy. It contains bioactive substances, including erinacines and hericenones, that promote neurogenesis, increase NGF, and offer anti-inflammatory properties. While modest clinical trials suggest possible cognitive benefits, especially in moderate cognitive impairment, preclinical research has shown an increase in memory and learning. It is more appealing as an adjuvant strategy due to its generally positive safety profile. Lion's mane can boost dementia care by accelerating neuroprotection and bolstering cognitive resilience.

REFERENCES

Al-Azzawi, R. R. A., & Bandr, L. K. (2023). Effect of Adding Lion's Mane (*Hericium erinaceus*) and Reishi Mushroom (*Ganoderma lucidum*) to Broiler Diets on the Productive Performance. *IOP Conference Series: Earth and Environmental Science*, *1262*(7), 072034. https://doi.org/10.1088/1755-1315/1262/7/072034

Allegra, M. (2019). Antioxidant and Anti-Inflammatory Properties of Plant Extracts. *Antioxidants*, *8*(11), 549. https://doi.org/10.3390/antiox8110549

Almjalawi, B. S. A., Chechan, R. A., & Alhesnawi, A. S. M. (2024). Some Applications of *Hericium erinaceus* Mushrooms: A Review. *South Asian Research Journal of Agriculture and Fisheries*, *6*(5), 92–100. https://doi.org/10.36346/sarjaf.2024.v06i05.002

Alves, L. M. G. (2024). *Hericium Erinaceus Polyphenols in the Strive against Alzheimer's Disease*. http://hdl.handle.net/10400.5/100897

Alzheimer's Association. (2023). *Alzheimer's Disease Facts and Figures*. https://www.alz.org/alzheimers-dementia/facts-figures.

Asuku, A. O., Arinde, E., Saliu, S. B., Oyerinde, T., Olajide, T. S., Afolabi, H., & Kehinde, A. (2025b). Neuroplasticity: Harnessing the Brain Ability to Heal Itself. In Advancing Medical Research through Neuroscience. IGI Global Publisher. https://doi.org/10.4018/979-8-3693-5464-3.ch005

Asuku, A. O., Ayinla, M. T., Gbonjubola, O. O., Babatunde, S. S., Olajide, T. S., & Oyerinde, T. O. (2024). Neurocognitive Aspects of Dementia. In: Nutrition in Brain Aging and Dementia. Nutritional Neurosciences, Springer, August 2024, https://doi.org/10.1007/978-981-97-4117-5_5

Asuku, A. O., Ogungbangbe, G. O., & Olajide, T. S. (2025a). Effect of Happiness and Positive Emotions on Dementia. In Neuroscience of Feelings. Progress in Brain Research. Elsevier. https://doi.org/10.1016/bs.pbr.2025.03.001

Banerjee, S., Gupta, S., Raj, R., Gupta, M., Kumari, S., & Kuar, G. (2024). Unlocking the Potential of Lion's Mane Mushroom (*Hericium erinaceus*). *Journal of Applied and Natural Science*, *16*(1), 39–50. https://doi.org/10.31018/jans.v16i1.5224

Blagov, A. V., Summerhill, V. I., Sukhorukov, V. N., Zhigmitova, E. B., Postnov, A. Y., & Orekhov, A. N. (2024). Potential Use of Antioxidants for the Treatment of Chronic Inflammatory Diseases. *Frontiers in Pharmacology*, *15*. https://doi.org/10.3389/fphar.2024.1378335

Contato, A. G., Aranha, G. M., Filho, B. A. A., Peralta, R. M., & de Souza, C. G. M. (2021). Evaluation of the Antioxidant and Antimicrobial Activity of the Culinary-Medicinal Mushroom *Lentinula boryana* (Agaricomycetes) from Brazil. *International Journal of Medicinal Mushrooms*, *23*(10), 1–7. https://doi.org/10.1615/IntJMedMushrooms.2021039966

Contato, A. G., Brugnari, T., Sibin, A. P. A., Buzzo, A. J. R., de Sá-Nakanishi, A. B., Bracht, L., Bersani-Amado, C. A., Peralta, R. M., & de Souza, C. G. M. (2020). Biochemical Properties and Effects on Mitochondrial Respiration of Aqueous Extracts of Basidiomycete Mushrooms. *Cell Biochemistry and Biophysics*, *78*(1), 111–119. https://doi.org/10.1007/s12013-020-00901-w

Contato, A. G., & Conte-Junior, C. A. (2025). Lion's Mane Mushroom (*Hericium erinaceus*): A Neuroprotective Fungus with Antioxidant, Anti-Inflammatory, and Antimicrobial Potential—A Narrative Review. *Nutrients*, *17*(8), 1307. https://doi.org/10.3390/nu17081307

Corana, F., Cesaroni, V., Mannucci, B., Baiguera, R. M., Picco, A. M., Savino, E., Ratto, D., Perini, C., Kawagishi, H., Girometta, C. E., & Rossi, P. (2019). Array of Metabolites in Italian *Hericium erinaceus* Mycelium, Primordium, and Sporophore. *Molecules*, *24*(19), 3511. https://doi.org/10.3390/molecules24193511

Doar, E., Meyer, K. W., Bair, Z. J., Nally, R., McNalley, S., Davis, R., & Beathard, C. (2025). Influences of Substrate and Tissue Type on Erinacine Production and Biosynthetic Gene Expression in *Hericium erinaceus*. *Fungal Biology and Biotechnology*, *12*(1), 4. https://doi.org/10.1186/s40694-025-00194-9

Freund, N., & Croughan, M. (2018). A Simple Method to Reduce Both Lactic Acid and Ammonium Production in Industrial Animal Cell Culture. *International Journal of Molecular Sciences*, *19*(2), 385. https://doi.org/10.3390/ijms19020385

Galla, R., Ferrari, S., Miletto, I., Mulè, S., & Uberti, F. (2025). In Vitro Neuroprotective Effects of a Mixed Extract of Bilberry, *Centella asiatica*, *Hericium erinaceus*, and Palmitoylethanolamide. *Foods*, *14*(15), 2678. https://doi.org/10.3390/foods14152678

Gálvez, C., Pérez-Puig, H., & Elorriaga-Verplancken, F. R. (2022). Northernmost Habitat Range of Guadalupe Fur Seals (*Arctocephalus townsendi*) in the Gulf of California, México. *Aquatic Mammals*, *48*(3), 223–233. https://doi.org/10.1578/AM.48.3.2022.223

Ge, F., Chen, Y., Wang, B., Zhou, W., Du, B., & Hou, L. (2025). Bioactive Polysaccharides from *Hericium erinaceus*: Extraction, Structure, Bioactivities, and Applications. *Molecules*, *30*(8), 1850. https://doi.org/10.3390/molecules30081850

Jahedi, A., Ahmadifar, S., & Mohammadigoltapeh, E. (2024). Revival of Wild Edible-Medicinal Mushroom (*Hericium erinaceus*) Based on Organic Agro-Industrial Waste-Achieving a Commercial Protocol with the Highest Yield; Optimum Reuse of Organic Waste. *Scientia Horticulturae*, *323*, 112510. https://doi.org/10.1016/j.scienta.2023.112510

Juszczyk, G., Mikulska, J., Kasperek, K., Pietrzak, D., Mrozek, W., & Herbet, M. (2021). Chronic Stress and Oxidative Stress as Common Factors of the Pathogenesis of Depression and Alzheimer's Disease: The Role of Antioxidants in Prevention and Treatment. *Antioxidants*, *10*(9), 1439. https://doi.org/10.3390/antiox10091439

Khare, T., Anand, U., Dey, A., Assaraf, Y. G., Chen, Z.-S., Liu, Z., & Kumar, V. (2021). Exploring Phytochemicals for Combating Antibiotic Resistance in Microbial Pathogens. *Frontiers in Pharmacology*, *12*. https://doi.org/10.3389/fphar.2021.720726

Lee, Y. T., Yunus, M. H. M., Ugusman, A., & Yazid, M. D. (2022). Natural Compounds Affecting Inflammatory Pathways of Osteoarthritis. *Antioxidants*, *11*(9), 1722. https://doi.org/10.3390/antiox11091722

Leyane, T. S., Jere, S. W., & Houreld, N. N. (2022). Oxidative Stress in Ageing and Chronic Degenerative Pathologies: Molecular Mechanisms Involved in Counteracting Oxidative Stress and Chronic Inflammation. *International Journal of Molecular Sciences*, *23*(13), 7273. https://doi.org/10.3390/ijms23137273

Li, H., Zhao, H., Liu, W., Feng, Y., Zhang, Y., Yuan, F., & Jia, L. (2021). Liver and Brain Protective Effect of Sulfated Polysaccharides from Residue of Lion's Mane Medicinal Mushroom, *Hericium erinaceus* (Agaricomycetes), on D-Galactose-Induced Aging Mice. *International Journal of Medicinal Mushrooms*, *23*(5), 55–65. https://doi.org/10.1615/IntJMedMushrooms.2021038241

Mendoza, C., Perez-Urrutia, N., Alvarez-Ricartes, N., Barreto, G. E., Pérez-Ordás, R., Iarkov, A., & Echeverria, V. (2018). Cotinine Plus Krill Oil Decreased Depressive Behavior, and Increased Astrocytes Survival in the Hippocampus of Mice Subjected to Restraint Stress. *Frontiers in Neuroscience*, *12*. https://doi.org/10.3389/fnins.2018.00952

Menon, A., Jalal, A., Arshad, Z., Nawaz, F. A., & Kashyap, R. (2025). Benefits, Side Effects, and Uses of *Hericium erinaceus* as a Supplement: A Systematic Review. *Frontiers in Nutrition*, *12*. https://doi.org/10.3389/fnut.2025.1641246

Neganova, M., Liu, J., Aleksandrova, Y., Klochkov, S., & Fan, R. (2021). Therapeutic Influence on Important Targets Associated with Chronic Inflammation and Oxidative Stress in Cancer Treatment. *Cancers*, *13*(23), 6062. https://doi.org/10.3390/cancers13236062

Picone, P., Girgenti, A., Buttacavoli, M., & Nuzzo, D. (2024). Enriching the Mediterranean Diet Could Nourish the Brain More Effectively. *Frontiers in Nutrition*, *11*. https://doi.org/10.3389/fnut.2024.1489489

Popoola, O. A., Adetunji, C. O., Oyewole, O. A., Inobeme, A., & Israel, A. (2025). Discovery and Relevance of Novel Pharmacological Substances from Beneficial Mushrooms. In Mushroom Biotechnology for Improved Agriculture and Human Health (pp. 183–207). Wiley. https://doi.org/10.1002/9781394212699.ch8

Rodríguez-Yoldi, M. J. (2021). Anti-Inflammatory and Antioxidant Properties of Plant Extracts. *Antioxidants*, *10*(6), 921. https://doi.org/10.3390/antiox10060921

Rybczyńska-Tkaczyk, K., Grenda, A., Jakubczyk, A., Kiersnowska, K., & Bik-Małodzińska, M. (2023). Natural Compounds with Antimicrobial Properties in Cosmetics. *Pathogens*, *12*(2), 320. https://doi.org/10.3390/pathogens12020320

Saitsu, Y., Nishide, A., Kikushima, K., Shimizu, K., & Ohnuki, K. (2019). Improvement of Cognitive Functions by Oral Intake of *Hericium erinaceus*. *Biomedical Research*, *40*(4), 125–131. https://doi.org/10.2220/biomedres.40.125

Smith, A. V., Zhu, H., Mats, L., & Bozzo, G. (2025). Polysorbate 80 Differentially Impacts Erinacine Production Profiles in Submerged Cultures of Hericium. *Molecules*, *30*(13), 2823. https://doi.org/10.3390/molecules30132823

Spangenberg, E. T., Moneypenny, A., Bozzo, G. G., & Perreault, M. L. (2025). Unveiling the Role of Erinacines in the Neuroprotective Effects of *Hericium erinaceus*: A Systematic Review in Preclinical Models. *Frontiers in Pharmacology*, *16*. https://doi.org/10.3389/fphar.2025.1582081

Sun, W., & Shahrajabian, M. H. (2023). Therapeutic Potential of Phenolic Compounds in Medicinal Plants—Natural Health Products for Human Health. *Molecules*, *28*(4), 1845. https://doi.org/10.3390/molecules28041845

Surendran, G., Saye, J., Binti Mohd Jalil, S., Spreadborough, J., Duong, K., Shatwan, I. M., Lilley, D., Heinrich, M., Dodd, G. F., & Surendran, S. (2025). Acute Effects of a Standardised Extract of *Hericium erinaceus* (Lion's Mane Mushroom) on Cognition and Mood in Healthy Younger Adults: A Double-Blind Randomised Placebo-Controlled Study. *Frontiers in Nutrition*, *12*. https://doi.org/10.3389/fnut.2025.1405796

Szućko-Kociuba, I., Trzeciak-Ryczek, A., Kupnicka, P., & Chlubek, D. (2023). Neurotrophic and Neuroprotective Effects of *Hericium erinaceus*. *International Journal of Molecular Sciences*, *24*(21), 15960. https://doi.org/10.3390/ijms242115960

Valu, M.-V., Soare, L. C., Ducu, C., Moga, S., Negrea, D., Vamanu, E., Balseanu, T.-A., Carradori, S., Hritcu, L., & Boiangiu, R. S. (2021). *Hericium erinaceus* (Bull.) Pers. Ethanolic Extract with Antioxidant Properties on Scopolamine-Induced Memory Deficits in a Zebrafish Model of Cognitive Impairment. *Journal of Fungi*, *7*(6), 477. https://doi.org/10.3390/jof7060477

World Health Organization. (2019). Risk Reduction of Cognitive Decline and Dementia: WHO Guidelines. World Health Organization. https://www.who.int/publications/i/item/9789241550543

World Health Organization. (2021). *Dementia*. https://www.who.int/news-room/fact-sheets/detail/dementia

Yao, F., Gao, H., Yin, C.-M., Shi, D.-F., Lu, Q., & Fan, X.-Z. (2021). Evaluation of In Vitro Antioxidant and Antihyperglycemic Activities of Extracts from the Lion's Mane Medicinal Mushroom, *Hericium erinaceus* (Agaricomycetes). *International Journal of Medicinal Mushrooms*, *23*(3), 55–66. https://doi.org/10.1615/IntJMedMushrooms.2021037646

Zaitseva, O., Sergushkina, M., Popyvanov, D., Nazarova, Y., Polezhaeva, T., Solomina, O., & Khudyakov, A. (2024). The Influence of Polysaccharide Fractions of the Lion's Mane Medicinal Mushroom *Hericium erinaceus* (Agaricomycetes) on the Immunomodulatory and Antioxidant Activity of Neutrophils Exposed to Stress of Different Durations. *International Journal of Medicinal Mushrooms*, *26*(11), 11–25. https://doi.org/10.1615/IntJMedMushrooms.2024054937

14 *Ganoderma lucidum* for Managing and Treating Dementia

Akintayo Christopher Oloruntoba, Olaleye Tobi Opeyemi, Olowomeye Mary Oluwakemi, Omoruyi Gloria Osayamen, Onyekweli Samuel, Akinsomisoye Olumide Stephen, and Olaniyi Kehinde Samuel

14.1 INTRODUCTION

14.1.1 Dementia and Historical Discovery of *Ganoderma lucidum*

Dementia is usually the broad phenomenon said to have symptom diversification, which affects an individual's way of thinking and ability to remember things (basically immediate events/occurrences), which in a way has a negative outcome on social functioning and day-to-day activities. There are underlying factors in dementia responsible for brain cell damage, and this hinders affected individuals' communication (World Health Organization, 2023). Most of the people who come down with dementia have difficulty in language, mood change/behavioral change, loss of memory, and most times confusion as their clinical presentation. However, affected persons tend to present with forgetfulness of recent events and tend to struggle to find words. With progression of this condition, the symptoms tend to get worse, thereby losing communication abilities and becoming unable to recognize family and friends and to cater for themselves (Alzheimer's Association, 2024). It has been said that age is the strongest known risk factor responsible for dementia, especially in persons above the age of 65 (Asuku et al., 2022a, 2022b). Other possible factors identified are lifestyle, genetics, cardiovascular health/disease, and educational background, which has been said to affect the risk (Livingston et al., 2020). Other conditions that also tend to lead to the development of dementia are social isolation, cigarette smoking, obesity, high blood pressure, pressure, and diabetes. This can be diagnosed via clinical evaluation, cognitive investigations, and imaging studies like magnetic resonance imaging/computed tomography (MRI/CT) scans. There is no cure for dementia at the moment, but there are ongoing research studies regarding medications that could slow the progression and improve associated symptoms (Alzheimer's Association, 2024). There are non-pharmacological strategies, including regular exercise, healthy eating, and cognitive stimulation therapy, that could help in achieving a better life quality (World Health Organization, 2023). However, an optimal care level should be instituted for people living with dementia, as this will further cause depreciation in cognition, as this will affect their living conditions significantly (World Health Organization, 2023). *Ganoderma lucidum*, a medicinal mushroom, shows promise in managing dementia due to its neuroprotective properties. It contains bioactive compounds like triterpenoids and polysaccharides, which exhibit antioxidant and anti-inflammatory activities. These actions may reduce neuronal damage and oxidative factors implicated in dementia progression. Additionally, *G. lucidum* has been found to enhance cognitive function by modulating neurotransmitters and promoting neurogenesis. Its multifaceted effects suggest potential as a therapeutic agent for dementia management (Wang et al., 2018a). Among the bioactive substances present in mushrooms that are not yet fully used are

DOI: 10.1201/9781003657385-14

glycoproteins, polysaccharides (mostly β-glucans), and secondary metabolites. These physiologically active components may be used as immunopotentiating, hepatoprotective, antiviral, anticancer, and hypocholesterolemic medications. Mushrooms were considered the "miraculous king of herbs" by the Chinese (221–206 BC); they were only served during great feasts. Native Americans used mushrooms in ancient rituals (such as magical hallucinogens) to break through mental and physical barriers. Numerous studies conducted, which started in the West over three decades ago, have revealed intriguing biological characteristics of *G. lucidum*. The earliest of the more than 250 species in the *Ganoderma* genus, *G. lucidum*, was discovered in 1831 by the Italian mycologist Carlo Vittadini, who examined a lot of fungal species (Ekiz et al., 2023). In many Asian countries, such as China, Vietnam, and India, this fungus has been used as traditional medicine for centuries. As a tropical plant, *G. lucidum* is predominant in tropical and subtropical rainforests of Asia, Africa, and the United States (Loyd et al., 2018). However, it is now being cultivated in China, Korea, Japan, and several nations in South America and Southeast Asia, using intensive farming technologies. In southern Vietnam, it germinates naturally on dầu lim trees. Mainly used to promote longevity and good health, it was seen as a source of spiritual strength and immortality (Money, 2016). Traditionally, it was used to treat conditions like bronchial cough in the elderly, "deficiency fatigue," carcinoma, and chronic hepatitis. More recently, however, it has been shown to have anti-tumor potential (El Sheikha, 2022).

14.1.2 A Brief Overview of *Ganoderma lucidum*

This is a member of the Ganodermataceae family, a suborder of the Polyporales, and is frequently referred to as Reishi or Ling Zhi. It is an enormous black mushroom with a glossy appearance and a woody touch. "*Ganoderma*," its scientific name, is a combination of the Greek words "ganos," meaning "brightness, sheen," and "derma," meaning "skin." The word "*lucidum*" comes from the Latin "lucidus," which means "light, bright, clear." It is well recognized by names such as "lacquered bracket" and "varnished conk" and is renowned for its glossy appearance.

14.1.3 Description of *Ganoderma lucidum*

G. lucidum typically grows on decaying hardwood, hemlock. However, it can also grow on conifer stumps (Figure 14.1). With a woody feel and a glossy surface, the mushroom is covered in a layer of fungal metabolites. It is large and black. Fresh mushrooms are squishy, flat, and cork-like, and they release their spores through small pores. They also have no underside gills. The bottom pores of the mushroom can be either brown or white, depending on how old it is.

It consists of the morphological elements as follows: cap (pileus), which gives it a semi-circular, kidney-like shape. It can have a diameter of 3–35 cm (1.2–13.8 in.). It comes in a range of colors and has a surface that is obviously lacquered or polished. It appears reddish-brown with circular growth zones that become yellow, orange, or maroon as it grows. Little, round pores, approximately 4–5 mm apart, are visible on the bottom, or pores, which have a white to light brown surface. It produces brown spores that are ellipsoid and have two walls. Stem (stipe): often eccentric, sporadically missing, or lateral. It has a hard, polished, woodsy appearance (Figure 14.1). Since only the rich used to eat *G. lucidum*, there was no need to expand its cultivation. This opinion has changed, though, as mushroom production has increased in response to growing demand from a range of socioeconomic levels. Because they boost exports and hold promise for the food and pharmaceutical industries, products made from *G. lucidum* are vital to the Chinese economy. Natural chemicals obtained from it have numerous biological characteristics, such as cosmetic and nutraceutical/therapeutic effects. Its mycelial, spore, and fruit body components are commercially available and sold as finished commodities. Its palatability may be enhanced, nonetheless, by turning it into manufactured goods, including powders and dietary supplements.

FIGURE 14.1 *Ganoderma lucidum* growing in the soil and on a Tree Stump.

14.1.4 Nutritional Profile of *Ganoderma lucidum*

This is one of the untapped natural medicinal products that is benefiting growers and consumers globally (Chang & Wasser, 2017; Adetunji et al., 2022). According to Cör et al. (2018), it contains around 400 physiological ingredients, such as alkaloids, fatty acids, triterpenoids, and nucleosides, among others. *G. lucidum* contains a significant 19.5 g per 100 g of mushroom (weight for weight) of water-soluble protein.

A substantial amount of water-soluble protein has been reported in *G. lucidum*, with a concentration of 19.5 g per 100 g of mushroom (weight for weight). Eighteen different amino acids have been identified, with leucine being the most prevalent amino acid (Zhang et al., 2018). Furthermore, there are 3.5 g of dietary fiber for every 100 g of dry weight mushrooms. It contains high levels of sulfur and trace elements (including copper), crucial in maintaining cellular structure and function, as well as activating enzymes. Potassium (K), one of the essential elements, is necessary for muscle contraction, neuronal transmission, and preserving the fluid balance. Calcium (Ca) is vital in both teeth/bone strength, the release of enzymes and hormones, the contraction of muscle fibers, and the conduction of nerve impulses. Immunological activities, including immunoglobulin production, antibody-mediated cytolysis, lymphocyte-IgM interaction, macrophage reactivity to lymphokines, and immunological cell adhesion, all depend on magnesium (Mg). A vital component of hemoglobin, iron (Fe) is responsible for carrying oxygen throughout the circulation and is essential for all metabolic functions. Protein synthesis and/or repair of cells and tissues depend on phosphorus (P). Sulfur (S) is in charge of blood coagulation, immunological response, and transfer across cell membranes. Electrolyte balance and neuron activity are supported by sodium (Na). Iron metabolism and erythropoiesis are mediated by copper (Cu). Furthermore, the mushroom contains a variety of vitamins, including riboflavin, niacin, thiamin, β-carotene, and vitamins C, D, and E, among others (Ahmad, 2020). Certain metalloproteases, which slow down the clotting process, are also present in *G. lucidum.*

14.2 *GANODERMA LUCIDUM* MECHANISMS OF ACTION

14.2.1 Antioxidant Mechanism

Oxidation is a biological process that generates harmful compounds, which causes a reduction in the nutritional content of food, shortens a product's shelf life, induces oxidative stress, and contributes to the development of chronic diseases (Bishop et al., 2015). It occurs as a result of imbalance in the redox status, causing damage in the central nervous system. Accumulation of reactive oxygen species (ROS) in cells also leads to lipid peroxidation, protein denaturation, apoptosis, and DNA damage. Additionally, ROS can directly bind to the mitochondrial membrane, leading to defects in mitochondrial structure and function. A decrease in dopamine and acetylcholine levels results in behavioral impairment, and a reduction in the activity of neurons is one of the events that are triggered by these abnormalities. By promoting the body's antioxidant defense mechanism through a redox cycle, the polysaccharides exhibit effective defense through a variety of mechanisms (Ferreira et al., 2015; Kou et al., 2022). *G. lucidum* possesses polysaccharides that prevent the respiratory burst and reduce ROS to help/promote the anti-aging process. Its triterpenes, leucine, and glycoproteins—which are produced by the covalent bonding of polysaccharides with peptides or proteins—also contribute to its antioxidant properties (Ferreira et al., 2015; El Sheikha, 2022). By preventing t-butyl hydroperoxide (t-BOOH) from inducing necrosis in macrophages, *G. lucidum* peptidoglycan has been demonstrated to prevent the hampering effects of ROS and protect from damaging macrophage microvilli, endoplasmic reticulum, and mitochondria. Ekiz et al. (2023) report that ethanolic extracts of *G. lucidum* contain the phenolic components.

14.2.2 Effect on Mitochondrial Dysfunction

Mitochondria serve as important energy producers and are also involved in the structure and neuronal functions. While mitochondrial deleterious outcome usually results in a chaotic cellular redox environment, insufficient adenosine triphosphate (ATP) supply, and triggers inflammation, progressive harm to mitochondria leads to a severe inflammatory outcome, apoptosis, and ultimately, neurodegenerative diseases (Islam, 2017). Neurodegenerative diseases, in which cells specifically destroy defective or senescent mitochondria, have also been strongly associated with mitochondrial autophagy, a process that involves the autophagolysosome pathway. Thus, by scavenging free radicals, *G. lucidum* protects mitochondria from oxidative stress and may have an impact on mitochondrial development.

14.2.3 Effect on Inflammatory Mechanism

Triterpenoids from *G. lucidum* can stop important inflammatory processes because they impede the nuclear factor-kappa B (NF-κB) signaling mechanism (Ma et al., 2024). This pathway decreases chronic inflammation (Meng et al., 2023), lowers the production of proinflammatory cytokines (TNF-α, IL-1β, and IL-6), and, finally, protects against diseases of neurodegenerative/multiple sclerosis (Hapuarachchi et al., 2016; Plosca et al., 2025).

14.3 *GANODERMA LUCIDUM* IN TREATING AND MANAGING DEMENTIA

A hallmark of neurodegenerative disorders, including Huntington's disease, Parkinson's disease, Alzheimer's, and amyotrophic lateral sclerosis (ALS), is the loss of neuronal structure and function (Asuku et al., 2022a). These conditions affect motor, behavioral, and cognitive abilities of the elderly and are characterized by malfunctioning protein accumulation, dysfunctional mitochondria, and oxidation (Lian et al., 2024). Dementia encompasses a group of neurodegenerative disorders characterized by cognitive decline, memory impairment, and behavioral changes, with Alzheimer's disease (AD) being the most common subtype (Asuku et al., 2024, 2025). It is a neurological condition that affects daily functioning by causing gradual memory loss/language impairment (Giau et al. 2018). Currently, there is no known cure or preventative strategy for AD (Toledo et al., 2017; Yu et al., 2020; Adetunji et al., 2023; Olaniyan et al., 2023a, 2023b). WHO revealed that AD causes 60–70% of dementia cases globally, and its incidence is expected to quadruple by 2050 (Luo et al., 2025; Rani et al., 2025). In clinical settings, N-methyl-D-aspartate (NMDA) receptor antagonists and acetyl choline esterase (AChE) inhibitors are commonly used for patients to reduce symptoms without halting or reversing the degenerative process (Mullane & Williams, 2018). This has spurred research into the use of natural products for AD. Traditional treatments often offer limited efficacy, prompting the exploration of complementary therapies such as *G. lucidum*, a medicinal mushroom long used in East Asian medicine for its neuroprotective properties.

G. lucidum contains various bioactive compounds, including triterpenoids, polysaccharides, and peptidoglycans, which exhibit antioxidant, anti-inflammatory, and immunomodulatory effects. These properties are critical because oxidative stress and neuroinflammation are central contributors to the pathogenesis of dementia.

Research indicates that *G. lucidum* may mitigate cognitive decline by reducing neuronal damage through its antioxidative activity. Animal studies have demonstrated that administration of *G. lucidum* extracts enhances learning and memory performance, likely through modulation of oxidative stress markers and inflammatory cytokines in the brain (Wang et al., 2018b). Additionally, polysaccharides extracted from *G. lucidum* have been shown to promote neurogenesis and synaptic plasticity, suggesting potential for improving cognitive functions (Liu et al., 2024).

Beyond antioxidation, *G. lucidum* may inhibit acetylcholinesterase activity, increasing acetylcholine levels in the brain, a neurotransmitter critical for memory and learning impaired in dementia (Chen et al., 2019). This cholinergic enhancement aligns with current therapeutic targets for AD, where acetylcholinesterase inhibitors are standard treatment.

Despite promising preclinical data, human clinical trials remain limited. Small-scale studies suggest improvements in cognitive function and quality of life among dementia patients treated with *G. lucidum* supplements, but further randomized controlled trials are necessary to establish efficacy, safety, and optimal dosing regimens (Zhao et al., 2019).

A report by Huang et al. (2017) indicated that oral administration of *G. lucidum*'s polysaccharides and water extract stimulates pathways involving stress kinases like phosphorylated c-Jun N-terminal kinase (JNK), and p38 mitogen-activated protein kinase (MAPK) in transgenic AD mice. The resultant effect of this is enhanced neurogenesis, increased proliferation of neural progenitor cells (NPCs), and a decrease in cognitive deficits in the mice. Additionally, Huang et al. (2017) showed that polysaccharides produced from *G. lucidum* can improve cognitive performance and brain progenitor proliferation in a mouse model of AD. Conclusively, *G. lucidum* triterpenoids prevented AD in mice by reducing apoptosis rate and the expression levels of Bax, Bcl-2, and caspase-3/cleaved caspase-3 in the neuronal and hippocampus tissues via the Rho-associated coiled-coil kinase (ROCK) signaling pathway (Yu et al., 2020). In summary, *G. lucidum* exhibits multiple neuroprotective effects that could be beneficial in the management of dementia, primarily through antioxidative, anti-inflammatory, and cholinergic mechanisms. However, clinical validation is essential before it can be recommended as a standard complementary therapy.

14.4 CONCLUSION

Owing to its complex etiology, AD continues to be a widespread, incurable illness that places a heavy burden on individuals who have it and their loved ones. The complexity of this problem makes the advancement of AD treatment approach a herculean task. However, due to the neuroprotective benefits contained in bioactive components and their mode of action of *G. lucidum*, this traditional medicinal fungus has become a potentially significant choice in research on natural herbal medications. Although studies on *G. lucidum*'s potential as a therapy for AD are still in their early phases, the data suggest that it could be a good option. According to some studies, *G. lucidum* may help patients with AD manage their behavioral alterations and enhance better life outcomes due to the numerous beneficial properties it possesses.

REFERENCES

Adetunji, C.O., Olugbemi, T.O., Adetunji, J.B., Osemwegie, O.O., & Ubi, B.E. (2022). African Mushrooms as Functional Foods and Nutraceuticals. *Fermentation and Algal Biotechnologies for the Food, Beverage, and Other Bioproduct Industries*, Publisher CRC Press, 233–251.

Adetunji, C.O., Olugbemi, T.O., Adeyomoye, O., Dare, A., Adeniyi, M.J., &Enoch, A. (2023). Classification of Neurodegenerative Disorders Using Machine Learning Techniques. *Artificial Intelligence for Neurological Disorders*, Publisher Academic Press, 261–273.

Ahmad, M.F. (2020). *Ganoderma lucidum*: A rational pharmacological approach to surmount cancer. *J. Ethnopharmacol.*, *260*, 113047.

Alzheimer's Association. (2024). Alzheimer's disease facts and figures. https://www.alz.org/media/Documents/alzheimers-facts-and-figures.pdf

Asuku, A.O., Adeyemo, M.B., Adeyemo, R.O., Agoro, S.A., & Odugbesan, O.A. (2022a). The physiology and neuroprotective role of melatonin in neurodegenerative diseases. *Int. J. Med. Sci. Res.*, *4*(1), 54–60.

Asuku, A.O., Adeyemo, M.B., Adeyemo, R.O., Agoro, S.A., Odugbesan, O.A., & Ayoola, S.A. (2022b). Applications of stem cell therapy in neurodegenerative diseases. *Eur. J. Biotechnol. Biosci.*, *10*(2), 17–24.

Asuku, A.O., Ayinla, M.T., Gbonjubola, O.O., Babatunde, S.S., Olajide, T.S., & Oyerinde, T.O. (2024). Neurocognitive Aspects of Dementia. *Nutrition in Brain Aging and Dementia. Nutritional Neurosciences*, Springer. https://doi.org/10.1007/978-981-97-4117-5_5

Asuku, A.O., Ogungbangbe, G.O., &Olajide, T.S. (2025). Effect of Happiness and Positive Emotions on Dementia. *Neuroscience of Feelings. Progress in Brain Research*, Elsevier. https://doi.org/10.1016/bs.pbr.2025.03.001

Bishop, K.S., Kao, C.H., Xu, Y., Glucina, M.P., Paterson, R.R.M., Ferguson, L.R., et al. (2015). From 2000 years of *Ganoderma lucidum* to recent developments in nutraceuticals. *Phytochemistry*, *114*, 56–65. https://doi.org/10.1016/j.phytochem.2015.02.015

Chang, S.T., & Wasser, S.P. (2017). The cultivation and environmental impact of mushrooms. *Oxford Res. Encycl. Environ. Sci.* Retrieved 26 Sep. 2025, from https://oxfordre.com/environmentalscience/view/10.1093/acrefore/9780199389414.001.0001/acrefore-9780199389414-e-231

Chen, X., Zhang, Y., & Wang, Z. (2019). Inhibitory effects of *Ganoderma lucidum* triterpenoids on acetylcholinesterase activity: Implications for Alzheimer's disease therapy. *J. Ethnopharmacol.*, *243*, 112123. https://doi.org/10.1016/j.jep.2019.112123

Cör, D., Knez, Ž., & Hrnčič, M.K. (2018). Antitumour, antimicrobial, antioxidant and antiacetylcholinesterase effect of *Ganoderma lucidum* terpenoids and polysaccharides: A review. *Molecules*, *23*, 649.

Ekiz, E., Oz, E., Abd El-Aty, A.M., Proestos, C., Brennan, C., Zeng, M., Tomasevic, I., Elobeid, T., Çadırcı, K., Bayrak, M., & Oz, F. (2023). Exploring the potential medicinal benefits of *Ganoderma lucidum*: From metabolic disorders to coronavirus infections. *Foods*, *12*(7), 1512. https://doi.org/10.3390/foods12071512

El Sheikha, A.F. (2022). Nutritional profile and health benefits of *Ganoderma lucidum* "Lingzhi, Reishi, or Mannentake" as functional foods: Current scenario and future perspectives. *Foods*, *11*, 1030.

Ferreira, I.C., Heleno, S.A., Reis, F.S., Stojkovic, D., Queiroz, M.J.R., Vasconcelos, M.H., & Sokovic, M. (2015). Chemical features of *Ganoderma* polysaccharides with antioxidant, antitumor, and antimicrobial activities. *Phytochemistry*, *114*, 38–55.

Giau, V.V., Wu, S.Y., Jamerlan, A., An, S.S.A., Kim, S.Y., & Hulme, J. (2018). Gut microbiota and their neuroinflammatory implications in Alzheimer's disease. *Nutrients*, *10*, 1765. https://doi.org/10.3390/nu10111765

Hapuarachchi, K., Wen, T., Jeewon, R., Wu, X., & Kang, J. (2016). Mycosphere Essays 15. *Ganoderma lucidum*-are the beneficial medical properties substantiated? *Mycosphere*, *7*, 687–715.

Huang, S., Mao, J., Ding, K., et al. (2017). Polysaccharides from *Ganoderma lucidum* promote cognitive function and neural progenitor proliferation in mouse model of Alzheimer's disease. *Stem Cell Rep.*, *8*(1), 84–94.

Islam, M.T. (2017). Oxidative stress and mitochondrial dysfunction-linked neurodegenerative disorders. *Neurol. Res.*, *39*, 73–82.

Kou, R.W., Xia, B., Wang, Z.J., Li, J.N., Yang, J.R., Gao, Y.Q., Yin, X., & Gao, J.M. (2022). Triterpenoids and meroterpenoids from the edible *Ganoderma resinaceum* and their potential anti-inflammatory, antioxidant and anti-apoptosis activities. *Bioorg. Chem.*, *121*, 105689.

Lian, W., Yang, X., Duan, Q., Li, J., Zhao, Y., Yu, C., He, T., Sun, T., Zhao, Y., &Wang, W. (2024). The biological activity of *Ganoderma lucidum* on neurodegenerative diseases: The interplay between different active compounds and the pathological hallmarks. *Molecules*, *29*(11), 2516. https://doi.org/10.3390/molecules29112516

Liu, J., Zhang, B., Wang, L., Li, S., Long, Q., & Xiao, X. (2024). Bioactive components, pharmacological properties and underlying mechanism of *Ganoderma lucidum* spore oil. A review. *Chin Herb Med*, *16*(3): 375–391. https://doi.org/10.1016/j.chmed.2023.09.007

Livingston, G., Huntley, J., Sommerlad, A., Ames, D., Ballard, C., Banerjee, S., &Mukadam, N. (2020). Dementia prevention, intervention, and care: 2020 report of the Lancet Commission. *Lancet*, *396*(10248), 413–446.

Loyd, A.L., Richter, B.S., Jusino, M.A., Truong, C., Smith, M.E., Blanchette, R.A., & Smith, J.A. (2018). Identifying the mushroom of immortality: Assessing the *Ganoderma* species composition in commercial Reishi products. *Front Microbiol.*, *9*, 1557.

Luo, Y., Qiao, L., Li, M., Wen, X., Zhang, W., & Li, X. (2025). Global, regional, national epidemiology and trends of Parkinson's disease from 1990 to 2021: Findings from the global burden of disease study 2021. *Front Aging Neurosci.*, *16*, 1498756.

Ma, Y., Han, J., Wang, K., Han, H., Hu, Y., Li, H., Wu, S., & Zhang, L. (2024). Research progress of *Ganoderma lucidum* polysaccharide in prevention and treatment of *Atherosclerosis*. *Heliyon*, *10*, e33307.

Meng, M., Wang, L., Yao, Y., Lin, D., Wang, C., Yao, J., Sun, H., & Liu, M. (2023). *Ganoderma lucidum* polysaccharide peptide (GLPP) attenuates rheumatic arthritis in rats through inactivating NF-κB and MAPK signaling pathways. *Phytomedicine*, *119*, 155010.

Money, N.P. (2016). Are mushrooms medicinal? *Fungal Biol.*, *120*, 449–453.

Mullane, K., &Williams, M. (2018). Alzheimer's disease (AD) therapeutics– 2: Beyond amyloid – Re-defining AD and its causality to discover effective therapeutics. *Biochem. Pharmacol.*, *158*, 376–401.

Olaniyan, T.O., Adetunji, C.O., Dare, A., Adeyomoye, O., Adeniyi, M.J., & Enoch, A. (2023a). Clinical Applications of Deep Learning in Neurology and Its Enhancements with Future Predictions. *Artificial Intelligence for Neurological Disorders*, Publisher Academic Press, 209–224.

Olaniyan, T.O., Adetunji, C.O., Dare, A., Adeyomoye, O., Adeniyi, M.J., & Enoch, A. (2023b). New Trends in Deep Learning for Neuroimaging Analysis and Disease Prediction. *Artificial Intelligence for Neurological Disorders*, Publisher Academic Press, 275–287.

Plosca, M.-P., Chis, M.S., Fărcas, A.C., &Păucean, A. (2025). *Ganoderma lucidum*—From ancient remedies to modern applications: Chemistry, benefits, and safety. *Antioxidants*, *14*, 513. https://doi.org/10.3390/antiox14050513

Rani, P., Shailja, Singh, B., & Jalwal, P. (2025). Neurotherapeutic effects of *Ganoderma lucidum*: Biochemical characterization and behavioral assessment in animal models. *J. Pharmacogn. Phytochem.*, *14*(3), 690–698. https://www.doi.org/10.22271/phyto.2025.v14.i3i.15426

Toledo, J.B., Arnold, M., Kastenmüller, G., et al. (2017). Metabolic network failures in Alzheimer's disease: A biochemical road map. *Alzheimers Dement.*, *13*(9), 965–984.

Wang, G.H., Wang, L.H., Wan, C., & Qin, L.H. (2018). Spore powder of Ganoderma lucidum for the treatment of Alzheimer's disease: A pilot study. *Medicine*, *97*(19), e0636. https://doi.org/10.1097/MD.0000000000010636

World Health Organization. (2023). Demetia. http://www.who.int/news-room/fact-sheets/detail/dementia

Yu, N., Huang, Y., Jiang, Y., Zou, L., Liu, X., Liu, X., Chen, F., Luo, J., & Zhu, Y. (2020). *Ganoderma lucidum* triterpenoids (GLTs) reduce neuronal apoptosis via inhibition of ROCK signal pathway in APP/PS1 transgenic Alzheimer's disease mice. *Oxid. Med. Cell. Longev.*2572, 1. https://doi.org/10.1155/2020/9894037

Zhang, H., Jiang, H., Zhang, X., &Yan, J. (2018). Amino acids from *Ganoderma lucidum*: Extraction optimization, composition analysis, hypoglycemic and antioxidant activities. *Curr. Pharm. Anal.*, *14*, 562–570.

Zhao, C., Zhang, C., Xing, Z., Ahmad, Z., Li, J.S., & Chang, M.W. (2019). Pharmacological effects of natural Ganoderma and its extract on neurological diseases: A comprehensive review. *Int J Biol Macromol. 121*, 1160–1178. https://doi.org/10.1016/j.ijbiomac.2018.10.076

15 Fucoidan for Managing and Treating Dementia

Mosunmola Busayo Oyeleke and Ahmed Olalekan Bakare

15.1 INTRODUCTION

Dietary fiber is an integral nutrient necessary to maintain normal homeostatic functions in the physiology of humans (Barber et al., 2020). It functions to reduce food intake, increasing satiety and promoting intestinal motility. Likewise, it controls the community of the intestinal flora in promoting the levels of beneficial intestinal bacteria, thereby preventing the advancement of various diseases in the host (Barber et al., 2020). Dietary fibers have been specifically known to modulate the gut bacteria environment, thus helping to prevent the occurrence of metabolic and neurodegenerative diseases such as Alzheimer's disease (AD). Functioning biomolecules such as short-chain fatty acids (SCFAs) are produced from the metabolism of the soluble dietary fibers by specialized gut microbiota; these SCFAs are responsible for the regulation of metabolism processes taking place in the host community (Guan et al., 2021, Wu et al., 2021). The SCFAs including propionate and butyrate carry out several functions in the physiological metabolism of energy, including anti-inflammatory properties, reformation of brain insulin sensitivity, thereby enhancing cognitive functions (Brahe et al., 2013, Li et al., 2017, Daily et al., 2021, Zhang et al., 2022a). Moreover, dietary fibers take part in modulatory changes in the reabsorption of acidic bile and metabolic activities within the gut microbiota environment thereby having an influence on the homeostasis of the brain via the gut–brain axis (Park et al., 2022, Zhang et al., 2022a, 2022b). Natural products have greatly contributed to the research leading to the discovery and development of drug over the past decades. Several natural products, especially from different plants parts (especially the roots, barks, flowers and their essential oils) are now deemed safe as effective alternatives to treat several diseases (Hans et al., 2019, Park et al., 2021, Yang et al., 2021).

15.2 CLASSIFICATION OF SEAWEED

Seaweeds are examples of natural products and they are sources of dietary fibers. Based on their pigmentation, seaweeds are grouped into three different types namely, the green seaweed aka chlorophyta, the red seaweed aka rhodophyta, and the brown seaweed aka phaeophyta, and this serves as the basis of sulfated polysaccharides (Arata et al., 2016, Jin et al., 2023). All of these have different species (*spp*) ranging from over 4000 *spp* of red to 900 *spp* of green and 1500 *spp* of brown seaweeds worldwide. Polysaccharides readily available in all living organisms serve as reservoirs of energy, providing anatomic integrity (Olasehinde et al., 2019a, Bauer et al., 2021). Most seaweeds have around 80% of polysaccharides (Ganesan et al., 2019), and they depend on the sulfated polysaccharides for strength to prevent against tidal damage (Freile-Pelegrín & Tasdemir, 2019). Sulfation of polysaccharides are majorly involved in the biological activity of polysaccharides (Bauer et al., 2021), and this is possibly because of the increase in the free radical scavenging capability of sulfated polysaccharides (Song et al., 2019). The sulfated phaeophyta are sources of dietary fibers and are consumed as appetizer and diets in some countries located in Asian continent such as Japan and Korea. Fucoidan (sulfated glycans) is the polysaccharide found naturally in brown seaweeds (Ganesan et al., 2019). It is the general name for algae polysaccharides containing sulfuric acids that are present in phaeophyta.

DOI: 10.1201/9781003657385-15

15.3 CATEGORIES AND GENERAL FUNCTIONS OF FUCOIDANS

Based on the species of seaweed these polysaccharides originate from, various types of fucoidan exist. Hence, there are variations in the chemical structures of fucoidans, but their general structure is almost identical (Zhong et al., 2019). Fucoidan could be categorized into two broad groups, one with a fucopyranosyl residuum and the other with glycosyl residuum (containing glucuronic acid) as their backbone structure (Deniaud-Bouët et al., 2017, Jin et al., 2023). Seaweeds, including algae, are enriched with vital bioactive compounds which serve as sources of natural products utilized in drug development employed in preventing and treating several disorders for several years, and these are still in use for health care services in several countries (Wells et al., 2017, Zhang et al., 2022b). It has been documented in literature that fucoidan, as a dietary fiber and natural product, possesses numerous activities biologically, including, but not limited to, antioxidant, anticancer, anti-inflammatory, antidiabetic, anti-apoptotic, and immunomodulatory properties (Wei et al., 2017, Ganesan et al., 2019). Fucoidan is a potential prebiotic that exerts its influence on the gut microbiota, although, fucoidan research is part of the uprising research field. The molecular weight of fucoidan varies; it could be either high or low. The high molecular weight (310–1614 kD) has been reported to elevate the levels of some probiotic bacteria such as *Lactobacillus* and *Akkermentia* and coupled with a decrease in pathogenic bacteria such as *Alistipes and Helicobacter rodentium* (Shang, 2020). However, the fucoidan with a low molecular weight of 185D to 2.67 kD performed the prebiotic function better than the one with a high molecular weight of 89–974 kD (Zhang et al., 2022b). Sulfated polysaccharides from the marine field including fucoidan have long been documented to possess antioxidants, antibacterial, anti-inflammatory, and modulatory effects on gut microbiota and their metabolic products (Zhang et al., 2022b). They enhance cognitive functions and prevent oxidative stress and inflammation both *in-vivo* in Aβ(1–40, 1–42)-injected rats and in vitro, thereby repressing neuronal damage (Liu et al., 2017, Myers et al., 2019).

15.4 ALZHEIMER'S DISEASE AND FUCOIDAN

AD is a continuous and escalating form of dementia that is irreversible (Olasehinde et al., 2019b). The major cause of AD is not well deciphered; however, the disease has several factors contributing to its progression (Olasehinde et al., 2019b). Some of these factors include oxidative stress, neuro-inflammation, mitochondrial dysfunction, apoptotic process, cholinergic system disturbance, metabolic derangement, and abnormal protein assemblage among others (Barbalace et al., 2019, Bertsch et al., 2021). The two major pathologic hallmarks of this disease include the phosphorylated tau proteins and the accumulation of neurotoxic amyloid-beta (Oyeleke et al., 2022, Asuku et al., 2024, 2025). Increased life expectancy as a result of emerging medical sciences leads to an increasing number of AD occurrences (Bertsch et al., 2021) with a speculated 80 million cases in 2040 and 100 million by 2050 (Alghazwi et al., 2020). The amyloid beta hypothesis proposed that an imbalance exists between amyloid beta production and its clearance in the central nervous system (Wang et al., 2018), leading to its accumulation in AD patients. Build-up of these Aβ and hyperphosphorylated tau proteins promotes a vicious cycle of different signaling pathways leading to dysfunctional neuronal synapse, oxidative stress, inflammation, synaptic degradation, neuronal cell loss, and invariably decline in cognitive capabilities (Wang et al., 2018). Presently, the only available medications for managing AD are the cholinesterase inhibitors and the newly discovered disease-modifying medications (Lecanemab and Donanemad) (Singh et al., 2024). These medications present with serious unwanted effects such as liver damage, brain swelling, angioedema and may worsen the AD progression in some cases (Poo, 2020, Bauer et al., 2021, Singh et al., 2024). Recently, research findings proposed that brown seaweeds polysaccharide extracts have reduced cytotoxic effect, are readily available with cost-effective production costs, hence it might have therapeutics benefits in AD drug research (Bauer et al., 2021). However, as at the time of writing this report, there is no positive report on its effectiveness in treating AD in clinical trials (Poo, 2020, Bauer et al., 2021). Past research articles have shown that treatment with fucoidan attenuated apoptosis induced by oxidative stress in PC12 cells (Wei et al., 2017) and inhibited neuronal cell death prompted by N-methyl-D-aspartate (NMDA) in a hippocampal organotypic culture slice (Wang et al., 2018). Likewise, fucoidan inhibited the generation of Aβ-induced ROS, and caspase activation in rats' basal forebrain nerve cell cultures, thereby decreasing the resultant neuronal death (Wang et al., 2018). Therefore, these studies and several others have proposed that fucoidans could be a prospective source of drug discovery responsible for treating and/or inhibiting AD based on its cognitive enhancing properties (Liu et al., 2017, Myers et al., 2019). The aim of this chapter is to explore the mechanisms of actions through which fucoidan performs its cognitive enhancing functions.

15.5 MECHANISMS OF ACTION OF FUCOIDAN

Fucoidan has been reported to boost the levels of glutathione peroxide, an important antioxidant enzyme, and inhibit the production of ROS (Schepers et al., 2020). Elevated levels of acetylcholinesterase are linked with cholinergic dysfunction (Ayinla et al., 2020, 2022), resulting in memory impairments in Alzheimer's patients. Aβ has been reported in literatures to alter cholinergic neurotransmission, leading to a decreased choline uptake and acetylcholine (AchE) release. Likewise, synaptic loss of cholinergic neurons and the generation of amyloid fibrils are linked with the neurotoxic effects of Aβ oligomers, and the interactions between acetylcholinesterase activities and amyloid beta peptides formation (Ferreira-Vieira et al., 2016, Hampel et al., 2018). Two classes of cholinesterases exist, the acetylcholinesterase and butyrylcholinesterase. Previous studies have reported that one of the major pathological factors causing AD is the acetylcholinesterase (AChE) which is found in the brain and neuronal synapses; therefore, AChE inhibition could be considered as a promising approach for the treatment of AD (Oyeleke & Owoyele, 2022). Previous research studies have documented that polysaccharide extracts from seaweed inhibit acetylcholinesterase and butyrylcholinesterase activities, thereby enhancing cholinergic functions in AD patients (Olasehinde et al., 2019a). Fucoidan extracts have also been documented to prevent ROS generation, inhibit the build-up of the toxic Aβ proteins, and significantly reduce the toxicity of hydrogen peroxide in mice. This action of fucoidan equally reduced apoptosis caused by oxidative damage (Alghazwi et al., 2020). The body has a natural defense mechanism to combat oxidative stress, and this is done mainly by antioxidants. Past research established the fact that Brown seaweed's sulfated polysaccharide present with greater antioxidant capability in comparison with any other known compound, and this could be as a result of their high level of sulfation (Jose & Kurup, 2016). These polysaccharides from seaweed help lower ROS production and help to discard reactive oxygen species (O2 and OH), which increases cell survival and reduces apoptotic cell death, and this invariably leads to better learning and memory ability (Li et al., 2017). These polysaccharide extracts from seaweed help reduce or prevent oxidative stress in the cells treated with them through an increased level of antioxidant enzymes, SOD1 and SOD2 (Lee et al., 2017).

Neuroinflammation has been identified as the third central pathological feature of AD, including Aβ deposition and hyperphosphorylation of tau proteins, and it begins decades before the clinical manifestation of AD (Hampel et al., 2020). It is a key factor that influences and controls the damage and repair of brain tissue (Schepers et al., 2020). In AD, inflammatory responses are often found near the Aβ plaques (Watson et al., 2017). Polysaccharide extracts from seaweed reduce inflammatory responses within the cells through a decreased production of lipid peroxidation and a decrease in hemolysis of erythrocytes (Li et al., 2017). Beta-site amyloid precursor protein cleaving enzyme 1 (BACE-1) is a protease enzyme that splits amyloid precursor protein (APP), generating Aβ peptides that build up to generate senile plaques (Mullard, 2017, Hampel et al., 2021). BACE-1 plays a key role in AD progression and pathophysiology; hence it is the main drug target for AD to reduce the production of Aβ (Hampel et al., 2021). Inhibition of BACE-1 activity resulted in reduced levels of toxic Aβ and a better synaptic activity (Hampel et al., 2021). Fucoidan extracts have been well documented in literatures to prevent BACE-1 production, resulting in reduced amyloid-beta generation

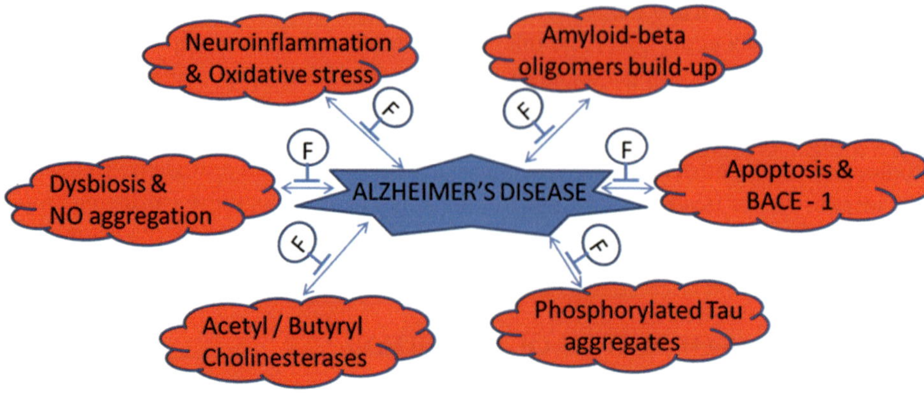

FIGURE 15.1 Mechanistic actions of fucoidan in relieving AD symptoms.

(Schepers et al., 2020). This action of fucoidan inhibits the aggregation of Aβ, thereby lowering the cytotoxic effect of Aβ protein (Olasehinde et al., 2019a) as shown in Figure 15.1. Likewise, fucoidan can improve memory and spatial learning deficits by limiting the generation of amyloid-beta plaques (Schepers et al., 2020).

15.6 CONCLUSION

Conclusively, the proposed mechanisms of action through which fucoidan extracts might help ameliorate AD are the following:

A. Possession of anti-inflammatory and antioxidant activities (Alghazwi et al., 2020);
B. Reactive oxygen species scavenging capabilities, including the prevention of nitric oxide and prostaglandins generation in the body (Ganesan et al., 2019);
C. Preventing the production of free radicals (Zhang et al., 2022);
D. Reducing the production and accumulation of mitochondrial protein (Alghazwi et al., 2020);
E. Directly inhibiting the accumulation of protein residues, thereby preventing the formation of Aβ oligomers and fibrils (Olasehinde et al., 2019a);
F. Preventing the accumulation of acetylcholinesterase at the neuronal synapses (Olasehinde et al., 2019a);
G. Limiting apoptotic cell death induced by Aβ through the JNK pathway (Bauer et al., 2021); and
H. Finally influencing the gut microbiome, which in turn affects neuroinflammation (Park et al., 2022).

REFERENCES

Alghazwi, M., Charoensiddhi, S., Smid, S., & Zhang, W. (2020). Impact of Ecklonia radiata extracts on the neuroprotective activities against amyloid beta (Aβ1-42) toxicity and aggregation. *Journal of Functional Foods*, *68*, 103893.

Arata, P.X., Quintana, I., Raffo, M.P., & Ciancia, M. (2016). Novel sulfated xylogalactoarabinans from green seaweed *Cladophora falklandica*: Chemical structure and action on the fibrin network. *Carbohydrate Polymers*, *154*, 139–150.

Asuku, A.O., Ayinla, M.T., Gbonjubola, O.O., Babatunde, S.S., Olajide, T.S., & Oyerinde, T.O. (2024). Neurocognitive Aspects of Dementia. In: Nutrition in Brain Aging and Dementia. Nutritional Neurosciences. Springer, Singapore.

Asuku, A.O., Ogungbangbe, G.O., & Olajide, T.S. (2025). Effect of Happiness and Positive Emotions on Dementia. In: Neuroscience of Feelings. Progress in Brain Research. Elsevier. https://doi.org/10.1016/bs.pbr.2025.03.001

Ayinla, M.T., Asuku, A.O., Bangwai, T.B., & Owoyele, B.V. (2022). Aqueous leaf extract of *Cola acuminata* (Malvaceae) alleviate scopolamine-induced memory impairment in wistar rats by modulating cholinergic neurotransmission. *Biokemistri*, *34*(3), 34248–34259.

Ayinla, M.T., Asuku, A.O., Bayo-Olugbami, A., Ayeni, A.O., Abiola, A.A., & Owoyele, B.V. (2020). Melatonin and vitamin C modulate cholinergic neurotransmission and oxidative stress in Scopolamine–induced rat model of memory impairment. *Journal of African Association of Physiological Sciences*, *8*(1), 20.

Barbalace, M.C., Malaguti, M., Giusti, L., Lucacchini, A., Hrelia, S., & Angeloni, C. (2019). Anti-inflammatory activities of marine algae in neurodegenerative diseases. *International Journal of Molecular Sciences*, *20*(12), 3061.

Barber, T.M., Kabisch, S., Pfeiffer, A.F., & Weickert, M.O. (2020). The health benefits of dietary fibre. *Nutrients*, *12*(10), 3209.

Bauer, S., Jin, W., Zhang, F., & Linhardt, R.J. (2021). The application of seaweed polysaccharides and their derived products with potential for the treatment of Alzheimer's disease. *Marine Drugs*, *19*(2), 89.

Bertsch, M., Franchi, B., Meschini, V., Tesi, M.C., & Tosin, A. (2021). A sensitivity analysis of a mathematical model for the synergistic interplay of amyloid beta and tau on the dynamics of Alzheimer's disease. *Brain Multiphysics*, *2*, 100020.

Brahe, L.K., Astrup, A., & Larsen, L.H. (2013). Is butyrate the link between diet, intestinal microbiota and obesity-related metabolic diseases? *Obesity Reviews*, *14*(12), 950–959.

Daily, J.W., Kang, S., & Park, S. (2021). Protection against Alzheimer's disease by luteolin: Role of brain glucose regulation, anti-inflammatory activity, and the gut microbiota-liver-brain axis. *Biofactors*, *47*(2), 218–231.

Deniaud-Bouët, E., Hardouin, K., Potin, P., Kloareg, B., & Hervé, C. (2017). A review about brown algal cell walls and fucose-containing sulfated polysaccharides: Cell wall context, biomedical properties and key research challenges. *Carbohydrate Polymers*, *175*, 395–408.

Ferreira-Vieira, T.H., Guimaraes, I.M., Silva, F.R., & Ribeiro, F.M. (2016). Alzheimer's disease: Targeting the cholinergic system. *Current Neuropharmacology*, *14*(1), 101–115.

Freile-Pelegrín, Y., & Tasdemir, D. (2019). Seaweeds to the rescue of forgotten diseases: A review. *Botanica Marina*, *62*(3), 211–226.

Ganesan, A.R., Tiwari, U., & Rajauria, G. (2019). Seaweed nutraceuticals and their therapeutic role in disease prevention. *Food Science and Human Wellness*, *8*(3), 252–263.

Guan, Z.W., Yu, E.Z., & Feng, Q. (2021). Soluble dietary fiber, one of the most important nutrients for the gut microbiota. *Molecules*, *26*(22), 6802.

Hampel, H., Caraci, F., Cuello, A.C., Caruso, G., Nisticò, R., Corbo, M., Baldacci, F., Toschi, N., Garaci, F., Chiesa, P.A., Verdooner, S.R., & Lista, S. (2020). A path toward precision medicine for neuroinflammatory mechanisms in Alzheimer's disease. *Frontiers in Immunology*, *11*, 456.

Hampel, H., Mesulam, M.M., Cuello, A.C., Farlow, M.R., Giacobini, E., Grossberg, G.T., Khachaturian, A.S., Vergallo, A., Cavedo, E., Snyder, P.J., & Khachaturian, Z.S. (2018). The cholinergic system in the pathophysiology and treatment of Alzheimer's disease. *Brain*, *141*(7), 1917–1933.

Hampel, H., Vassar, R., De Strooper, B., Hardy, J., Willem, M., Singh, N., Zhou, J., Yan, R., Vanmechelen, E., De Vos, A., Nisticò, R., Corbo, M., Imbimbo, B.P., Streffer, J., Voytyuk, I., Timmers, M., Tahami Monfared, A.A., Irizarry, M., Albala, B., Koyama, A., Watanabe, N., Kimura, T., Yarenis, L., Lista, S., Kramer, L., & Vergallo, A. (2021). The β-secretase BACE1 in Alzheimer's disease. *Biological Psychiatry*, *89*(8), 745–756.

Hans, N., Naik, S.N., & Malik, A. (2019). Platform Molecules from Algae by Using Supercritical CO2 and Subcritical Water Extraction. In: Handbook of Algal Technologies and Phytochemicals (pp. 229–243). CRC Press.

Jin, W., Lu, C., Zhu, Y., Zhao, J., Zhang, W., Wang, L., Linhardt, R.J., Wang, C., & Zhang, F. (2023). Fucoidans inhibited tau interaction and cellular uptake. *Carbohydrate Polymers*, *299*, 120176.

Jose, G.M., & Kurup, G.M. (2016). In vitro antioxidant properties of edible marine algae *Sargassum swartzii*, *Ulva fasciata* and *Chaetomorpha antennina* of Kerala coast. *Pharmaceutical Bioprocessing*, *4*(6), 100–108.

Lee, Y.J., Hwang, J.H., Kim, K.J., & Lee, B.Y. (2017). Ecklonia cava-derived polysaccharide prevent hydro peroxide-induced oxidative stress and neurotoxicity in human microglial HMO6 cells. *Journal of Food and Nutrition Research*, *5*, 187–190.

Li, H., Ding, F., Xiao, L., Shi, R., Wang, H., Han, W., & Huang, Z. (2017). Food-derived antioxidant polysaccharides and their pharmacological potential in neurodegenerative diseases. *Nutrients*, *9*(7), 778.

Liu, Y., Jiang, L., & Li, X. (2017). κ-Carrageenan-derived pentasaccharide attenuates Aβ25–35-induced apoptosis in SH-SY5Y cells via suppression of the JNK signaling pathway. *Molecular Medicine Reports*, *15*(1), 285–290.

Mullard, A. (2017). BACE inhibitor bust in Alzheimer trial. *Nature Reviews Drug Discovery*, *16*(3), 155–156.
Myers, M.J., Deaver, C.M., & Lewandowski, A.J. (2019). Molecular mechanism of action responsible for carrageenan-induced inflammatory response. *Molecular Immunology*, *109*, 38–42.
Olasehinde, T.A., Mabinya, L.V., Olaniran, A.O., & Okoh, A.I. (2019a). Chemical characterization of sulfated polysaccharides from *Gracilaria gracilis* and *Ulva lactuca* and their radical scavenging, metal chelating, and cholinesterase inhibitory activities. *International Journal of Food Properties*, *22*(1), 100–110.
Olasehinde, T.A., Olaniran, A.O., & Okoh, A.I. (2019b). Macroalgae as a valuable source of naturally occurring bioactive compounds for the treatment of Alzheimer's disease. *Marine Drugs*, *17*(11), 609.
Oyeleke, M.B., Oni, H.T., Arokoyo, O.L., & Owoyele, B.V. (2022). Therapeutic effects of crude extracts of *Bacopa floribunda* on beta-amyloid 1-42-induced Alzheimer's disease via suppression of dyslipidemia, systemic inflammation and oxidative stress in male Wistar Rats. *Heliyon*, *8*(4), e09255.
Oyeleke, M.B., & Owoyele, B.V. (2022). Saponins and flavonoids from *Bacopa floribunda* plant extract exhibit antioxidant and anti-inflammatory effects on amyloid beta 1-42-induced Alzheimer's disease in BALB/c mice. *Journal of Ethnopharmacology*, *288*, 114997.
Park, S., Kim, C.J., Ha, K.C., Baek, H.I., Yang, H.J., Kim, M.J., & Park, S.J. (2021). Efficacy and safety of aronia, red ginseng, shiitake mushroom, and nattokinase mixture on insulin resistance in prediabetic adults: A randomized, double-blinded, placebo-controlled trial. *Foods*, *10*(7), 1558.
Park, S., Zhang, T., Yue, Y., & Wu, X. (2022). Effects of bile acid modulation by dietary fat, cholecystectomy, and bile acid sequestrant on energy, glucose, and lipid metabolism and gut microbiota in mice. *International Journal of Molecular Sciences*, *23*(11), 5935.
Poo, M.M. (2020). New light on the horizon of Alzheimer's disease. *National Science Review*, *7*(5), 831–831.
Schepers, M., Martens, N., Tiane, A., Vanbrabant, K., Liu, H.B., Lütjohann, D., Mulder, M., & Vanmierlo, T. (2020). Edible seaweed-derived constituents: An undisclosed source of neuroprotective compounds. *Neural Regeneration Research*, *15*(5), 790–795.
Shang, Q. (2020). Revisit the effects of fucoidan on gut microbiota in health and disease: What do we know and what do we need to know? *Bioactive Carbohydrates and Dietary Fibre*, *23*, 100221.
Singh, B., Day, C.M., Abdella, S., & Garg, S. (2024). Alzheimer's disease current therapies, novel drug delivery systems and future directions for better disease management. *Journal of Controlled Release: Official Journal of the Controlled Release Society*, *367*, 402–424.
Song, J., Wu, Y., Jiang, G., Feng, L., Wang, Z., Yuan, G., & Tong, H. (2019). Sulfated polysaccharides from *Rhodiola sachalinensis* reduce d-gal-induced oxidative stress in NIH 3T3 cells. *International Journal of Biological Macromolecules*, *140*, 288–293.
Wang, X., Yi, K., & Zhao, Y. (2018). Fucoidan inhibits amyloid-β-induced toxicity in transgenic *Caenorhabditis elegans* by reducing the accumulation of amyloid-β and decreasing the production of reactive oxygen species. *Food & Function*, *9*(1), 552–560.
Watson, P.M.D., Kavanagh, E., Allenby, G., & Vassey, M. (2017). Bioengineered 3D glial cell culture systems and applications for neurodegeneration and neuroinflammation. *SLAS DISCOVERY: Advancing Life Sciences R&D*, *22*(5), 583–601.
Wei, H., Gao, Z., Zheng, L., Zhang, C., Liu, Z., Yang, Y., Teng, H., Hou, L., Yin, Y., & Zou, X. (2017). Protective effects of fucoidan on Aβ25-35 and d-Gal-induced neurotoxicity in PC12 cells and d-Gal-induced cognitive dysfunction in mice. *Marine Drugs*, *15*(3), 77.
Wells, M.L., Potin, P., Craigie, J.S., Raven, J.A., Merchant, S.S., Helliwell, K.E., Smith, A.G., Camire, M.E., & Brawley, S.H. (2017). Algae as nutritional and functional food sources: Revisiting our understanding. *Journal of Applied Phycology*, *29*(2), 949–982.
Wu, X., Kim, M.J., Yang, H.J., & Park, S. (2021). Chitosan alleviated menopausal symptoms and modulated the gut microbiota in estrogen-deficient rats. *European Journal of Nutrition*, *60*(4), 1907–1919.
Yang, H.J., Zhang, T., Wu, X.G., Kim, M.J., Kim, Y.H., Yang, E.S., Yoon, Y.S., & Park, S. (2021). Aqueous blackcurrant extract improves insulin sensitivity and secretion and modulates the gut microbiome in non-obese type 2 diabetic rats. *Antioxidants*, *10*(5), 756.
Zhang, T., Kim, M.J., Kim, M.J., Wu, X., Yang, H.J., Yuan, H., Huang, S., Yoon, S.M., Kim, K.N., & Park, S. (2022a). Long-term effect of porcine brain enzyme hydrolysate intake on scopolamine-induced memory impairment in rats. *International Journal of Molecular Sciences*, *23*(6), 3361.
Zhang, T., Wu, X., Yuan, H., Huang, S., & Park, S. (2022b). Mitigation of memory impairment with fermented fucoidan and λ-carrageenan supplementation through modulating the gut microbiota and their metagenome function in hippocampal amyloid-β infused rats. *Cells*, *11*(15), 2301.
Zhong, Q., Wei, B., Wang, S., Ke, S., Chen, J., Zhang, H., & Wang, H. (2019). The antioxidant activity of polysaccharides derived from marine organisms: An overview. *Marine Drugs*, *17*(12), 674.

16 Docohexanoic Acids (from Fish and Algae) for Managing and Treating Dementia

Olabisi Elizabeth Ayoola, Obafemi Samuel Olayinka, Victor Jesulayomi Bamisaye, Usman Okeme, and Ruqayyah Yetunde Ibiyeye

16.1 INTRODUCTION

Docosahexaenoic acid (DHA) is a long-chain omega-3 polyunsaturated fatty acid (PUFA) and an important structural and functional determinant of neuronal membranes. It is greatly concentrated in the gray matter of the human brain and retina where it makes up more than 90% of omega-3 fatty acids in the brain and almost 25% of all PUFAs (Bazinet & Layé, 2014). Although cardiovascular and anti-inflammatory properties have been widely investigated in DHA, its neuroprotective effects are becoming of increasing interest to the study of neurodegenerative disorders, specifically dementia. Marine ecosystems are extensive sources of bioactive compounds of high biomedical potential (Asuku, 2024; Asuku et al., 2024). Of these two, marine-derived omega-3 fatty acids, particularly DHA, have become of interest because of their wide range of influence on brain physiology and ameliorating VN (Calder, 2015). The marine bioactive compounds can readily penetrate the blood–brain barrier unlike terrestrial sources thus making them potential neuroprotective agents. The role of DHA in therapeutic argumentations about dementia is indicated by the rising preclinical and clinical evidence which reveal that DHA can participate in the fluidity of neuronal membranes, signaling, and neurogenesis (Cunnane et al., 2020). DHA is associated with systemic benefits and is relatively safe and uncomplicated in contrast to synthetic drugs which mainly have limited bioavailability or are prone to different side effects. Besides, other DHA derivatives have been found to involve neuroprotectin D1 and synaptamide in the pathways of neuronal survival/synaptic plasticity and inflammation control (Bazan, 2009).

16.2 SOURCES AND PRODUCTION OF DOCOSAHEXANOIC ACID (DHA)

16.2.1 DHA Derived From Fish Oil

Traditionally, DHA was provided by fish oil, the major source of which is fatty fish like salmon, mackerel, sardines, and anchovies. The oil is processed either by mechanical pressing or wet rendering and is then processed by a purification process (molecular distillation) by which contaminants such as mercury, dioxins, and polychlorinated biphenyls (PCBs) are removed (Shahidi & Ambigaipalan, 2018). DHA quality and concentration are affected by the methods of its refinement: newer techniques involve supercritical CO_2 extraction and have yield and purity advantages. Nevertheless, the manufacturing of fish oils cannot be free of constraints such as the threat of over-fishing, pollution, and oxidation during product production and storage. Figure 16.1 shows the extraction process of DHA from the fish oil.

DOI: 10.1201/9781003657385-16

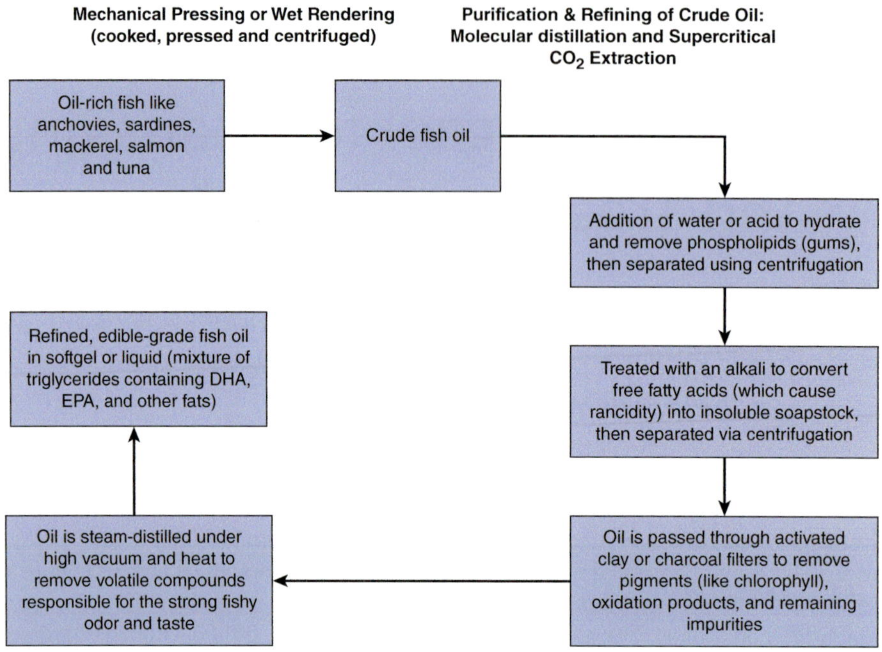

FIGURE 16.1 DHA extraction from fish oil.

16.2.2 Algae-Derived DHA

The most prominent non-animal sources of DHA are microalgae, *Schizochytrium* sp., and *Crypthecodinium cohnii*. The species are grown in bioreactors with controlled conditions, which are sterile, making the product consistent, and reducing an impact on the environment (Martins et al., 2013). DHA-producing algae require heterotrophic fermentation, under which the algae feed on glucose in the dark and cause the accumulation of the lipids. Once there is enough biomass, a solvent or enzymatic method is used to extract DHA. Algal DHA can be particularly applied to infant formula, vegan supplements, and pharmaceutical-grade products. Table 16.1 shows the differences between DHA from fish oil and algae oil.

16.2.3 Advances in DHA Production Technologies

Advancements in biotechnology have also enhanced high efficiencies and scales of DHA transport. The field is changing with genetic modification of microalgae and yeast species, enzymatic

TABLE 16.1
Comparative Characteristics of Fish Oil versus Algae Oil DHA

Parameter	Fish Oil DHA	Algae-Derived DHA
Original source	Marine fish (e.g., salmon, anchovies, sardines, mackerel, and tuna)	Microalgae (*Schizochytrium*, *Crypthecodinium cohnii*)
Extraction method	Mechanical pressing, molecular distillation	Fermentation, solvent, or enzyme extraction
Contaminants risk	High (unless refined)	Low
Sustainability	May be low (dependent on fishing practices)	High (cultivated under controlled conditions)
Vegetarian-friendly	No	Yes
Common use	Dietary supplements	Infant formula, vegan products, pharmaceuticals

conversion of shorter precursors (such as the 20-carbon 21-carbon-linolenic acid) and nanotechnology-enabled encapsulation techniques (Zhang et al., 2020). Not only are its production cost savings realized by these advances but also DHA bioavailability and stability are improved. In addition, there are also attempts to exploit biorefinery strategies where waste biomass is used as algal growth substrates that would make the production of DHA more environmentally and economically friendly.

16.3 CHEMISTRY AND BIOAVAILABILITY OF DHA

16.3.1 Chemical Structure and Molecular Forms

DHA is a long-chain PUFA consisting of 22 carbon atoms and six double bonds. Its distinctive molecular arrangement (Table 16.2) plays a crucial role in regulating membrane characteristics and biological activities (Hashimoto et al., 2017). DHA is classified as an omega-3 fatty acid because its first double bond is located at the third carbon from the methyl (omega) end. In three dimensions, DHA is highly flexible and dynamic, enabling it to assume multiple conformations that affect its positioning and function within lipid bilayers. This elevated fluidity regulates the activity of membrane-bound enzymes and receptors, ultimately influencing vital cellular processes.

16.3.2 Absorption, Distribution, and Brain Uptake

The human body has limited capacity for endogenous synthesis of DHA, making dietary intake the primary source (Ma et al., 2025). Plasma DHA concentrations reflect dietary consumption, and peripheral tissues such as muscles, heart, and kidneys mirror plasma phospholipid-DHA levels (Stark, 2008). In contrast, brain DHA concentrations remain relatively stable and are not markedly influenced by fluctuations in blood DHA (Tu et al., 2013; Ma et al., 2025). Conventional DHA supplements, including fish and krill oil, have shown little therapeutic benefit in neurological disorders such as Alzheimer's, Huntington's, and schizophrenia (Phillips et al., 2015). This limited efficacy is thought to result from their inability to substantially increase lysophosphatidylcholine-DHA.

16.3.3 Factors Influencing Bioavailability

DHA, an omega-3 PUFA (C22:6n-3), serves as a vital structural and functional nutrient with well-established therapeutic benefits, yet the bioavailability remains a critical public health challenge (Couëdelo et al., 2024).

Thus, studies in rats reveal that DHA bioavailability varies significantly by molecular form:

- Triglyceride (TG)-DHA (e.g., from microalgal oil) exhibits 23% higher absorption than ethyl ester (EE)-DHA due to more efficient lipolysis (Couëdelo et al., 2024).

TABLE 16.2
Molecular Forms, Structures, and Primary Sources of DHA

Form	Structure	Primary Sources
TAG-DHA (triacylglycerol)	DHA esterified to glycerol backbone (CH_2-O-CO-DHA in sn-1/2/3 positions)	Fish oils, algal oils, fatty fish
PL-DHA (phospholipid)	DHA bound to phospholipids (e.g., phosphatidylcholine)	Krill oil, eggs, marine lecithin
EE-DHA (ethyl ester)	DHA linked to ethanol (DHA-CO-O-CH_2CH_3)	Synthetic supplements
FFA-DHA (free fatty acid)	Unesterified DHA (COOH terminal)	Digestive products, some supplements

- Phospholipid (PL)-blended formulations may further enhance delivery to neural tissues compared to EE or TG forms (Couëdelo et al., 2024).
- Omega-3 supplements exhibit pronounced bioavailability differences due to chemical binding form (EE versus TG versus PL), supramolecular structure (e.g., emulsions versus bulk oils), or co-ingested nutrients that alter digestive kinetics (Schuchardt & Hahn, 2013).

16.4 MECHANISMS OF ACTION IN THE CENTRAL NERVOUS SYSTEM

16.4.1 Neuroprotection and Neurogenesis Activity

DHA exerts its neurodevelopmental and neuroprotective effects through dual mechanisms:

- Modulation of membrane-associated signaling pathways and
- Generation of bioactive metabolites.

These processes collectively enhance neuronal survival, synaptic plasticity, and inflammation-resisting responses within the central nervous system (Kim et al., 2022).

16.4.2 Anti-inflammatory and Antioxidant Signaling Pathways

DHA, with its six cis double bonds, integrates into phospholipid membranes, where its polyunsaturated structure modulates fluidity and signaling (Sung et al., 2020). However, this high unsaturation renders DHA susceptible to free radical oxidation, generating lipid hydroperoxides that propagate oxidative chain reactions (Borgonovi et al., 2023). Such oxidation can compromise membrane integrity and yield toxic byproducts (e.g., malondialdehyde) (Borgonovi et al., 2023). To counterbalance this, DHA activates endogenous antioxidant defenses such as the following:

- Enzymatic systems: Increases the activities of superoxide dismutase, catalase (CAT), and glutathione peroxidases (Aranda-Rivera et al., 2022).
- Non-enzymatic scavengers: Boosts glutathione and thioredoxin activity (Mirończuk-Chodakowska et al., 2018).

DHA also attenuates inflammation through metabolite-driven pathways by modulating nuclear receptors by binding to PPAR-γ, which suppresses pro-inflammatory NF-κB and cytokine production (Korbecki et al., 2019). It also regulates SREBP to limit cholesterol-driven inflammation (Borgonovi et al., 2023). DHA also attenuates inflammation through specialized pro-resolving mediators (SPMs) which are metabolized into neuroprotectin D1 (NPD1) via lipoxygenase, which inhibits IL-6 and TNF-α release (Ferreira et al., 2022), blocking oxidative stress in neural tissues in the process (Raghunath et al., 2018). These dual antioxidant and anti-inflammatory actions of DHA help to preserve membrane integrity under oxidative stress (Bie et al., 2020), downregulate inflammation-supporting transcription factors, and enhance cellular resilience via Nrf2-driven antioxidant synthesis (Raghunath et al., 2018).

16.4.3 Anti-Apoptotic Signaling Activity

DHA demonstrates selective incorporation into brain phospholipid membranes, with preferential accumulation in two key glycerophospholipid classes [phosphatidylethanolamine (PE) and phosphatidylserine (PS)], within the inner membrane leaflet (Kim et al., 2022). This enrichment creates DHA-PS microdomains that facilitate neuroprotective signaling by recruiting and activating kinases like Akt, Raf-1, and PKC (Kim et al., 2022). PS synthesis is enhanced only with DHA and requires serine (Kim et al., 2022). DHA-PS enrichment prevents apoptosis by Akt Activation. PS binds Akt's PH domain (residues R15/K20) and regulatory domain (K419/K420), enabling conformational changes for phosphorylation at T308 (by PDK1) and S473 (by mTORC2) (Sarbassov et al., 2005).

16.4.4 DHA-Derived Bioactive Metabolites (e.g., Neuroprotectin D1, Synaptamide)

DHA is enzymatically transformed in the brain and retina into biomolecules that mediate neuroprotection, neurogenesis, and anti-inflammatory effects. These include the following:

- *Neuroprotectin D1 (NPD1)*: Formed via lipoxygenase pathways (Kim et al., 2022).
- *Synaptamide (N-docosahexaenoylethanolamine)*: Produced via N-acylation of DHA (Kevala et al., 2020).
- *Elovanoids (ELV-N32, ELV-N34)*: Synthesized from very-long-chain DHA derivatives in the retina (Jun et al., 2017).

Neuroprotectin D1 (NPD1) is synthesized during ischemia-reperfusion in glial cells (Kim et al., 2022), exhibiting anti-inflammatory (inhibits NF-κB, COX-2, and IL-1β) (Hong et al., 2003), apoptotic-inhibiting (upregulates Bcl-2, Bcl-xl, and Bfl-1) (Asatryan & Bazan, 2017), neuroprotective (counters amyloid-β42-induced inflammation) activities (Lukiw et al., 2005).

Figure 16.2 provides an illustration of synaptamide-activated GPR110 signaling, whereby at low dose (nM), synaptamide acts as a promising candidate for treating neurodevelopmental disorders and neurodegenerative diseases by stimulating neurite growth, promoting synaptogenesis, and regulating synaptic protein expression (e.g., synapsins, NR2B) (Kim et al., 2011; Kim et al., 2022), inducing neural stem cell differentiation via cAMP/PKA/CREB signaling (Rashid et al., 2013), suppresses LPS-induced NF-κB and cytokine release in microglia (Tyrtyshnaia et al., 2021), enhances recovery after optic nerve injury (Kwon et al., 2021), and improves outcomes of post-traumatic brain injury (Tyrtyshnaia et al., 2020).

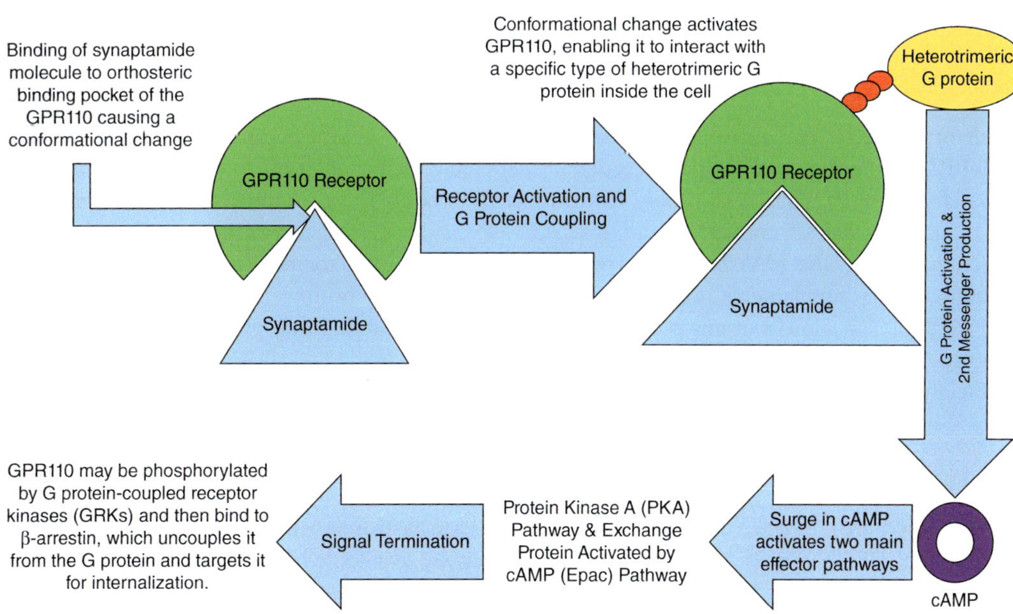

FIGURE 16.2 Synaptamide-activated GPR110 signaling.

16.5 CLINICAL EVIDENCE AND THERAPEUTIC APPLICATIONS OF DHA

16.5.1 *In Vitro* Studies: Animal Models of Dementia

In cultured neurons and animal models, DHA exerts multiple neuroprotective actions. It is the precursor to anti-inflammatory mediators (resolvins and neuroprotectin D1) and supports membrane fluidity and synaptic integrity (Teng et al., 2015). In vitro, DHA can reduce Aβ generation and aggregation and attenuate oxidative stress in stressed neural cells. In vivo, DHA supplementation typically increases brain DHA and ameliorates AD-related pathology. 3×Tg-AD mice were fed DHA-enriched diets for four to six months and observed shorter water-maze escape latencies, longer platform search times, and significantly lower hippocampal Aβ plaque and nerve fiber burdens in DHA-fed mice (Xiao et al., 2022). Similarly, 450 mg/kg daily DHA in aged Tg2576 mice enhanced contextual fear memory and increased markers of synaptogenesis versus controls (Badesso et al., 2023). Animal research also implicates DHA in vascular neuroprotection. In rodent stroke or vascular dementia models, its metabolites reduce infarct size, preserve white matter, and improve recovery (Zhang et al., 2017).

16.5.2 Clinical Trials in Alzheimer's Disease and Other Dementias

Human trials of DHA have focused mainly on Alzheimer's disease (AD) and mild cognitive impairment (MCI), with few studies in vascular or Lewy body dementias. A landmark randomized trial tested high-dose DHA (2.15 g/day) with B vitamins in cognitively normal older adults over six months. The treatment significantly raised cerebrospinal fluid (CSF) DHA (+28%) and EPA levels versus placebo, though APOE-ε4 carriers showed less increase than noncarriers. Despite bioavailability, there were no short-term cognitive or brain-volume benefits (Arellanes et al., 2020). In patients with MCI or AD, trials have been disappointing. A 12-month RCT of DHA-rich fish oil (1491 mg DHA + 351 mg EPA daily) versus placebo in 60 elders with MCI. There was no significant difference in cognitive outcomes between groups. However, DHA did lower systolic blood pressure overall, and among APOE-ε4 carriers it reduced depression and anxiety scores (Mengelberg et al., 2022). Notably, no RCTs of omega-3 DHA in non-Alzheimer dementias (vascular or Lewy body) have been published (Burckhardt et al., 2016).

16.6 SAFETY, TOLERABILITY, AND REGULATORY CONSIDERATIONS

Regulatory agencies recognize fish oil and algal DHA as safe dietary components, i.e., in the US, DHA/EPA supplements are regulated as dietary supplements, not as drugs for dementia. The FDA allows claims on EPA/DHA (but only for heart health), stating, e.g., "consuming EPA and DHA may reduce the risk of coronary heart disease". No health claims are authorized for cognitive protection. In fact, the FDA has issued warnings about products illegally claiming to treat AD, including dietary supplements marketed with DHA for memory noting such claims are unproven and potentially unsafe (Li et al., 2021). In practice, no DHA-containing pharmaceutical is FDA-approved for dementia; the only approved omega-3 drugs (e.g., Lovaza™, Vascepa™) are for hypertriglyceridemia.

16.6.1 Safety Profile and Adverse Effects

Events in DHA are similar to placebo. Reported effects are gastrointestinal: e.g., mild nausea, heartburn, abdominal discomfort (Gould et al., 2021), which are avoidable by taking enteric-coated formulations. High-dose fish oil did not increase bleeding or major metabolic issues when oxidative stability was ensured, but extremely high intakes (g/day) may prolong clotting time via anti-platelet effects, so mega-doses (>5 g/day) warrant caution (Skulas-Ray et al., 2019). Blood pressure is minimally lowered by fish oils, which is beneficial for cardiovascular co-morbidities. In pregnancy and breastfeeding, moderate fish oil intake is considered safe and recommended to support fetal neurodevelopment (Patted et al., 2024)

16.6.2 Drug Interactions and Contraindications

Anticoagulants and anti-platelets: Up to 3 g/day DHA does not alter INR in warfarin users; bleeding guidance recommends monitoring, but discontinuation of anticoagulants is rarely necessary (Green et al., 2017).

Antihypertensives: DHA can lower blood pressure by ~2–4 mmHg (Khan et al., 2019), which, while usually beneficial, may require monitoring in patients with multiple antihypertensives.

Prescribers should ask about bleeding symptoms, monitor BP, and be aware of the FDA upper limit of 3 g/day from all sources. Focus DHA as a supportive adjunct, not a replacement for standard dementia therapies. Thus, DHA supplementation is recommended as part of a comprehensive strategy for dementia prevention and early cognitive support, rather than as a treatment for established dementia (Wei et al., 2023) For older adults, a reasonable approach is encouraging dietary omega-3 intake (e.g., fatty fish twice weekly) or a daily supplement providing approximately 500–1000 mg combined EPA + DHA (McMullan et al., 2025). Given its low risk and potential benefit, DHA supplements can be integrated into dementia care with minimal concern.

16.7 FUTURE PROSPECTS AND CHALLENGES

16.7.1 Innovative Formulations

Because dementia is multifactorial, DHA is frequently combined with other nutrients or drugs to enhance efficacy. Nutraceutical combinations, such as DHA with luteolin and urolithin A, have shown an additive neuroprotective effect in cell models exposed to amyloid-β, suggesting possible synergistic benefit in designer nutrient blends (Jayatunga et al., 2021). Co-supplementation with B vitamins (folate, B6, B12) is common in DHA trials aimed at reducing homocysteine levels, with evidence that individuals with high baseline DHA benefit more cognitively (Van Soest et al., 2022). The MIND or Mediterranean diets, rich in fish, fruits, vegetables, and olive oil, consistently correlate with lower dementia incidence illustrating the value of holistic dietary integration (Fekete et al., 2025). Lifestyle factors such as physical exercise and cognitive training may also synergize with DHA by promoting neuroplasticity though empirical data remain limited.

16.7.2 Limitations of DHA as a Therapy for Dementia

Numerous human trials in MCI and early AD have failed to demonstrate significant improvements in cognitive function with omega-3 supplementation (Kalamara et al., 2025). Multiple randomized trials of Souvenaid, a DHA-containing multi-nutrient formulation, yielded only minimal benefits on standard cognitive outcomes in mild-to-moderate AD (Soininen et al., 2017; Shah et al., 2023a, 2023b). Additional barriers include metabolic competition with other fatty acids and high rates of turnover and oxidation. DHA is highly susceptible to lipid peroxidation, and without protective delivery mechanisms or antioxidant co-supplementation, much of it may degrade before reaching neuronal membranes (Valenzuela et al., 2019).

Achieving and sustaining high tissue levels of DHA require daily intake in gram doses for multiple years, raising concerns around tolerability, cost, and adherence (Swanson et al., 2012). While side effects are typically mild (e.g., gastrointestinal discomfort), high-dose fish oil may slightly increase the risk of bleeding or LDL oxidation in some individuals (Wang et al., 2023). This doesn't negate its benefits, but it limits how aggressively DHA can be dosed, especially in frail elderly populations. Environmental factors, such as pollution and poor diet, also reduce how well DHA works in the body, as these can make it harder for DHA to protect the brain. Furthermore, genes play a role, in that some people may have certain genes, like APOE4, which affect how their body uses DHA, meaning they may not benefit much from it even with high doses (Yassine et al., 2016). Thus, until precision medicine tools can stratify responders from non-responders, DHA will likely remain an adjunct, rather than a core treatment, for dementia. Finally, because of all these limits, DHA is not strong enough to be the main treatment for dementia.

16.8 CONCLUSION

Findings from this study indicate that DHA is an important nutrient for the brain whereby it supports memory, protects brain cells, and reduces inflammation. Research in animals indicates very good results, but it's inconsistent, especially when treatment begins after dementia has already developed. Furthermore, DHA is safe, easy to get from fish or supplements, and works better when taken as part of a healthy diet and lifestyle. While it may not cure dementia, it can play a helpful role in protecting the brain and slowing down memory loss when used early and together with other treatments. Thus, because of its safety, low cost, and benefits, DHA remains an important option for supporting brain health, though more observational studies on patients are needed to know the best dose, time, and combinations for real impact in dementia care.

REFERENCES

Aranda-Rivera, A. K., Cruz-Gregorio, A., Arancibia-Hernández, Y. L., Hernández-Cruz, E. Y., & Pedraza-Chaverri, J. (2022). RONS and oxidative stress: An overview of basic concepts. *Oxygen*, *2*(4), 437–478. https://doi.org/10.3390/oxygen2040030

Arellanes, I. C., Choe, N., Solomon, V., He, X., Kavin, B., Martinez, A. E., … & Yassine, H. N. (2020). Brain delivery of supplemental docosahexaenoic acid (DHA): A randomized placebo-controlled clinical trial. *EBioMedicine*, *59*, 102883. https://doi.org/10.1016/j.ebiom.2020.102883

Asatryan, A., & Bazan, N. G. (2017). Molecular mechanism of signaling via the docosanoid neuroprotectin D1 for cellular homeostasis and neuroprotection. *Journal of Biological Chemistry*, *292*(30), 12390–12397. https://doi.org/10.1074/jbc.R117.783076

Asuku, A. O. (2024). Pharmacological Applications of Marine Metabolites: Anticancer, Antimicrobial, Anti-Inflammatory, Antiviral, and Other Therapeutic Properties. In: Marine Metabolites in Disease Management. IGI Publisher. https://doi.org/10.4018/979-8-3693-5878-8.ch003

Asuku, A. O., Ayinla, M. T., Ajibare, A. J., & Krishnamurthy, R. (2024). Modulatory and Neuroprotective Roles of Marine Green Spirulina on the Brain and Neurodegeneration. In: Marine Greens: Environmental, Agricultural, Industrial, and Biomedical Applications. CRC Press, Taylor and Francis group, Boca Raton, 17, 89. https://doi.org/10.1201/9781003369738

Badesso, S., Cartas-Cejudo, P., Espelosin, M., Santamaría, E., Cuadrado-Tejedor, M., & García-Osta, A. (2023). Docosahexaenoic acid ameliorates contextual fear memory deficits in the Tg2576 Alzheimer's disease mouse model: Cellular and molecular correlates. *Pharmaceutics*, *15*(1), 82. https://doi.org/10.3390/pharmaceutics15010082

Bazan, N. G. (2009). Cell survival matters: Docosahexaenoic acid signaling, neuroprotection and photoreceptors. *Trends in Neurosciences*, *29*(5), 263–271. https://doi.org/10.1016/j.tins.2006.03.005

Bazinet, R. P., & Layé, S. (2014). Polyunsaturated fatty acids and their metabolites in brain function and disease. *Nature Reviews Neuroscience*, *15*(12), 771–785. https://doi.org/10.1038/nrn3820

Bie, N., Han, L., Meng, M., Yan, Z., & Wang, C. (2020). The immunomodulatory effect of docosahexaenoic acid (DHA) on the RAW264.7 cells by modification of the membrane structure and function. *Food & Function*, *11*(3), 2603–2616.

Borgonovi, S. M., Iametti, S., & Di Nunzio, M. (2023). Docosahexaenoic acid as master regulator of cellular antioxidant defenses: A systematic review. *Antioxidants*, *12*(6), 1283. https://doi.org/10.3390/antiox12061283

Burckhardt, M., Herke, M., Wustmann, T., Watzke, S., Langer, G., & Fink, A. (2016). Omega-3 fatty acids for the treatment of dementia. *Cochrane Database of Systematic Reviews*, *2016*(4), CD009002. https://doi.org/10.1002/14651858.CD009002.pub3

Calder, P. C. (2015). Marine omega-3 fatty acids and inflammatory processes: Effects, mechanisms and clinical relevance. Biochimica et Biophysica Acta (BBA). *Molecular and Cell Biology of Lipids*, *1851*(4), 469–484. https://doi.org/10.1016/j.bbalip.2014.08.010

Couëdelo, L., Lennon, S., Abrous, H., Chamekh, I., Bouju, C., Griffon, H., Vaysse, C., Larvol, L., & Breton, G. (2024). In vivo absorption and lymphatic bioavailability of docosahexaenoic acid from microalgal oil according to its physical and chemical form of vectorization. *Nutrients*, *16*(7), 1014. https://doi.org/10.3390/nu16071014

Cunnane, S. C., Courchesne-Loyer, A., St-Pierre, V., Vandenberghe, C., Pierotti, T., Fortier, M., …, & Castellano, C.-A. (2020). Brain energy rescue: An emerging therapeutic concept for neurodegenerative disorders of ageing. *Nature Reviews Drug Discovery*, *19*(9), 609–633. https://doi.org/10.1038/s41573-020-0072-x

Fekete, M., Varga, P., Ungvari, Z. et al. (2025). The role of the Mediterranean diet in reducing the risk of cognitive impairment, dementia, and Alzheimer's disease: A meta-analysis. *GeroScience*, *47*, 3111–3130. https://doi.org/10.1007/s11357-024-01488-3

Ferreira, I., Falcato, F., Bandarra, N., & Rauter, A. P. (2022). Resolvins, protectins, and maresins: DHA-derived specialized pro-resolving mediators, biosynthetic pathways, synthetic approaches, and their role in inflammation. *Molecules*, *27*(5), 1677. https://doi.org/10.3390/molecules27051677

Green, P., Glozman, S., Kamensky, B., & Yavin, E. (2017). Developmental changes in rat brain membrane lipids and fatty acids. The preferential prenatal accumulation of docosahexaenoic acid. *Journal of Lipid Research*, *40*(5), 960–966. https://doi.org/10.1016/S0022-2275(20)32132-5

Gould, J. F., Roberts, R. M., & Makrides, M. (2021). The influence of omega-3 long-chain polyunsaturated fatty acid, docosahexaenoic acid, on child behavioral functioning: A review of randomized controlled trials of DHA supplementation in pregnancy, the neonatal period and infancy. *Nutrients*, *13*(2), 415. https://doi.org/10.3390/nu13020415

Hashimoto, M., Hossain, S., Al Mamun, A., Matsuzaki, K., & Arai, H. (2017). Docosahexaenoic acid: One molecule diverse functions. *Critical Reviews in Biotechnology*, *37*(5), 579–597. https://doi.org/10.1080/07388551.2016.1207153

Hong, S., Gronert, K., Devchand, P. R., Moussignac, R.-L., & Serhan, C. N. (2003). Novel docosatrienes and 17S-resolvins generated from docosahexaenoic acid in murine brain, human blood and glial cells: Autacoids in anti-inflammation. *Journal of Biological Chemistry*, *278*(17), 14677–14687. https://doi.org/10.1074/jbc.M300218200

Jayatunga, D. P. W., Hone, E., Fernando, W. M. A. D. B., Garg, M. L., Verdile, G., & Martins, R. N. (2021). A synergistic combination of DHA, luteolin, and urolithin A against Alzheimer's disease. *Frontiers in Aging Neuroscience*, *14*, e780602. https://doi.org/10.3389/fnagi.2022.780602

Jun, B., Mukerjee, P. K., Asatryan, A., Kautzmann, M.-A., Heap, J., Gordon, W. C., Bhattacharjee, S., Yang, R., Petasis, N. A., & Bazan, N. G. (2017). Elovanoids are novel cell-specific lipid mediators necessary for neuroprotective signaling for photoreceptor cell integrity. *Scientific Reports*, *7*(1), 5279. https://doi.org/10.1038/s41598-017-05433-7

Kalamara, T. V., Dodos, K., Georgakopoulou, V. E., Fotakopoulos, G., Spandidos, D. A., & Kapoukranidou, D. (2025). Cognitive efficacy of omega-3 fatty acids in Alzheimer's disease: A systematic review and meta-analysis. *Biomedical Reports*, *22*, 62. https://doi.org/10.3892/br.2025.1940

Kevala, K., Lagarde, M., Spector, A. A., & Kim, H.-Y. (2020). Biosynthesis of N-docosahexaenoylethanolamine from unesterified docosahexaenoic acid and docosahexaenoyl-lysophosphatidylcholine in neuronal cells. *International Journal of Molecular Sciences*, *21*(22), 8768. https://doi.org/10.3390/ijms21228768

Khan, S. U., Khan, M. U., Riaz, H., Valavoor, S., Zhao, D., Vaughan, L., Okunrintemi, V., Riaz, I. B., Khan, M. S. et al., (2019). Effects of nutritional supplements and dietary interventions on cardiovascular outcomes. *Annals of Internal Medicine*. *171*(3), 190–198. https://doi.org/10.7326/M19-0341

Kim, H.-Y., Huang, B. X., & Spector, A. A. (2022). Molecular and signaling mechanisms for docosahexaenoic acid-derived neurodevelopment and neuroprotection. *International Journal of Molecular Sciences*, *23*(9), 4635. https://doi.org/10.3390/ijms23094635

Kim, H.-Y., Moon, H. S., Cao, D., Lee, J., Kevala, K., Jun, S. B., Lovinger, D., Akbar, M., & Huang, B. X. (2011). N-docosahexaenoylethanolamide promotes development of hippocampal neurons. *Biochemical Journal*, *435*(2), 327–336. https://doi.org/10.1042/BJ20102118

Korbecki, J., Bobiński, R., & Dutka, M. (2019). Self-regulation of the inflammatory response by peroxisome proliferator-activated receptors. *Inflammation Research*, *68*(6), 443–458. https://doi.org/10.1007/s00011-019-01231-1

Kwon, H., Kevala, K., Xin, H., Patnaik, S., Marugan, J., & Kim, H.-Y. (2021). Ligand-induced GPR110 activation facilitates axon growth after injury. *International Journal of Molecular Sciences*, *22*(7), 3386. https://doi.org/10.3390/ijms22073386

Li, J., Pora, B. L. R., Dong, K., & Hasjim, J. (2021). Health benefits of docosahexaenoic acid and its bioavailability: A review. *Food Science & Nutrition*, *9*(9), 5229–5243. https://doi.org/10.1002/fsn3.2299

Lukiw, W. J., Cui, J.-G., Marcheselli, V. L., Bodker, M., Botkjaer, A., Gotlinger, K., Serhan, C. N., & Bazan, N. G. (2005). A role for docosahexaenoic acid-derived neuroprotectin D1 in neural cell survival and Alzheimer's disease. *Journal of Clinical Investigation*, *115*(10), 2774–2783. https://doi.org/10.1172/JCI25420

Ma, K., Xie, S., Zhang, Y., Liu, H., Yu, W., & Chen, P. (2025). Novel approach to enhancing brain DHA uptake: The role of Nannochloropsis microalgae extract. *Frontiers in Nutrition*, *12*, 1573310. https://doi.org/10.3389/fnut.2025.1573310

Martins, D. A., Custódio, L., Barreira, L., Pereira, H., Ben-Hamadou, R., Varela, J., & Abreu, M. H. (2013). Alternative sources of n-3 long-chain polyunsaturated fatty acids in marine microalgae. *Marine Drugs*, *11*(7), 2259–2281. https://doi.org/10.3390/md11072259

McMullan, J. E., Kumar, R. A., Yeates, A. J., Allsopp, P. J., Mulhern, M. S., van Wijngaarden, E., Strain, J. J., & McSorley, E. M. (2025). Influence of fish consumption and ω-3 supplementation on the ω-3 index of young adults: A 2×2 factorial randomized controlled trial (YouFish Study). *The Journal of Nutrition*, *155*(12), 4345–4355. https://doi.org/10.1016/j.tjnut.2025.10.010

Mengelberg, A., Leathem, J., Podd, J., Hill, S., & Conlon, C. (2022). The effects of docosahexaenoic acid supplementation on cognition and well-being in mild cognitive impairment: A 12-month randomised controlled trial. *International Journal of Geriatric Psychiatry*, *37*(5), e5707. https://doi.org/10.1002/gps.5707

Mirończuk-Chodakowska, I., Witkowska, A. M., & Zujko, M. E. (2018). Endogenous non-enzymatic antioxidants in the human body. *Advances in Medical Sciences*, *63*(1), 68–78. https://doi.org/10.1016/j.advms.2017.05.005

Patted, P. G., Masareddy, R. S., Patil, A. S. et al. (2024). Omega-3 fatty acids: A comprehensive scientific review of their sources, functions, and health benefits. *Future Journal of Pharmaceutical Sciences*, *10*, e94. https://doi.org/10.1186/s43094-024-00667-5

Phillips, M. A., Childs, C. E., Calder, P. C., & Rogers, P. J. (2015). No effect of omega-3 fatty acid supplementation on cognition and mood in individuals with cognitive impairment and probable Alzheimer's disease: A randomised controlled trial. *International Journal of Molecular Sciences*, *16*(10), 24600–24613. https://doi.org/10.3390/ijms161024600

Raghunath, A., Sundarraj, K., Nagarajan, R., Arfuso, F., Bian, J., Kumar, A. P., Sethi, G., & Perumal, E. (2018). Antioxidant response elements: Discovery, classes, regulation and potential applications. *Redox Biology*, *17*, 297–314. https://doi.org/10.1016/j.redox.2018.05.002

Rashid, M. A., Katakura, M., Kharebava, G., Kevala, K., & Kim, H.-Y. (2013). N-docosahexaenoylethanolamine is a potent neurogenic factor for neural stem cell differentiation. *Journal of Neurochemistry*, *125*(6), 869–884. https://doi.org/10.1111/jnc.12255

Sarbassov, D. D., Guertin, D. A., Ali, S. M., & Sabatini, D. M. (2005). Phosphorylation and regulation of Akt/PKB by the rictor-mTOR complex. *Science*, *307*(5712), 1098–1101. https://doi.org/10.1126/science.1106148

Schuchardt, J. P., & Hahn, A. (2013). Bioavailability of long-chain omega-3 fatty acids. *Prostaglandins, Leukotrienes and Essential Fatty Acids*, *89*(1), 1–8. https://doi.org/10.1016/j.plefa.2013.03.010

Shah, R. C., Kamphuis, P. J., Melis, R. J., de Wilde, M. C., Maquet, D., & Scheltens, P. (2023a). Nutritional interventions in mild cognitive impairment and Alzheimer's: Findings from the LipiDiDiet trial. *The Lancet Healthy Longevity*, *4*(2), e126–e137.

Shah, R. C., Kamphuis, P. J. G. H., Leurgans, S., Swinkels, S. H. N., Sadowsky, C. H., Bongers, A., Scheltens, P. et al. (2023b). Long-term Souvenaid use and cognitive decline in Alzheimer's disease. *Research & Therapy*, *15*(1), 1–9.

Shahidi, F., & Ambigaipalan, P. (2018). Omega-3 polyunsaturated fatty acids and their health benefits. *Annual Review of Food Science and Technology*, *9*, 345–381. https://doi.org/10.1146/annurev-food-111317-095850

Skulas-Ray, A. C., Wilson, P. W. F., Harris, W. S., Brinton, E. A., Kris-Etherton, P. M., Richter, C. K., …, & Welty, F. K.. (2019). Omega-3 fatty acids for the management of hypertriglyceridemia: A science advisory from the American Heart Association. *Circulation*, *140*(12). https://doi.org/10.1161/CIR.0000000000000709

Soininen, H., Solomon, A., Visser, P. J., Hendrix, S. B., Blennow, K., Kivipelto, M., & Hartmann, T. (2017). 24-month intervention with a specific multinutrient in people with prodromal Alzheimer's disease (LipiDiDiet): A randomized, double-blind, controlled trial. *The Lancet Neurology*, *16*(12), 965–975. https://doi.org/10.1016/S1474-4422(17)30332-0

Stark, K. D. (2008). The percentage of n-3 highly unsaturated fatty acids in total HUFA as a biomarker for omega-3 fatty acid status in tissues. *Lipids*, *43*(1), 45–53. https://doi.org/10.1007/s11745-007-3128-3

Sung, N. J., Kim, N. H., Bae, N. Y., Jo, H. S., & Park, S. A. (2020). DHA inhibits Gremlin-1-induced epithelial-to-mesenchymal transition via ERK suppression in human breast cancer cells. *Bioscience Reports, 40*(8), BSR20200164. https://doi.org/10.1042/BSR20200164

Swanson, D., Block, R., & Mousa, S. A. (2012). Omega-3 fatty acids EPA and DHA: Health benefits throughout life. *Advances in Nutrition, 3*(1), 1–7. https://doi.org/10.3945/an.111.000893

Teng, M. W. L., Galon, J., Fridman, W.-H., & Smyth, M. J. (2015). From mice to humans: Developments in cancer immunoediting. *Journal of Clinical Investigation, 125*(9):3338–3346. https://doi.org/10.1172/JCI83871

Tu, W. C., Mühlhäusler, B. S., Yelland, L. N., & Gibson, R. A. (2013). Correlations between blood and tissue omega-3 LCPUFA status following dietary ALA intervention in rats. *Prostaglandins, Leukotrienes and Essential Fatty Acids, 88*(1), 53–60. https://doi.org/10.1016/j.plefa.2012.04.005

Tyrtyshnaia, A., Konovalova, S., Bondar, A., Ermolenko, E., Sultanov, R., & Manzhulo, I. (2021). Anti-inflammatory activity of N-docosahexaenoylethanolamine and N-eicosapentaenoylethanolamine in a mouse model of lipopolysaccharide-induced neuroinflammation. *International Journal of Molecular Sciences, 22*(19), 10728. https://doi.org/10.3390/ijms221910728

Tyrtyshnaia, A. A., Egorova, E. L., Starinets, A. A., Ponomarenko, A. A., Ermolenko, E., & Manzhulo, I. V. (2020). N-docosahexaenoylethanolamine attenuates neuroinflammation and improves hippocampal neurogenesis in rats with chronic constrictive injury. *Marine Drugs, 18*(10), 516. https://doi.org/10.3390/md18100516.

Valenzuela, R., Echeverría, F., Ortiz, M., Espinosa, A., Illesca, P., Videla, L. A. et al. (2019). Molecular adaptations in liver and brain induced by a high-fat diet and n-3 PUFA supplementation: Effects on mitochondrial bioenergetics and oxidative stress. *Progress in Lipid Research, 76*, 101–115.

Van Soest, A. P. M., Van de Rest, O., Witkamp, R. F. et al. (2022). DHA status influences effects of B-vitamin supplementation on cognitive ageing: A post-hoc analysis of the B-proof trial. *European Journal of Nutrition, 61*, 3731–3739. https://doi.org/10.1007/s00394-022-02924-w

Wang, T., Zhao, H., Zhang, Y., & Liu, J. (2023). Dose-response associations between omega-3 polyunsaturated fatty acid intake and low-density lipoprotein cholesterol: A meta-analysis. *Journal of the American Heart Association, 12*(11), e029512. https://doi.org/10.1161/JAHA.123.029512

Wei, B. Z., Li. L. Dong. C. W. et al. (2023). The relationship of omega-3 fatty acids with dementia and cognitive decline: A systematic review and meta-analysis. *The American Journal of Clinical Nutrition, 117*(6), 1096–1109. https://doi.org/10.1016/j.ajcnut.2023.04.001

Xiao, M., Xiang, W., Chen, Y., Peng, N., Du, X., Lu, S., … & Li, X. (2022). DHA ameliorates cognitive ability, reduces amyloid deposition, and nerve fiber production in Alzheimer's disease. *Frontiers in Nutrition, 9*, 852433. https://doi.org/10.3389/fnut.2022.852433.

Yassine, H. N., Rawat, V., Mack, W. J., Quinn, J. F., Yurko-Mauro, K., Bailey, R., & Cummings, J. (2016). APOE genotype modulates the effects of omega-3 fatty acids on biomarkers of Alzheimer's disease. *Journal of Alzheimer's Disease, 50*(3), 647–658.

Zhang, Y., Chen, J., Qiu, J., Li, Y., Wang, J., & Jiao, J. (2017). Effects of omega-3 fatty acid supplementation on cognitive function in older adults: A meta-analysis of randomized controlled trials. *Clinical Nutrition, 36*(5), 234–241. https://doi.org/10.1016/j.clnu.2016.12.001

Zhang, Y., Ma, X. N., Chen, F., & Jiang, Y. (2020). Biotechnology for high-value algal metabolites: From molecular biology to bioprocessing. *Bioresource Technology, 289*, 121619. https://doi.org/10.1016/j.biortech.2019.121619

17 Chitosan for Managing and Treating Dementia

Olugbemi Tope Olaniyan and Duyilemi Chris Ajonijebu

17.1 INTRODUCTION

Chitosan has been widely utilized as a natural biomaterial for the development of several pharmaceutical delivery systems due to the fact that chitosan-based drug delivery is cheap, bioavailable, biocompatible, biodegradable, and widely available (Asuku et al., 2024a, 2024b). Chitosan-based drug delivery system is also non-reactogenic and nontoxic; it is an excellent pharmaceutical-drug-based carrier for polysaccharides (Adetunji et al., 2023a, 2023b). Chitosan is described to contain N-acetyl, D-glucosamine, and D-glucosamine, and the units are selectively distributed throughout its entire structure, thus posing chelating and mucoadhesive properties. Chitosan is made by deacetylating chitin, a carbohydrate generally found in crab and snail shells, insects, fungi, crustaceans, and mollusks. Using biotechnological tools and approach, this chitin is subjected to deproteinization, deacetylation, decolorization, and demineralization, with all sequential steps necessary (biological and chemical) in the technique for the extraction of chitin. In recent years, increased attention toward the utilization of chitosan for various industrial applications and drug delivery systems has increased significantly (Olaniyan et al., 2023a, 2023b). Studies have revealed that chitosan was discovered in the nineteenth century; since then, it has rapidly grown as a multifunctional agent in drug delivery and biomedical applications. Chitosan is not soluble in water and is a weak base, but is soluble in solutions like diluted aqueous acidic solutions with a pH less than 6.5. Chitosan can be found in the market as dry flakes, fine powder, and a solution with low toxicity, good biocompatibility, and biodegradability. The breakdown of chitosan is through the cell lysosomal hydrolytic enzymes into nontoxic, noncarcinogenic, immunoadjuvant, anti-cholesteremic, antithrombogenic, and nonimmunogenic amino sugars, which can be absorbed by the body (Olaniyan et al., 2023a, 2023b). Generally, the biomedical applications of chitosan include drug delivery, dressings and wound-healing ointments, artificial membranes, immobilization of enzymes, eye bandages and contact lenses, ophthalmology, gene delivery, regenerative therapies, tissue engineering, and living cells. The modification and techniques involved in chitosan derivatives are carboxyalkylation, sulfation, thiolation, phosphorylation, graft copolymerization, carboxymethylation, cross-linking, N-trimethylation, and esterification strategies. These strategies confer novel physiochemical properties on chitosan derivatives, such as solubility, rheological, and thermal stability properties. The aim of this chapter is to give an overview of the management of dementia using chitosan.

17.2 CHITOSAN FOR MANAGING AND TREATING DEMENTIA

Iyer et al. (2024) reported the role of chitosan in the management of neurobiological disorders such as dementia, which has become a global public health issue. The treatment or management of dementia using medication is often confronted with issues of the blood–brain barrier difficulties. In the past, many therapeutic agents have been utilized, such as cholinesterase inhibitors, molecular chaperones, metal chelators, and antibodies, as direct drug delivery to the brain has not yielded any positive solutions in the management of neurodegenerative disorders. In recent years, chitosan as a natural multifunctional polymer has emerged as a useful compound in the pharmaceutical and medical industries for several biological applications (Olaniyan et al., 2023b). Chitosan and its derivatives can be utilized for drug delivery in a stable form with fewer adverse effects in

 DOI: 10.1201/9781003657385-17

the management of dementia among patients (Adetunji et al., 2023a; Olaniyan et al., 2023a). The authors demonstrated that recent advancement in polymer drug delivery leads to improved therapeutic use of chitosan for neurodegenerative disorders, warranting more investigations.

Wilson et al. (2010) reported that chitosan loaded with tacrine nanoparticles produced good drug-loading capacity for the treatment of neurodegenerative diseases. The authors prepared the chitosan loaded with tacrine nanoparticles through a spontaneous emulsification process. These particles were later subjected to investigation to determine the zeta potential and size using Zetasizer and scanning probe microscopy, respectively. The authors revealed that these particles produced high biodistribution after intravenous injection. The authors described the chitosan used as a natural polysaccharide with copolymers of N-acetylglucosamine and glucosamine with the ability to cross the blood–brain barrier due to the presence of tacrine. Ojeda-Hernández et al. (2020) revealed that the regenerative capacity of the central nervous system is very limited; thus, the blood–brain barrier prevents the penetration of many compounds and molecules. The emerging role of biomolecules and biomaterials in medical and pharmaceutical industries for diverse therapeutic applications is gaining more attention to combat the limitation of the blood–brain barrier (Adetunji et al., 2023b). In addition, in recent years, chitosan has been revealed to possess many advantages in its application as a biomaterial for biological applications due to the beneficial properties such as high versatility, hydrogel formation capacity, mucoadhesiveness, porous scaffolds formation, and biocompatibility. Furthermore, the application of chitosan-based drug delivery has been utilized in the therapeutic treatment of various conditions across the gene and cell delivery vehicles. The authors concluded that chitosan-based biomaterials offer important therapeutic usefulness in drug delivery, tissue engineering, regenerative medicine, and gene therapy for neurodegenerative diseases (Adetunji et al., 2023f).

Atai et al. (2017) also reported the important role of low-molecular-weight chitosan in the treatment of fungal infections. In addition, Alqahtani et al. (2024) showed that chitosan positively impacts memory and enhances neurobiological parameters in the rat model. In their study, it was discovered that chitosan had cognitive-enhancing effects through the suppression of oxidative biomarkers. Aarti and Preeti (2014) showed that dementia is associated with cognitive and memory damage across some selective brain regions, such as the neocortex, amygdala, basal forebrain, and hippocampus, involved in cholinergic neurotransmission. The authors further explained that the effects of cognitive dysfunction were caused by amyloid plaque deposition, disruption in the cholinergic activity, increased oxidative stress, and inflammatory response. Chitosan as chitin can be found in insects, crustaceans, and fungi as a polysaccharide for pharmaceutical and medical applications. The authors revealed that chitin can suppress amyloid, cholinesterase, oxidative stress, inflammation, and promote antioxidant properties. Yamaguch et al. (2004) reported that chitosan can be effectively used to treat vascular dementia as a natural herbal medicine. In their study, it was revealed that chitosan modulates cognitive functions and brain activity through improving electrophysiological parameters among neurodegenerative patients.

Kerch (2022) reported that chitosan oligosaccharide can stimulate Nrf2 and prevent NF-κB activation, an inflammatory marker in the induction of dementia. In dementia, astrocytoma cells are stimulated by interleukin 1β and amyloid beta peptide, which subsequently inhibit AMP-activated protein kinase. Chitosan has been shown to block this pathway, promote Nrf2 nuclear translocation, tumor necrosis factor TNF-α, cytokinesinterleukin-1β, inhibit phosphorylation of p65, IKKβ, IKBα, and suppress NF-κB-mediated inflammatory response. All these responses make chitosan a potential candidate for effective management of neurodegenerative agents. Saleem (2020) showed that neurodegenerative diseases can be ameliorated using chitosan nanoparticles. In their study, the authors utilized zebrafish dementia Aβ model to investigate the protective role of chitosan. From the results, it was discovered that the therapeutic intervention chrysin–chitosan nanoparticles restored and protected the synapses through the disintegration of amyloid protein in vivo. Bashir et al. (2023) reported that dementia involves loss of cognitive functions, which can be treated using chitosan through inhibition of acetylcholinesterase and oxidative stress. In their study, the authors utilized magnoflorine-loaded chitosan collagen nanocapsules as an active compound to treat dementia. The compound possesses antioxidant, anti-inflammatory, and the ability to cross the blood–brain barrier to enhance the cognitive functions of the brain.

Manek and Petroianu (2021) revealed that natural polymers such as cellulose from plants and animal chitin from fungi, protozoa, nematodes, and arthropods are nontoxic polysaccharides can be utilized as nanoparticles due to their biocompatibility, low immunogenicity, and physiochemical properties. Simar et al. (2011) revealed that chitosan nanoparticles for drug delivery could enhance drug efficacy, reduce toxicity, and improve patient compliance and bioavailability in patients with dementia. The nanoparticles target tissues and cells, inducing the therapeutic mechanisms. The authors revealed that through molecular investigation using scanning electron microscopy, differential scanning calorimetry, photon correlation spectroscopy, and transmission electron microscopy, the nanoparticle shape is spherical, 258 nm, and 0.261 for particle size and polydispersity index, respectively. The authors revealed that the nanoparticle chitosan drug was released in a sustained manner. Kerch (2015) revealed that neurodegenerative diseases affect elderly individuals, resulting in age-related dysfunctions; thus, the search for a treatment strategy that can penetrate the blood–brain barrier with little adverse effects. They revealed that the derivatives of chitosan and chito-oligosaccharides possess the ability to ameliorate dementia in neurodegenerative diseases. Chitosan can reduce the generation of oxidative stress, tissue stiffness, low-density lipoprotein oxidation, aging-associated chronic inflammation, and protein conformational changes, which are all implicated in the pathogenesis of dementia.

Clementino et al. (2021) demonstrated that for the treatment of neurodegenerative diseases using neuroprotective drugs, nasal delivery is one of the alternative routes for the management of the condition. Through the use of nanocarriers, nanotherapeutic nanoparticles can be delivered into the nasal epithelia. The authors showed that hybrid lecithin/chitosan nanoparticles as a drug delivery nanoparticle were delivered to the nasal epithelia to induce cytokine suppression and bioavailability in nose-to-brain delivery. The authors showed that this route is highly effective in the management of neurodegenerative diseases. Khemariya and Khemariya (2016) demonstrated that polysorbate-80-coated chitosan nanoparticles formation through nanotechnology-associated drug delivery can improve drug efficacy, reduce toxicity, and improve patient compliance and bioavailability of the drug for the management of neurodegenerative diseases like dementia (Adetunji et al., 2023c, 2023d). The nanoparticle size, zeta potential, and drug–polymer compatibility and shape were investigated using transmission electron microscopy, differential scanning calorimetry, and microscopy.

Cunha et al. (2021) demonstrated that the quality of patients' lives in neurodegenerative diseases like dementia can be improved using nasal route drug delivery, such as lipid-based nanosystems (nanostructured lipid carriers and nanoemulsions), due to the ability to cross the blood–brain barrier. The nanodrug in situ nanohydrogel delivered through the nose-to-brain had reduced side effects and high therapeutic effects like enhanced cognitive functions in neurodegeneration. Tamilselvan and Raghavan (2015) demonstrated that nanoparticle chitosan loaded with drug was evaluated for its neuroprotective role in neurodegenerative diseases. The authors formulated a hydrophilic nanodrug chitosan with a size of 200 nm using a simple ionic gelation approach. The authors discovered that nanodrug chitosan improved the cell viability, was nontoxic, and had greater efficiency. Ouyang et al. (2017) reported that neurodegenerative disease, such as dementia, involves the death of cholinergic neurons, metal ion dynamic equilibrium disorder, acetylcholine deficit, and deposition of tau and amyloid protein. These conditions result in oxidative stress, neuroinflammation, and apoptosis. The authors revealed that the current therapeutic management is deficient with a lot of side effects; thus, there is a need to search for an alternative solution. They revealed that several marine-based bioactive products can be developed, such as chitosan-oligosaccharide, chitosan, and their derivatives, with many physiological mechanisms, such as immunomodulatory and neuroprotective activities and anti-inflammatory and antioxidant pathways (Sharifi-Rad et al., 2023). From their study, it was demonstrated that chitosan and its derivatives are neuroprotective agents as an AChE inhibitor and a nontoxic secretase.

Manek et al. (2020) revealed that the rate of neurodegenerative diseases is increasing at an alarming rate globally, presenting a social, medical, and economic burden if not quickly addressed. The authors revealed that several pharmacological drugs have high adverse effects with less therapeutic intervention; thus, there is a search for alternatives. Polymeric chitosan-based nanoparticles are shown to have the ability to cross the blood–brain barrier with fewer side effects and are biocompatible, nontoxic, and biodegradable (Wike et al., 2023). From their study, it was discovered that the polymeric chitosan-based nanoparticle has intrinsic bioactivity and mucoadhesive characteristics, thus promoting brain penetration. Abbas et al. (2022) reported that the new luteolin-loaded chitosan nanoparticle can be utilized for the management of neurodegenerative diseases like dementia. The mechanisms of action were revealed to be anti-inflammatory, amyloidogenic, and antioxidant pathways. The nanoparticles were delivered through the non-invasive intranasal route to improve spatial memory and learning, increase neuronal survival rate, and decrease amyloid plaques. Georgieva et al. (2023) reported that galantamine can be delivered using chitosan-based nanoparticles in the treatment of dementia. Galantamine is a natural alkaloid noted for its significant physiological role in improving cognitive functions through anticholinesterase activity. Mohammadbaghban et al. (2024) demonstrated that catechin-loaded chitosan-alginate nanoparticles significantly improve aluminum-chloride-induced neurodegenerative diseases such as dementia. The authors utilized an ionotropic gelation approach to synthesize the nanoparticles, and characterization was done using the scanning electron microscope, Fourier transform infrared spectroscopy, and Zetasizer Nano system. From the results of the study, the authors showed that the nanoparticles suppressed oxidative stress and increased antioxidant capacity in the hippocampus.

Saleh et al. (2023) reported that neurodegenerative diseases are one of the leading causes of age-related dementia globally. These are further compounded by the limited number of pharmaceutical agents to cross the blood–brain barrier. The development of memantine-loaded chitosan nanocrystals using an ionic gelation approach has proved to be an effective drug for the management of dementia with little side effects as compared to other orthodox medications. ElMosbah et al. (2024) showed that the management of neurodegeneration is a global challenge due to lower bioavailability, short half-life, and lower concentration of orthodox drugs. Recently, it was discovered that phytochemicals like nasal nanoparticles rivastigmine-loaded chitosan can significantly suppress the detrimental effect of heavy metals on the brain. From the study, nanoparticles were able to suppress tau expression, downregulate cytokines such as NF-κB, Nrf-2, and Caspase-3, 7, and 9, leading to improvement in neuronal viability. Etxebeste-Mitxeltorena et al. (2024) reported that Tomafran chitosan nanoparticles can be utilized for their health-promoting effect in the management of neurodegenerative diseases like dementia. Using picrocrocin and crocin, a low-cost and efficient chitosan compound containing apocarotenoids with increased bioactivity was developed as a neuroprotective strategy, thus preventing dementia and rescuing the neuroblastoma cell line.

Darwish et al. (2024) demonstrated the development of chitosan lipid nanoparticles for the treatment of cognitive deficit. This nanoparticle promotes a drug delivery system and serves as an adjuvant to enhance bioavailability for ameliorating the oxidative stress and inflammation associated with neurodegeneration (Adetunji et al., 2023e; Alquahtani et al., 2024). Norouzi et al. (2025) demonstrated that dementia can be managed using saffron (*Crocus sativus* L.), which contains anti-inflammatory and antioxidant properties with fewer side effects. Chitosan-based herbal therapeutic agents can potentially be utilized for the treatment of dementia. Heidari et al. (2024) demonstrated that dementia and amnesia are increasing globally, with little progress in the management of the condition. There is a search for alternative solutions using medicinal plants containing phytochemicals for the development of chitosan-based nanoparticles. Nano-chitosan encapsulation that is injected intraperitoneally was found to improve memory and learning in mice. Table 17.1 shows the neuroprotective functions and the applications of chitosan for the management of dementia.

TABLE 17.1
Neuroprotective Functions of Chitosan and Its Use for the Management of Dementia

SN	Mechanism	Functions	Authors
1.	Antithrombogenic Immunoadjuvant Regenerative therapy Tissue engineering	Drug delivery Biomedical application	Olaniyan et al. (2023a)
2.	Metal chelators Chaperones	Neuroprotective	Iyer et al. (2024)
3.	Regenerative capacity	Neuroprotective	Wilson et al. (2010) Ojeda-Hernández et al. (2020)
4.	Antioxidant Anti-inflammation	Dementia	Atai et al. (2017) Alqahtani et al. (2024) Yamaguch et al. (2004)
5.	Anti-inflammation Beta amyloid reduction	Dementia	Kerch (2022) Saleem (2020)
6.	Anti-acetylcholinesterase Drug delivery	Dementia	Bashir et al. (2023) Simar et al. (2011) Clementino et al. (2021)
7.	Nanosystem	Neuroprotective	Cunha et al. (2021) Tamilselvan and Raghavan (2015)
8.	Apoptotic Antioxidant Anti-inflammatory Anticholinesterase	Neuroprotective	Ouyang et al. (2017) Sharifi-Rad et al. (2023) Wike et al. (2023) Manek et al. (2020)
9.	Drug delivery Antioxidant	Dementia	Georgieva et al. (2023) Mohammadbaghban et al. (2024)
10.	Biodistribution Blood–brain barrier Adjuvant	Dementia	Saleh et al. (2023) ElMosbah et al. (2024) Darwish et al. (2024)

17.3 CONCLUSION

Chitosan as a natural multifunctional biopolymer has numerous physiological applications due to its physiochemical properties, such as non-reactogenic, immunomodulatory, and neuroprotective activities, and anti-inflammatory, antioxidant, and nontoxic effects. The role of chitosan in the management of dementia using nanotechnology to increase the bioavailability and compatibility of the biopolymer is highlighted in the chapter. The physiological mechanisms of chitosan in the management of dementia include reduction in Aβ aggregation, suppression of oxidative stress and inflammation, senescent cell-clearing, immune system modulation, blockage of β-secretase, and inhibition of AMP-activated protein kinase, amyloidogenic, butyrylcholinesterase, and anticholinesterase.

REFERENCES

Aarti, S., & Preeti, K. (2014). Chitin and derivatives as anti-Alzheimer's agents. *Indo American Journal of Pharmaceutical Research* 4(9): 3689–3696.

Abbas, H., Sayed, N.S.E., Youssef, N., Gaafar, M.E., Mousa, P., Fayez, M.R., & Elsheikh, A.M. (2022). Novel luteolin-loaded chitosan decorated nanoparticles for brain-targeting delivery in a sporadic Alzheimer's disease mouse model: Focus on antioxidant, anti inflammatory, and amyloidogenic pathways. *Pharmaceutics* 14: 1003. https://doi.org/10.3390/pharmaceutics14051003

Adetunji, C.O., Ogundolie, F.A., Mathew, J.T., Inobeme, A., Olotu, T., Olaniyan, O.T., Ghazanfar, S., Ijabadeniyi, O.A., Ajiboye, M.D., Ajayi, O.O., Dauda, W., & Adetunji, J.B. (2023d). Nanotube platforms for effective drug delivery applications. Novel platforms for drug delivery applications. Pages 317–332. *Woodhead Publishing.*

Adetunji, C.O., Ogundolie, F.A., Mathew, J.T., Inobeme, A., Olotu, T., Olaniyan, O.T., Ijabadeniyi, O.A., Ajiboye, M.D., Ajayi, O.O., Dauda, W., Ghazanfar, S., & Adetunji, J.B. (2023f). Graphene-based nanomaterials for targeted drug delivery and tissue engineering. Novel platforms for drug delivery applications. Pages 277–288. *Woodhead Publishing.*

Adetunji, C.O., Olaniyan, O.T., Dare, A., Adeniyi, M.J., & Ajayi, O.O. (2023a). Significance of nanochitosan for improving bioavailability, stability and shelf life of sensitive ingredients in aquatic and animal-based foods. Next generation nanochitosan. Pages 421–430. *Academic Press.*

Adetunji, C.O., Olaniyan, O.T., Inobeme, A., Bodunrinde, R.E., Wike, N.Y., Dauda, P.W., Adetunji, J.B., Ozolua, P.O., Farid, A., Ghazanfar, S., Adeyomoye, O., Akram, M., Chukwu, C.V., & Yerima, M.B. (2023c). Application of nanotechnology for the bioengineering of useful metabolites derived from algae and their multifaceted applications. Next-generation algae. Volume 2. Pages 285–320. *John Wiley & Sons, Inc.*

Adetunji, C.O., Olaniyan, O.T., Rebezov, M., Shariati, M.A., Ijabadeniyi, O.A., Ajayi, O., Inobeme, A., Ajiboye, M.D., Dauda, W.P., Ghazanfar, S., Adetunji, J.B., & Odoh, U.E. (2023e). Roles of nutrigenomics in drug discovery and development. Role of nutrigenomics in modern-day healthcare and drug discovery. Pages 277–299. *Elsevier.*

Adetunji, C.O., Singh, R.B.K., Inobeme, A., Olaniyan, O.T., Oladosu, T., Dauda, W., Ghazanfar, S., Singh, J., & Singh, R.P. (2023b). Biological nanocarriers for nutrient delivery. Nanobiotechnology for the livestock industry. Pages 99–110. *Elsevier.*

Alqahtani, M., Virk, P., Elfaki, K., Alqahtani, W., & Albasher, G. (2024). The role of chitosan in mitigating tert-butylhydroquinone induced cognitive impairment and neurotoxicity in a female rat model. *European Review for Medical and Pharmacological Sciences* 28: 3318–3329.

Asuku, A.O., Ayinla, M.T., Aransiola, S.A., Ajibare, A.J., Adeyemo, M.B., & Adeyemo, R.O. (2024a). Chitosan-based nanoparticles to bypass the blood–brain barrier for the treatment of neurological diseases: A review. *Pacific Journal of Medical Sciences* 25(1), 2072–1625.

Asuku, A.O., Ogungbangbe, G.O., Ajibare, A.J., & Babatunde, S.S. (2024b). Chitosan nano-formulations for the treatment of alopecia: A review. *Pacific Journal of Medical Sciences* 25(2), 2072–1625.

Atai, Z., Mohammad, A., Javid, A., & Negar, S. (2017). In vivo study of antifungal effects of low-molecular-weight chitosan against *Candida albicans*. *Journal of Oral Science* 59(3): 425–430.

Bashir, D.J., Saliha, M., Sarfaraj, M., Shekh, M.A., Masarat, B., NidhiShweta, R., Indu, A., & Mohammed, S. (2023). Magnoflorine-loaded chitosan collagen nanocapsules ameliorate cognitive deficit in scopolamine-induced Alzheimer's disease-like conditions in a rat model by down regulating IL-1β, IL-6, TNF-α, and oxidative stressand upregulating brain-derived neurotrophic factor and DCX expressions. *ACS Omega* 8: 2227–2236.

Clementino, A.R., Marchi, C., Pozzoli, M., Bernini, F., Zimetti, F., & Sonvico, F. (2021) Anti-inflammatory properties of statin-loaded biodegradable lecithin/chitosan nanoparticles: A step toward nose-to-brain treatment of neurodegenerative diseases. *Frontiers in Pharmacology* 12: 716380. https://doi.org/10.3389/fphar.2021.716380

Cunha, S., Ben, F., José, M.S.L., & Ana, C.S. (2021). Improving drug delivery for Alzheimer's disease through nose-to-brain delivery using nano-emulsions, nanostructured lipid carriers (NLC) and in situ hydrogels. *International Journal of Nanomedicine* 16: 4373–4390.

Darwish, A.B., Amira, M.M., Shaimaa, E., Rania, E., & Mostafa, M.Y. (2024). Development of chitosan lipid nanoparticles to alleviate the pharmacological activity of piperine in the management of cognitive deficit in diabetic rats. *Scientific Reports* 14: 8247. https://doi.org/10.1038/s41598-024-58601-x

ElMosbah, D.E., Marwa, S.K., Marwa, A.I., Mona, I.E., & El Miniawy, H.M.F. (2024). Preclinical efficacy of oral and nasal rivastigmine-loaded chitosan nano-particles on AlCl3-induced Alzheimer's-like disease in rats. *Inflammopharmacology* 32: 3943–3952. https://doi.org/10.1007/s10787-024-01541-9

Etxebeste-Mitxeltorena, M., Enrique, N., Cristián, M.F., Carmen, G., Gómez-Gómez, L., Martinez, A., & Ahrazem, O. (2024). Neuroprotective properties of exosomes and chitosan nanoparticles of Tomafran, a bioengineered tomato enriched in crocins. *Natural Products and Bioprospecting*: 14–19. https://doi.org/10.1007/s13659-023-00425-9

Georgieva, D., Nikolova, D., Vassileva, E., & Kostova, B. (2023). Chitosan-based nanoparticles for targeted nasal galantamine delivery as a promising tool in Alzheimer's disease therapy. *Pharmaceutics* 15: 829. https://doi.org/10.3390/pharmaceutics15030829

Heidari, M., Azami, N., Ebadi, M., Dardmeh, F., Sepahi Charmi, S., & Alipour, H. (2024). Chitosan nanoparticle encapsulation improves the effect of donepezil on impaired memory in an amnesia mouse model. *Discover Medicine* 1, Article 109. https://doi.org/10.1007/s44337-024-00093-9

Iyer, M., Ajay, E., Ramya, S., Harysh, W., Suresh, B., Saranya, T., Jayalakshmi, K., Mukesh, K.Y., Abilash, V.G., Arul, N., & Balachandar, V. (2024). Chitosan – An alternative drug delivery approach for neurodegenerative diseases. *Carbohydrate Polymer Technologies and Applications* 7(2024): 100460.

Kerch, G. (2015). The potential of chitosan and its derivatives in prevention and treatment of age-related diseases. *Marine Drugs* 13: 2158–2182. https://doi.org/10.3390/md13042158

Kerch, G. (2022). Role of extracellular matrix stiffness inhibitors in Alzheimer's disease. *Alzheimer's Disease & Dementia* 6(1): 140–144.

Khemariya, R.P., & Khemariya, P.S. (2016). New-fangled approach in the management of Alzheimer by Formulation of Polysorbate 80 Coated Chitosan Nanoparticles of Rivastigmine for brain delivery and their in vivo evaluation. *International Journal of Current Research in Medical Sciences* 2(2): 18–29.

Manek, E., Ferenc, D., & Georg, A.P. (2020). Use of biodegradable, chitosan-based nanoparticles in the treatment of Alzheimer's disease. *Molecules* 25: 4866. https://doi.org/10.3390/molecules25204866

Manek, E., & Petroianu, G.A. (2021). Chitosan-based nanoparticles in Alzheimer's disease: Messenger or message? *Neural Regeneration Research* 16(11): 2204–2205.

Mohammadbaghban, E., Ali, T., Hossein, K., Najafzadehvarzi, H., & Fatemeh, T. (2024). Oral administration of encapsulated catechin in chitosan- alginate nanoparticles improves cognitive function and neurodegeneration in an aluminum chloride- induced rat model of Alzheimer's disease. *Physiological Reports* 12: e16095. https://doi.org/10.14814/phy2.16095

Norouzi, A., Armin, J., Nadia, D., Mosaddeghi-Heris, R., & Hesam, S.A.N. (2025). Saffron's promise: A systematic review of its role in Alzheimer's treatment. *Egyptian Journal of Neurology, Psychiatry and Neurosurgery* 61: 21–11. https://doi.org/10.1186/s41983-025-00950-z

Ojeda-Hernández, D.D., Canales-Aguirre, A.A., Matias-Guiu, J., Gomez-Pinedo, U., & Mateos-Díaz, J.C. (2020). Potential of chitosan and its derivatives for biomedical applications in the central nervous system. *Frontiers in Bioengineering and Biotechnology* 8: 389. https://doi.org/10.3389/fbioe.2020.00389.

Olaniyan, O.T., Adetunji, C.O., Dare, A., Adeniyi, M.J., & Ajayi, O.O. (2023a). Application of nanochitosan and polymeric chitosan as antibacterial, antivirus and antifungal activities when incorporated into aquatic and animal-based food materials. Next generation nanochitosan. Pages 401–420. *Academic Press.*

Olaniyan, O.T., Adetunji, C.O., Dare, A., Ajayi, O.O., & Adeniyi, M.J. (2023b). Roles of nanochitosan and polymeric chitosan for the protection of farmed fish against disease-causing pathogens. Next generation nanochitosan. Pages 287–294. *Academic Press.*

Ouyang, Q., Shannon, Z., Si-Dong, L., & Cai, S. (2017). Application of chitosan, chitooligosaccharide, and their derivatives in the treatment of Alzheimer's disease. *Marine Drugs* 15: 322. https://doi.org/10.3390/md15110322.

Saleem, S. (2020). Chitosan nanoparticles alleviate Alzheimer's disease in zebrafish by adopting a dual strategy. *Alzheimer's & Dementia* 16(9): e037238.

Saleh, M.A., Mohamed, J.M.M., Ruby, J.J., Kanthiah, S., Alanazi, Y.F., Majrashi, K., Alshahrani, S.M., Eladl, M.A., Alaryani, F.S., El-Sherbiny, M., & Menaa1, F. (2023) Preparation of memantine-loaded chitosan nanocrystals: In vitro and ex vivo toxicity analysis. *Crystals* 13: 21. https://doi.org/10.3390/cryst13010021.

Sharifi-Rad J., Naoual, E., Alibek, Y., Naoufal, E., Nasreddine, E., Afaf, A.A., Farukh, S., Abdelhakim, B., Živković, J., Adetunji, C.O., Olaniyan, O.T., Miquel, M., Sönmez, G.E.D.A., Christophe, H., & Daniela, C. (2023). Bioactive compounds from Prosopis species as potential oxidative stress and inflammation modulators: An update on mechanisms. *Minerva Biotechnology & Biomolecular Research* 35(2). 127–144. https://doi.org/10.23736/S2724-542X.23.02977-2

Simar, P.K., Rekha, R., Afzal, H., & Sarita, K. (2011). Preparation and characterization of rivastigmine loaded chitosan nanoparticles. *Journal of Pharmaceutical Sciences and Research* 3(5): 1227–1232.

Tamilselvan, N., & Raghavan, C.V. (2015). Formulation and characterization of anti-Alzheimer's drug loaded chitosan nanoparticles and its In vitro biological evaluation. *Journal of Young Pharmacists* 7(1): 28–35. https://doi.org/10.5530/jyp.2015.1.6

Wike, N.Y., Akinbo, O., Olaniyan, O.T., Adetunji, C.O., & Adetunji, J.B. (2023). The role of nanochitosan for effective delivery of nutrients and drugs including hormones and vaccines in cattle. Next generation nanochitosan. Pages 171–202. *Academic Press.*

Wilson, B., Malay, K.S., Kumaraswamy, S., Sampath, K.K.P., Muthu, R., & Bhojraj, S. (2010). Chitosan nanoparticles as a new delivery system for the anti-Alzheimer drug tacrine. *Nanomedicine: Nanotechnology, Biology, and Medicine* 6(2010): 144–152. https://doi.org/10.1016/j.nano.2009.04.001

Yamaguch, S., Miwa, M., & Shetai, K. (2004). Event related brain potential changes after Chotosan administration in stroke patients with mild cognitive impairments. *Psychopharmacology* 171: 241–249.

18 Challenges and Future Perspectives on the Applications of Bioresources for Managing and Treating Dementia

Kafilat Temidayo Aransi-ola, Oyaname Miracle Adejumo, Precious Adeoye Oyedokun, Fabunmi A. Oyesanmi, and Ayodeji Johnson Ajibare

18.1 INTRODUCTION

Dementia cases would be rising to about 152.8 million in 2050, unless therapeutic interventions with well-tolerated and long-term effects are discovered (GBD, 2019). Despite the increasing prevalence and substantial advances made in understanding the pathophysiology of dementia, little progress has been made toward developing well-tolerated and efficacious treatment regimens (Rahman et al., 2021; Bhat et al., 2022). Hence, attention has moved toward the utilization of bioresources – natural products constituting bioactive phytonutrients with documented pharmacological activities (Bhat et al., 2022). However, despite the successful application of these bioresources for dementia treatment, they have also been found to be limited by several growing challenges, thereby making them insufficient as clinical interventions for dementia.

This chapter examines research findings on the application of bioresources as therapeutic interventions for dementia. Specifically, it explores the new technological innovations that can proffer solutions to such challenges, as well as the future directions of these bioresources applications in dementia treatment.

18.1.1 Overview of the Application of Bioresources for Managing and Treating Dementia

The bioresources are diverse with a near-exhaustive collection of naturally present remedies such as nutraceuticals, dietary supplements, and organic products containing phytocompounds, which show promising outcomes in the multifactorial pathogenesis of dementia (Bhat et al., 2022). This may also include *Ginkgo biloba*, with its bioactive constituents ginkgolides reported to reduce oxidative stress and improve cognitive impairment, and green tea-derived (–)-epigallocatechin-3-O-gallate (EGCG), which inhibits inflammatory responses (Asuku et al., 2024a). The naturally occurring products have offered a rich source of neuroprotective therapeutic agents (Alzobaidi et al., 2021; Asuku et al., 2024b; Ayinla & Asuku, 2025). Their mechanisms of action are consistent with the dementia characteristic (Alzobaidi et al., 2021), thereby making them a rather interesting subject of research.

 DOI: 10.1201/9781003657385-18

18.1.2 Overview of Bioresources Limitations, Challenges, and Future Perspectives for Managing and Treating Dementia

Despite the aforementioned synergistic effects of these bioresources on dementia, their limitations in terms of bioavailability, blood–brain barrier (BBB) permeability, preclinical studies, ethical issues, etc. have resulted in failed clinical trials (Passeri et al., 2022). However, in order to prevent their proven excellent potentials as therapeutic interventions for dementia from getting wasted, researchers are now steering toward the development of innovative approaches such as nanotechnology, advanced characterization techniques, gene precision medicine, and artificial intelligence (AI) utilization (Passeri et al., 2022). The combination of the various innovative approaches could be the link to future directions of bioresources transformation in managing and treating dementia.

18.2 CHALLENGES FACED IN THE APPLICATION OF BIORESOURCES FOR DEMENTIA

18.2.1 Pharmacokinetic and Bioavailability Challenges

Changes in pharmacokinetics and drug bioavailability remain a clinical challenge faced by patients living with dementia (Becker et al., 2008). Most of the bioresources used in dementia have failed because of inadequate trial design, with one of the major limitations being BBB impermeability (Becker et al., 2008) (Figure 18.1). BBB is reported to possess features, including transporters and

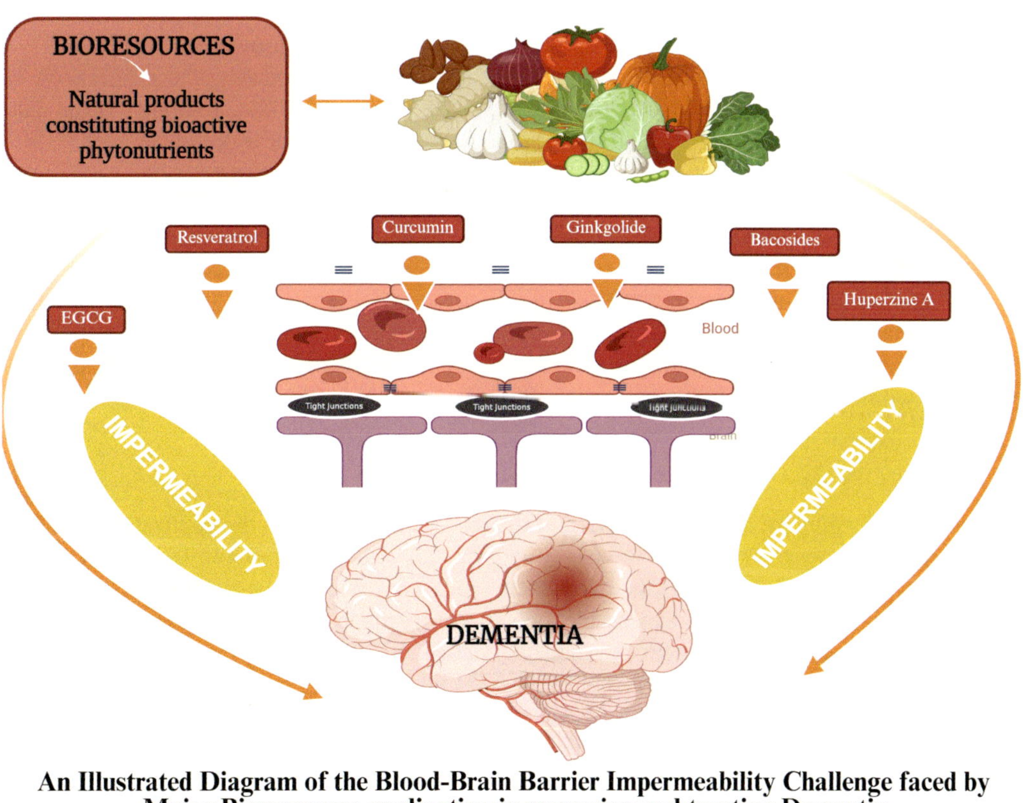

FIGURE 18.1 An illustrated diagram of the blood–brain barrier impermeability challenge faced by major bioresources application in managing and treating dementia.

tight junctions, which serve as physical and biochemical barriers restricting the permeation of most therapeutics across the blood–brain interface (Abbott et al., 2010; van de Haar et al., 2016; Asuku et al., 2024c).

BBB changes are regarded as a pharmacokinetic mechanism in drug delivery and bioavailability in the dementia (Cheng et al., 2010; Waller et al., 2022; Iqbal et al., 2024). Nevertheless, the perturbation of the BBB in therapeutic delivery cannot be allowed to cause serious complications due to the occurrence of sensitization of many classes of drugs and the development of a greater exposure to adverse drug reactions (ADRs) and the achievement of a reduced therapeutic index in most drugs (Sönnerstam et al., 2018).

There are several contributors to weight loss in patients with dementia, such as modifications in the gastrointestinal microbiome, changes in the gastric emptying rate, alterations in albumin and functional levels (Reeve et al., 2017), and small intestinal metabolizing enzymes and transporters (Liu et al., 2019; Wang et al., 2020). The pharmacokinetic-related vascular alterations are related to vascular changes, such as drug–drug interaction (DDI) and the hospitalization of older adults with dementia (Gustafsson et al., 2016).

18.2.2 Chemistry and Standardization Challenges

The constraints surrounding the analytical methods have added to the challenges of deconjugating the chemical complexity present in medicinal plants and their different preparations, which are influenced by genetic, cultural, and environmental factors that the medicinal plant is present in, a critical barrier to integrating herbal medicines into modern medical practice (Alzobaidi et al., 2021). In addition, there are obstructive compound classes in bioresources that may cause false-positive or false-negative outcomes in the application of dementia (Atanasov et al., 2015).

Reduced bioavailability, a consequence of inadequate solubility, prevents the progression of many bioresources to clinical trials (Coltescu et al., 2020; Das et al., 2022). Solubility is the analytical composition of a saturated solution expressed as a proportion of a designated solute in a designated solvent at equilibrium (van den Anker et al., 2018). Drug dissolution is one of the rate-controlling steps that determines the rate and degree of absorption of compounds (Das et al., 2022).

Poorly soluble molecules present challenges during *in vitro* and *in vivo* evaluations, thereby complicating the drug development process (Das et al., 2022). Incorrect solubility prediction can lead to faulty interpretation of results in numerous in vitro assays (Coltescu et al., 2020). Thus, optimizing an adequate analytical approach, such as *in-silico* models, will play a pivotal role, especially when targeting treatments for conditions related to dementia, to ensure effective administration and absorption within the body (Huuskonen et al., 2008; Bayliss et al., 2016; Das et al., 2022).

18.2.3 Preclinical and Clinical Translation Challenges

It remains challenging to validate and interpret the results obtained from the application of bioresources in complex systems, such as in vivo animal models of dementia (Table 18.1). They do not fully recapitulate neurodegenerative diseases due to species-specific genetic and morphological differences, although they remain indispensable tools in the drug discovery process for validating and interpreting the results obtained with cell models (Xicoy et al., 2017; Fisher & Bannerman, 2019). Since in vitro cell models are not affected by internal bodily factors such as the circulatory and immune systems, they cannot accurately predict the pharmacokinetics of bioresource-based treatments for dementia. However, employing high-quality cell models can help minimize the need for animal testing, thereby lowering ethical concerns and reducing costs during the preclinical stage of drug development (Cetin et al., 2022).

Some of the notable barriers around preclinical and clinical translation of bioresources applications in dementia are linked to regulatory burden, risk-averse trial designs, and sometimes a lack of urgency and not accounting for the costs of inaction, leading to clinical trial delays and

TABLE 18.1
Major Challenges Faced in the Applications of Bioresources for Managing and Treating Dementia

Pharmacokinetics and Bioavailability Challenges	Chemistry and Standardization Challenges	Preclinical and Clinical Translation Challenges	Regularization and Commercialization Challenges
• Blood–brain barrier impermeability	• Chemical complexity	• Risk-averse trial designs	• Lack of proper scientific database
• Gastric emptying rate	• Compositional variability of bioactive compounds	• Clinical trial entry barriers	• Intellectual property characteristics issues
	• Poor characterization	• Validation issues	• Limited sustainable sourcing
	• Inadequate solubility	• Species-specific genetic and morphological differences	• Quality control issues
	• Reduced bioavailability	• *In vivo* study and *in vitro* study imbalances	

higher overall costs (Dokholyan et al., 2022). Hence, building clinical ecosystems that reduces barriers to clinical trial success and provides cost-effective experimental research environment is significant.

Moreover, appropriate prescribing in older adults presents a significant challenge, and optimizing drug therapy for those with dementia is even more complex (Pfister et al., 2017). Elderly patients with dementia are especially susceptible to drug-related problems and face an increased risk of adverse effects, particularly due to potential DDIs and ADRs (Doraiswamy et al., 2002; Bunn et al., 2014). With this, frequent exposure to polymedication serves as a deterrent to the application of bioresources in dementia.

18.2.4 Regulatory and Commercialization Challenges

The application of findings from preclinical studies to clinical drug development is dependent on strict regulatory frameworks. However, bioresources have been shown to be limited in terms of their lack of proper scientific database, which precedes all their reported regulatory frameworks and commercialization challenges, all of which hinders their development into approved drugs for sales in the pharmaceutical industries (Heinrich et al., 2022).

According to Heinrich et al. (2022), prominent regulatory and commercialization challenge that facing bioresources application in dementia is due to their complicated classification as they usually fall into either dietary supplements, nutraceuticals, or pharmaceutical categories. Their complex classification has also been linked to their lack of specificity in terms of identification and quantification of their constituting bioactive compounds, chemical structures, mechanism of action, and their substandard characterization when compared with standard drug (Heinrich et al., 2022); all of these makes it hard to perform clinical trials and get regulatory organizations to design required taxonomic and approved frameworks.

Intellectual property limitation is another major commercialization challenge of bioresources application for managing dementia. This is because existing intellectual property rights (IPR) frameworks such as patency, trademarking, and copyrighting have been reported to center more specifically on single-person or specific group ownership and novel (Lal et al., 2024). However, these contradict with the long existing and shared generational characteristic of bioresources thus getting them considered public entities (Lal et al., 2024). Thus, bioresources recognition as public entities deters their patency and commercialization, as the possibility of outsiders claiming ownership on them or their constituting compounds without consequences discourages investors and drug development companies (Lal et al., 2024).

Another challenge facing the commercialization of the bioresources applied for managing and treating dementia is limited sustainable sourcing and quality control. The utilization of these bioresources as raw materials for drug development by pharmaceutical companies translates to increased demand for large amounts of these naturally occurring products (Kurnaz & Aksan Kurnaz, 2021). However, the constant environmental changes such as disrupted weather and climate conditions, seasonal loss of soil water, deforestation, and global warming reduce the availability and quality of these plants and their bioactive compounds, thereby making them useless for pharmaceutical companies to use for dementia drug development (Kurnaz & Aksan Kurnaz, 2021). Due to the effects of these environmental factors on the bioresources active compounds, quality control is limited as their consistently changing chemical composition deter their accurate dosing and effectiveness (Kurnaz & Aksan Kurnaz, 2021), an important requirement for dementia drug development in the pharmaceutical industry.

The resolution of these regulatory and commercialization challenges is very crucial for the successful translation of preclinical findings on bioresources application in dementia into effective therapeutic interventions. Hence, the combined effort of research institutes, regulatory organizations, and pharmaceutical industries is very important in achieving this goal.

18.3 EMERGING TECHNOLOGIES TO REPLACE LIMITED BIORESOURCES FOR DEMENTIA TREATMENT

18.3.1 Advanced Formulation and Delivery Approaches

The integration of nanotechnology has significantly advanced the longstanding obstacle in the pharmacological management of dementia by enabling nanoscale drug systems to bridge the blood–brain barrier and improve therapeutic efficacy (Fonseca-Santos & Gremião, 2015; Saraiva et al., 2016). This is due to the ability of the nanoscale drug delivery systems to enhance therapeutic molecules' bioavailability, enable reduced dosing, and minimize systemic toxicity (Tiwari et al., 2012; Saraiva et al., 2016). Several nanoparticles (NPs) including liposomes, solid lipid NPs, and polymeric NPs that encapsulate neuroprotective compounds such as curcumin, resveratrol, and therapeutic peptides have been reported (Wu et al., 2023). NPs are often functionalized with ligands like transferrin, lactoferrin, or monoclonal antibodies to enable receptor-mediated transcytosis across the BBB (Lim *et al.,* 2024). Once delivered to the brain, these nanocarriers facilitate targeted drug release in regions most affected by neurodegeneration, such as the hippocampus and cortex (Masserini, 2013). Nanoformulations overcome physicochemical barriers related to poor solubility and permeability to expand the therapeutic application of bioresources with limited pharmacokinetic profiles and providing synthetic alternatives in cases where natural supply chains are inadequate (Poudel & Park, 2022).

Traditional systemic drug delivery through oral and intravenous routes has also been reported to be partially deficient for the treatment of dementia due to several factors, such as inability to evade the BBB and first-pass metabolism (Pardridge, 2020). Other delivery routes, such as transdermal, intranasal, and intrathecal, have demonstrated more effective, targeted delivery of drugs to the brain without undergoing systemic degradation and with increased local bioavailability (Keller et al., 2021). Rawal et al. (2022) revealed that the intranasal route acts as a non-invasive direct route for insulin, oxytocin, and neurotrophic factors into the brain through the olfactory and trigeminal nerves to enhance cognitive reasoning and neuroprotection. Intranasal administration has also been reported to deliver several therapeutic constituents, including peptides, growth factors, or phytochemicals entrapped in nanoemulsions or gels to the Central Nervous System (CNS) in a rapid fashion.

In situ activation strategies provide site-specific activation of drugs in the brain with the lowest systemic toxicity and specificity of therapy. These sensors activate prodrugs or inactive precursors, which can be changed into the active form by local enzymatic activity or pH, or external forces like light fields or magnetic fields (Noble et al., 2021). When pathological microenvironments are identified, these systems release drugs, which maximize the efficacy of therapy and minimize off-target exposure (Noble et al., 2021). On the same note, near-infrared light-activated nanocarriers or photoactivatable agents have been produced to enable on-demand and targeted controlled delivery of drugs to the brain, thereby providing temporal specificity (Zhao et al., 2019).

18.3.2 Biotechnological Innovations

Synthetic biology and metabolic engineering modify cellular or microbial systems for the effective biosynthesis of bioactive molecules for pharmaceutical use (Yan et al., 2023). Both fields enhance the controlled and scalable biosynthesis of complex natural products that are seasonally, geographically, or ecologically constrained. Metabolic engineering involves the genetic modification of host organisms such as *Saccharomyces cerevisiae*, *Escherichia coli*, or *Streptomyces* spp. to synthesize required chemicals through biosynthetic pathways. Key pathways for dementia treatment include modifying neuroprotective alkaloids biosynthesis such as flavonoids, terpenoids, and polyketides that exhibit antioxidant, anti-amyloid, or anti-inflammatory effects (Navale et al., 2021). Synthetic biology modifies these systems through the integration of regulatory circuits, orthogonal gene expression modules, and feedback control (Navale et al., 2021).

Natural bioresources are often utilized as pharmacophores for drug development, but their natural unrefined form might lack optimal pharmacokinetics, target selectivity, or BBB permeability (Atanasov et al., 2021). Zanforlin et al. (2017) revealed that the development of semi-synthetic drugs through the chemical modification of naturally derived molecules optimizes their therapeutic benefits for more effective treatment of dementia (Newman & Cragg, 2020). Upon the discovery of neuroactive natural products, such as galantamine or *huperzine A* which is an inhibitor of acetylcholinesterase, chemists can selectively modify them in order to improve vital therapeutic features such as solubility, metabolic stability, or CNS selectivity (Rodrigues et al., 2016). Semi-synthetic derivatives ensure the retention of core features of natural products by preserving bioactive cores while enhancing drug-like properties (Rodrigues et al., 2016).

To guarantee purity, selectivity into targets, structural alterations, and functional activities in the neural systems, it is important to develop and validate biotechnological engineered compounds through sophisticated methods of characterization (Scheres, 2020). Chen et al. (2024) explained that innovative technologies play a critical role in developing sufficient structures and forming interactions with molecular targets that are linked to dementia, including tau protein, amyloid-beta (Aβ), or cholinergic receptors: nuclear magnetic resonance (NMR) spectroscopy, high-resolution mass spectrometry (HR-MS), and cryo-electron microscopy (cryo-EM).

18.3.3 Clinical Development Innovations

Flexible clinical trial proceedings, possible with adaptive trial designs, are characterized by alteration of key trial parameters, e.g. sample size, treatment conditions, and randomization ratios. With such trial designs, a few trial experiments can be conducted using interim data without compromising statistical analyses. The report of Burnett et al. (2020) describes the application of pre-specified decision rules, and Bayesian or frequentist statistical models in order to make real-time decisions in adaptive trials support it. For instance, certain experimental trials have compared multiple interventions to a common control at once, while other types have tested treatment across subtypes of dementia or biomarkers (Hsiao et al., 2023). The adaptive designs save nearly fifty percent of the cost and the trial time and allow the early detection of potential treatments. By minimizing the application of the natural products with the ineffective treatments, adaptive trials are used to eliminate the wastage of natural products in rare bioresource cases. They also allow testing synthetic derivatives in a short time, thereby leading to the development of more feasible replacement schemes (Park et al., 2020). Adaptive trials help to avoid wasting natural products since ineffective treatments are not performed in the case of rare bioresources. They also enable quick testing of synthetic derivatives, and this results in more effective replacement strategies (Park et al., 2020).

In Cummings and Kinney (2022), it was reported that the biomarkers and surrogate markers are measurable early markers of disease pathogenesis or of the response to treatment that are linked to clinical outcomes. The use of biomarkers involves several clinical assessments that are carried out at different points throughout the clinical trials to enhance the level of diagnosis, control the disease pathogenesis, and test the therapeutic efficacy (Cummings & Kinney, 2022). The important biomarkers used in dementia are Amyloid-beta, Tau Levels, Neurofilament Light Chain (NfL), Functional MRI (fMRI), and Volumetric MRI.

Efficiency of powerful biomarkers generates surrogate endpoints, which can be used to make early decisions during an experimental trial and perhaps to substitute long-term clinical endpoints like score on cognition or functional decline (Cummings et al., 2025). Surrogate endpoints are viewed as highly significant in the testing of synthetic/engineered therapies that aim to mimic or substitute medicines based on bioresources, offering a quicker and safer way to regulatory acceptance (Cummings et al., 2025).

Jameson et al. (2015) identified the use of a number of methods, including genetic composition, biomarkers, and phenotypic data, to classify patients into smaller, more homogeneous groups that should respond to specific treatment as patient stratification. Marques et al. (2024) further explained this stratification to make the trials more efficient and minimized the number of requirements related to the sample size to minimize the usage of bioresources. It was also highlighted as particularly significant in improving the development of customized synthetic or semi-synthetic therapeutics that focus on particular biological pathways in particular subgroups of patient (Schork, 2020).

18.3.4 Convergent Technology Integration

AI and ML have emerged as crucial components in speeding up drug discovery, optimizing therapeutic targets, and integrating new interventions without the utilization of scarce bioresources (Sharma & Mandal, 2022). The technologies utilize massive data sets such as proteomics, genomics, imaging, and clinical phenotypes to detect complex disease pathogenesis and the suitable approaches required to diagnose them (Yi et al., 2021). Sharma and Mandal (2022) published that ML algorithms analyze high-dimensional data to detect several features associated with certain diseases and subtypes of patients (Li et al., 2022). In cases of dementia, diagnostic techniques such as neuroimaging data (MRI, PET) have been reported to be effective for predicting disease pathogenesis, patient stratification, and early biomarker discovery (Wen et al., 2020).

Digital health technological inventions such as mobile health applications, wearable sensors, and telemedicine platforms ensure continuous observation, early detection, and remote interventions in cases of dementia care (Kourtis et al., 2020). Shaik et al. (2025) published that these inventions offer several benefits especially in low-resource settings where access to modern diagnosis and therapies derived from bioresources is inadequate. Digital equipment records real-time physiological, behavioral, and cognitive data and ensures adequate observation of disease pathogenesis.

The integration of digital innovations into clinical care models promotes a preventive and individualized model, thus, decreasing the over dependence on currently available bioresources to cure dementia (Yi et al., 2021).

18.4 FUTURE PERSPECTIVES AND EMERGING DIRECTIONS

18.4.1 Precision Medicine Applications

Precision medicine in dementia strives to modify treatment to individual genetic, environmental, and lifestyle factors in dementia. This strategy entails risk stratification, early identification of pathophysiological mechanism, and intervention targeting molecular disease causes (Devi, 2023). While precision medicine has made progress in some disorders, its utilization in Alzheimer's disease (AD) and other dementias is still in early stages.

In genotype-guided approaches, several studies have connected genes to the etiology of several forms of dementia, including vascular dementia and AD (Zhou et al., 2021). In advancing our understanding of the mechanisms underlying dementia, and designing targeted treatment plans, it is significant to identify associated genetic markers. Due to the significant impact of genetic variables on the diseases risk, progression, and treatment response, enabling tailored therapies targeting presenting symptoms, underlying pathophysiological processes, and their associated polymorphisms could be achieved through the genetic screening of markers such as Apolipoprotein E (ApoE), Tar DNA-Binding Protein 43 (TDP-43), amyloid precursor protein gene (*APP*), Microtubules-Associated Protein Tau (MAPT), Chromosome 9 Open Reading Frame 72 (C90RF72), and others (Caiado et al., 2024).

Specification of therapies to individual molecular causes of disease, patient grouping based on risk variables, and early detection of physiological changes are all microbiome-informed techniques (Shukla et al., 2024). The utilization of the microbiome's ability to influence health, disease progression, and treatment response also termed precision medicine (Zhao et al., 2023) has been shown to combine microbiota composition studies with factors such as cognitive testing, biochemical analyses, and genetic profiling to improve patient classification and treatment success in dementia (Yuan et al., 2025).

It is important to identify related genetic biomarkers in the advancement of knowledge of the mechanisms of dementia and in formulating specific treatment regimens. These genetic variables have a strong contribution to the risk, progression, and response to treatment. The genetic screening of such markers, including ApoE, Tar DNA-Binding Protein 43 (TDP-43), amyloid precursor protein gene (APP), MAPT, Chromosome 9 Open Reading Frame 72 (C90RF72), and others, could enable a tailored approach of the treatment based on presenting symptoms, underlying pathophysiological processes and their associated polymorphisms.

Enabling tailored therapies targeting presenting symptoms, underlying pathophysiological processes, and their associated polymorphisms could be achieved through the genetic screening of markers such as ApoE, Tar DNA-Binding Protein 43 (TDP-43), amyloid precursor protein gene (*APP*), MAPT, and Chromosome 9 Open Reading Frame 72 (C90RF72).

Individual molecular etiologies of disease, grouping of patients according to risk factors, and early detection of physiological alterations are all microbiome-informed strategies (Shukla et al., 2024). Combining the microbiota composition studies with factors like cognitive testing, biochemical analyses, and genetic profiling to enhance patient classification and treatment response in dementia has been shown to utilize the ability of the microbiome to influence health, disease progression, and respond to treatment termed precision medicine (Zhao et al., 2023).

18.4.2 Multi-Target and Network Pharmacology

In recent times, research has demonstrated the potential advantages of multi-target and network pharmacology in tackling dementia, with focus on the modification of several nodes in the pathogenic network to halt its progression (Hossain & Hussain, 2025). The rise of network pharmacology and AI as valuable techniques for studying the multi-target processes of different drug effects on neurodegenerative disorders through detection of bioactive substances, molecular targets, and linked pathways aids drug discovery and enhancement for dementia (Chen et al., 2025).

The development of network-based multi-target pharmaceuticals derived from medicinal herbs has been reported to be enabled by the integration of systems biology platforms and computational technology (Li et al., 2021). However, the improvement of reliability and efficacy of these approaches are hindered by obstacles, occurring in the regulation of procedures and integration of multiple '-omics' databases (Mohr et al., 2024).

18.4.3 Preventive and Early Intervention Application

Preventive measures center on alterable risk factors, like cardiovascular disease and education level. Early detection approaches, such as ultrasound and biomarkers identification of molecular disease signature and structural vascular changes, allow for early care of vascular risk factors, thereby slowing or reversing disease progression (Gautam & Singh, 2024). Current research aims to create risk stratification models and preventive strategies for midlife intervention, which could lead to more effective second-line prevention trials (Anderson et al., 2023).

Recent research highlights the potential of multimodal lifestyle interventions in preventing and managing dementia. These interventions target multiple modifiable risk factors simultaneously, including physical activity, diet regimentation, cognitive training, vascular management, and social engagement (Chong, 2025). The Finnish Geriatric Intervention Study to Avoid Cognitive Impairment and Disability showed that incorporating exercise, diet, social interaction, and cognitive training can help elderly at risk of cognitive decline maintain everyday functioning (Chong, 2025).

18.4.4 Emerging Bioresources Frontiers

A global approach for studying dementia is critical, considering society diverse attitude, neurological symptoms, and associated risk profiles. While there are no interventions that can definitively prevent dementia, addressing modifiable risk factors is recommended.

Recent research shows an important relationship between gut microbiota and dementia; the gut microbiota can influence cognitive function via the microbiome–gut–brain axis, causing neuroinflammation (Martel et al., 2022). Dysbiosis of gut microbiota could damage gut epithelial barrier, causing chronic inflammation (Martel et al., 2022). New research suggests that modification of microbiome using prebiotics, synbiotics, or probiotics could be effective in managing dementia by enhancing cognition and improving metabolic activity.

Exosomes use is another frontier in dementia management, as they can serve as potential biomarkers for the progression of dementia due to their transport of harmful proteins like amyloid beta and tau (Joo et al., 2023). These exosomes are also being explored as possible drug delivery vehicles due to their BBB permeability.

The use of extremophile-derived compounds in the management and treatment of dementia is another cutting-edge frontier; efficacious treatment of several neurological events associated with dementia such as neural inflammation, oxidative stress, amyloid beta accumulation, and tau phosphorylation has been shown to be exhibited by various marine-derived bioactive compounds (Kabir et al., 2021). Improved bioavailability, efficacy, and milder side effects have been reported to be demonstrated by specific marine-derived medications, like homotaurine, anabaseine, and bryostatin-1 derivatives (Kabir et al., 2021). However, more studies are still needed to further the knowledge of these frontiers in developing novel therapeutic interventions for dementia.

18.5 CONCLUSION

The application of bioresources as therapeutic interventions is a significant progress in the management of dementia; however, multiple challenges such as limited bioavailability, analytical barriers, clinical translation, and commercialization issues exist for them, thus making their translation into well-tolerated and efficacious standard impractical. Utilizing new technologies such as advanced delivery approaches, synthetic innovations, innovative trial designs, and AI has been shown to possibly lead to a step forward for the standard usage of bioresources in managing and treating dementia. The significance of genotype-guided approaches, early detection techniques, multi-target pharmacology and new frontiers such as extremophile-derived compound therapies in the improvement of bioresources application in dementia are prospective directions that cannot be overlooked. This is because their combination with the emerging technologies improves the chances of successful translation of bioresources preclinical findings into efficacious therapeutic interventions.

Successful clinical application of bioresource for dementia treatment will improve dementia care and reduce its global prevalence; however, this is only achievable with steadfast collaboration between research institutions, regulatory organizations, and pharmaceutical industry stakeholders in integrating bioresources application with innovative technologies.

REFERENCES

Abbott, N. J., Patabendige, A. A. K., Dolman, D. E. M., Yusof, S. R., & Begley, D. J. (2010). Structure and function of the blood-brain barrier. *Neurobiology of Disease*, *37*(1), 13–25. https://doi.org/10.1016/j.nbd.2009.07.030

Alzobaidi, N., Quasimi, H., Emad, N. A., Alhalmi, A., & Naqvi, M. (2021). Bioactive compounds and traditional herbal medicine: Promising approaches for the treatment of dementia. *Degenerative Neurological and Neuromuscular Disease*, *11*, 1–14. https://doi.org/10.2147/dnnd.s299589

Anderson, C., Bucholc, M., McClean, P. L., & Zhang, S. D. (2023). The potential of a stratified approach to drug repurposing in Alzheimer's disease. *Biomolecules*, *14*(1), 11.

Asuku, A. O., Ajibare, A. J., Ajagun, E. J., Babatunde, S. S., Olajide, T. S., Oyerinde, T. O., Ogungbangbe, G. O., & Odugbesan, O. A. (2024a). Herbal Extracts in Anti-Aging Products. In: Eco-Friendly Skin Solutions for Natural Cosmeceuticals. IGI Publisher. http://doi.org/10.4018/979-8-3693-5473-5.ch012

Asuku, A. O., Ayinla, M. T., Ajibare, A. J., & Krishnamurthy, R. (2024b). Modulatory and Neuroprotective Roles of Marine Green Spirulina on the Brain and Neurodegeneration. In: Marine Greens: Environmental, Agricultural, Industrial, and Biomedical Applications. CRC Press, Taylor and Francis Group, Boca Raton, 17: 89. https://doi.org/10.1201/9781003369738

Asuku, A. O., Ayinla, M. T., Aransiola, S. A., Ajibare, A. J., Adeyemo, M. B., & Adeyemo, R. O. (2024c). Chitosan-based nanoparticles to bypass the blood-brain barrier for the treatment of neurological diseases: A review. *Pacific Journal of Medical Sciences*, *25*(1), 45–65.

Atanasov, A. G., Waltenberger, B., Pferschy-Wenzig, E. M., Linder, T., Wawrosch, C., Uhrin, P., Temml, V., Wang, L., Schwaiger, S., Heiss, E. H., Rollinger, J. M., Schuster, D., Breuss, J. M., Bochkov, V., Mihovilovic, M. D., Kopp, B., Bauer, R., Dirsch, V. M., & Stuppner, H. (2015). Discovery and resupply of pharmacologically active plant-derived natural products: A review. *Biotechnology Advances*, *33*(8), 1582–1614. https://doi.org/10.1016/J.BIOTECHADV.2015.08.001

Atanasov, A. G., Zotchev, S. B., Dirsch, V. M., & Supuran, C. T. (2021). Natural products in drug discovery: Advances and opportunities. *Nature Reviews Drug Discovery*, 20(3), 200–216. https://doi.org/10.1038/s41573-020-00114-z

Ayinla, M. T., & Asuku, A. O. (2025). The neurotoxic effects of lead acetate and the abrogating actions of 6-gingerol-rich extract of ginger via modulation of antioxidant defence system, pro-inflammatory markers, and apoptotic cascade. *Naunyn-Schmiedeberg's Archives of Pharmacology*, *398*, 9341–9356. https://doi.org/10.1007/s00210-025-03873-x

Bayliss, M. K., Butler, J., Feldman, P. L., Green, D. V. S., Leeson, P. D., Palovich, M. R., & Taylor, A. J. (2016). Quality guidelines for oral drug candidates: Dose, solubility and lipophilicity. *Drug Discovery Today*, *21*(10), 1719–1727. https://doi.org/10.1016/j.drudis.2016.07.007

Becker, R. E., Greig, N. H., & Giacobini, E. (2008). Why do so many drugs for Alzheimer's disease fail in development? Time for new methods and new practices? *Journal of Alzheimer's Disease*, *15*(2), 303–325. https://doi.org/10.3233/JAD-2008-15213

Bhat, B. A., Almilaibary, A., Mir, R. A., Aljarallah, B. M., Mir, W. R., Ahmad, F., & Mir, M. A. (2022). Natural therapeutics in aid of treating Alzheimer's disease: A green gateway toward ending quest for treating neurological disorders. *Frontiers in Neuroscience*, *16*, 884345. https://doi.org/10.3389/fnins.2022.884345

Bunn, F., Burn, A. M., Goodman, C., Rait, G., Norton, S., Robinson, L., Schoeman, J., & Brayne, C. (2014). Comorbidity and dementia: A scoping review of the literature. *BMC Medicine*, *12*(1). https://doi.org/10.1186/S12916-014-0192-4

Burnett, T., Mozgunov, P., Pallmann, P., Villar, S. S., Wheeler, G. M., & Jaki, T. (2020). Adding flexibility to clinical trial designs: An example-based guide to the practical use of adaptive designs. *BMC Medicine*, *18*(1), 352.

Caiado, F. L., Morii, K. Y., de Biasi Alves, M. L., Destefani, A. C., & Destefani, V. C. (2024). Precision medicine in geriatric care: Harnessing the power of genetic profiling for personalized therapies in older adults. *Revista Ibero-Americana de Humanidades, Ciências e Educação*, *10*(8), 2730–2740.

Cetin, S., Knez, D., Gobec, S., Kos, J., & Pišlar, A. (2022). Cell models for Alzheimer's and Parkinson's disease: At the interface of biology and drug discovery. *Biomedicine & Pharmacotherapy*, *149*, 112924. https://doi.org/10.1016/J.BIOPHA.2022.112924

Chen, Q., Chen, G., & Wang, Q. (2025). Application of Network pharmacology in the treatment of neurodegenerative diseases with traditional Chinese medicine. *Planta Medica*, *91*(5): 226–237.

Chen, Z., Yu, J., Wang, H., Xu, P., Fan, L., Sun, F., Huang, S., Zhang, P., Huang, H., Gu, S., & Zhang, B. (2024). Flexible scaffold-based cheminformatics approach for polypharmacological drug design. *Cell*, *187*(9), 2194–2208.

Cheng, Z., Zhang, J., Liu, H., Li, Y., Zhao, Y., & Yang, E. (2010). Central nervous system penetration for small molecule therapeutic agents does not increase in multiple sclerosis- and Alzheimer's disease-related animal models despite reported blood-brain barrier disruption. *Drug Metabolism and Disposition*, *38*(8), 1355–1361. https://doi.org/10.1124/dmd.110.033324

Chong, W. F. W. (2025). Approaches to personalize multidomain lifestyle interventions for dementia prevention. *Heart and Mind*, *9*(1), 80–83.

Coltescu, A. R., Butnariu, M., & Sarac, I. (2020). The importance of solubility for new drug molecules. *Biomedical and Pharmacology Journal*, *13*(2), 577–583. https://doi.org/10.13005/BPJ/1920

Cummings, J., & Kinney, J. (2022). Biomarkers for Alzheimer's disease: Context of use, qualification, and roadmap for clinical implementation. *Medicina*, *58*(7), 952.

Cummings, J. L., Teunissen, C. E., Fiske, B. K., Le Ber, I., Wildsmith, K. R., Schöll, M., Dunn, B., & Scheltens, P. (2025). Biomarker-guided decision making in clinical drug development for neurodegenerative disorders. *Nature Reviews Drug Discovery*, *24*(8): 589–609.

Das, B., Baidya, A. T. K., Mathew, A. T., Yadav, A. K., & Kumar, R. (2022). Structural modification aimed for improving solubility of lead compounds in early phase drug discovery. *Bioorganic and Medicinal Chemistry*, *56*, 116614. https://doi.org/10.1016/j.bmc.2022.116614

Devi, G. (2023). A how-to guide for a precision medicine approach to the diagnosis and treatment of Alzheimer's disease. *Frontiers in Aging Neuroscience*, *15*, 1213968.

Dokholyan, N. V., Mohs, R. C., & Bateman, R. J. (2022). Challenges and progress in research, diagnostics, and therapeutics in Alzheimer's disease and related dementias. *Alzheimer's & Dementia: Translational Research & Clinical Interventions*, *8*(1), e12330. https://doi.org/10.1002/TRC2.12330

Doraiswamy, P. M., Leon, J., Cummings, J. L., Marin, D., & Neumann, P. J. (2002). Prevalence and impact of medical comorbidity in Alzheimer's disease. *Journals of Gerontology – Series A Biological Sciences and Medical Sciences*, *57*(3), M173–M177. https://doi.org/10.1093/GERONA/57.3.M173

Fisher, E. M. C., & Bannerman, D. M. (2019). Mouse models of neurodegeneration: Know your question, know your mouse. *Science Translational Medicine*, *11*(493). https://doi.org/10.1126/SCITRANSLMED.AAQ1818

Fonseca-Santos, B., Gremião, M. P. D., & Chorilli, M. (2015). Nanotechnology-based drug delivery systems for the treatment of Alzheimer's disease. *International Journal of Nanomedicine*, *10*, 4981–5003. https://doi.org/10.2147/IJN.S87148

Gautam, G., & Singh, H. (2024). Biomarkers in dementia research. In Nutrition in Brain Aging and Dementia. Springer Nature Singapore, Singapore, 93–107.

Global Burden of Diseases (2019). Dementia forecasting collaborators. Estimation of the global prevalence of dementia in 2019, and forecasted prevalence in 2050: An analysis of the Global Burden of Disease study, 2019, Lancet public health, 2022, *7* (2): e105–e125

Gustafsson, M., Sjölander, M., Pfister, B., Jonsson, J., Schneede, J., & Lövheim, H. (2016). Drug-related hospital admissions among old people with dementia. *European Journal of Clinical Pharmacology*, *72*(9), 1143–1153. https://doi.org/10.1007/s00228-016-2084-3

Heinrich, M., Jalil, B., Abdel-Tawab, M., Echeverria, J., Kulić, Ž., McGaw, L. J., Pezzuto, J. M., Potterat, O., & Wang, J.-B. (2022). Best practice in the chemical characterisation of extracts used in pharmacological and toxicological research—The ConPhyMP—Guidelines12. *Frontiers in Pharmacology*, *13*, 953205. https://doi.org/10.3389/fphar.2022.953205

Hossain, M. S., & Hussain, M. H. (2025). Multi-target drug design in Alzheimer's disease treatment: Emerging technologies, advantages, challenges, and limitations. *Pharmacology Research & Perspectives*, *13*(4), e70131.

Hsiao, J. J., Broglio, K. R., Quintana, M., & Grill, J. D. (2023). Increasing the efficiency of Alzheimer's disease clinical trials: A review of adaptive design methods. *Alzheimer's & Dementia: Translational Research & Clinical Interventions*, *9*(1), e12376.

Huuskonen, J., Livingstone, D. J., & Manallack, D. T. (2008). Prediction of drug solubility from molecular structure using a drug-like training set. *SAR and QSAR in Environmental Research*, *19*(3–4), 191–212. https://doi.org/10.1080/10629360802083855

Iqbal, I., Saqib, F., Mubarak, Z., Latif, M. F., Wahid, M., Nasir, B., Shahzad, H., Sharifi-Rad, J., & Mubarak, M. S. (2024). Alzheimer's disease and drug delivery across the blood–brain barrier: Approaches and challenges. *European Journal of Medical Research*, *29*(1), 313. https://doi.org/10.1186/S40001-024-01915-3

Jameson, J. L., & Longo, D. L. (2015). Precision medicine–personalized, problematic, and promising. *The New England Journal of Medicine*, *372*(23), 2229–2234. https://doi.org/10.1056/NEJMsb1503104

Joo, H. S., Jeon, H. Y., Hong, E. B., Kim, H. Y., & Lee, J. M. (2023). Exosomes for the diagnosis and treatment of dementia. *Current Opinion in Psychiatry*, *36*(2), 119–125.

Kabir, M. T., Uddin, M. S., Jeandet, P., Emran, T. B., Mitra, S., Albadrani, G. M., ... & Simal-Gandara, J. (2021). Anti-Alzheimer's molecules derived from marine life: Understanding molecular mechanisms and therapeutic potential. *Marine Drugs*, *19*(5), 251.

Keller, L. A., Merkel, O., & Popp, A. (2021). Intranasal drug delivery: Opportunities and toxicologic challenges during drug development. *Drug Delivery and Translational Research*, *12*(4), 735–757. https://doi.org/10.1007/s13346-020-00891-5

Kourtis, L. C., Regele, O. B., Wright, J. M., & Jones, G. B. (2020). Digital biomarkers for Alzheimer's disease: The mobile/wearable devices opportunity. *npj Digital Medicine*, *3*(1), 1–9

Kurnaz, M. L., & Aksan Kurnaz, I. (2021). Commercialization of medicinal bioeconomy resources and sustainability. *Sustainable Chemistry and Pharmacy*, *22*, 100484. https://doi.org/10.1016/j.scp.2021.100484

Lal, R., Baeraiya, B., Thakur, R., Gautam, R. K., Narayan, S., Yadav, D., Pathak, P., & Singha, S. (2024). Traditional knowledge in drug development and the rights of Indigenous peoples: A legal and ethical perspective. *Journal of Drug and Alcohol Research*, *13*.

Li, R., Zhang, W., Suk, H. I., Wang, L., Li, J., & Shen, D. (2022). Deep learning-based imaging data completion for improved brain disease diagnosis. *Medical Image Analysis*, *76*, 102296. https://doi.org/10.1016/j.media.2021.102296

Li, S., Ding, Q., & Wang, X. (2021). "Network target" theory and network pharmacology. In Network Pharmacology. Springer Singapore, Singapore, 1–34.

Lim, S. H., Lee, S., Park, J. H., & Lee, J. (2024). Nanoparticles for drug delivery across the blood–brain barrier: mechanisms and applications. *International Journal of Nanomedicine*, *19*, 2535–2553. https://doi.org/10.2147/IJN.S450853

Liu, P., Wu, L., Peng, G., Han, Y., Tang, R., Ge, J., Zhang, L., Jia, L., Yue, S., Zhou, K., Li, L., Luo, B., & Wang, B. (2019). Altered microbiomes distinguish Alzheimer's disease from amnestic mild cognitive impairment and health in a Chinese cohort. *Brain, Behavior, and Immunity*, *80*, 633–643. https://doi.org/10.1016/j.bbi.2019.05.008

Marques, L., Costa, B., Pereira, M., Silva, A., Santos, J., Saldanha, L., Silva, I., Magalhães, P., Schmidt, S., & Vale, N. (2024). Advancing precision medicine: A review of innovative in silico approaches for drug development, clinical pharmacology and personalized healthcare. *Pharmaceutics*, *16*(3), 332.

Martel, J., Chang, S. H., Ko, Y. F., Hwang, T. L., Young, J. D., & Ojcius, D. M. (2022). Gut barrier disruption and chronic disease. *Trends in Endocrinology & Metabolism*, *33*(4), 247–265.

Masserini, M. (2013). Nanoparticles for brain drug delivery. *ISRN Biochemistry*, *2013*, 238428. https://doi.org/10.1155/2013/238428

Mohr, A. E., Ortega-Santos, C. P., Whisner, C. M., Klein-Seetharaman, J., & Jasbi, P. (2024). Navigating challenges and opportunities in multi-omics integration for personalized healthcare. *Biomedicines*, *12*(7), 1496.

Navale, G. R., Dharne, M. S., & Shinde, S. S. (2021). Metabolic engineering and synthetic biology for isoprenoid production in Escherichia coli and Saccharomyces cerevisiae. *Applied Microbiology and Biotechnology*, *105*(2), 457–475.

Newman, D. J., & Cragg, G. M. (2020). Natural products as sources of new drugs over the nearly four decades from 01/1981 to 09/2019. *Journal of Natural Products*, *83*(3), 770–803. https://doi.org/10.1021/acs.jnatprod.9b01285

Noble, W., Burns, M. P., & Sigurdsson, E. M. (2021). Neurodegeneration editors' pick 2021. *Frontiers Media SA*. www.frontiersin.org/research-topics/21881/neurodegeneration-editor-pick-2021/magazine

Pardridge, W. M. (2020). Blood-brain barrier and delivery of protein and gene therapeutics to brain. *Frontiers in Aging Neuroscience*, *11*(373). https://doi.org/10.3389/fnagi.2019.00373

Park, J. J. H., Harari, O., Dron, L., Lester, R. T., Thorlund, K., & Mills, E. J. (2020). An overview of platform trials with a checklist for clinical readers. *Journal of Clinical Epidemiology*, *125*, 1–8. https://doi.org/10.1016/j.jclinepi.2020.04.025

Passeri, E., Elkhoury, K., Morsink, M., Rizzi, F., Baay, M. F., Cani, P. D., & Eeckhaut, V. (2022). Alzheimer's disease: Treatment strategies and their limitations. *International Journal of Molecular Sciences*, *23*(22), 13954. https://doi.org/10.3390/ijms232213954

Pfister, B., Jonsson, J., & Gustafsson, M. (2017). Drug-related problems and medication reviews among old people with dementia. *BMC Pharmacology and Toxicology*, *18*(1), 1–11. https://doi.org/10.1186/s40360-017-0157-2

Poudel, P., & Park, S. (2022). Recent advances in the treatment of Alzheimer's disease using nanoparticle-based drug delivery systems. *Pharmaceutics*, *14*(4), 835. https://doi.org/10.3390/pharmaceutics14040835

Rahman, M. H., Bajgai, J., Fadriquela, A., Sharma, S., Trinh, T. T., Akter, R., Jeong, Y. J., Goh, S. H., Kim, C.-S., & Lee, K.-J. (2021). Therapeutic potential of natural products in treating neurodegenerative disorders and their future prospects and challenges. *Molecules*, *26*(17), 5327. https://doi.org/10.3390/molecules26175327

Rawal, S. U., Patel, B. M., & Patel, M. M. (2022). New drug delivery systems developed for brain targeting. *Drugs*, *82*(7), 749–792

Reeve, E., Trenaman, S. C., Rockwood, K., & Hilmer, S. N. (2017). Pharmacokinetic and pharmacodynamic alterations in older people with dementia. *Expert Opinion on Drug Metabolism & Toxicology*, *13*(6), 651–668. https://doi.org/10.1080/17425255.2017.1325873

Rodrigues, T., Reker, D., Schneider, P., & Schneider, G. (2016). Counting on natural products for drug design. *Nature Chemistry*, 8(6), 531–541. https://doi.org/10.1038/nchem.2479

Saraiva, C., Praça, C., Ferreira, R., Santos, T., Ferreira, L., & Bernardino, L. (2016). Nanoparticle mediated brain drug delivery: Overcoming blood–brain barrier to treat neurodegenerative diseases. *Journal of Controlled Release*, *235*, 34–47. https://doi.org/10.1016/j.jconrel.2016.05.044

Scheres, S. H., Zhang, W., Falcon, B., & Goedert, M. (2020). Cryo-EM structures of tau filaments. *Current Opinion in Structural Biology*, *64*, 17–25

Schork, N. J. (2020). Artificial intelligence and personalized medicine: Hopes and limitations. *Trends in Genetics*, 36(8), 552–559. https://doi.org/10.1016/j.tig.2020.05.003

Shaik, M. A., Anik, F. I., Hasan, M. M., Chakravarty, S., Ramos, M. D., Rahman, M. A., Ahamed, S. I. & Sakib, N. (2025). Advancing remote monitoring for patients with Alzheimer disease and related dementias: Systematic review. *JMIR Aging*, *8*(1), e69175.

Sharma, S., & Mandal, P.K. (2022). A comprehensive report on machine learning-based early detection of Alzheimer's disease using multi-modal neuroimaging data. *ACM Computing Surveys (CSUR)*, *55*(2), 1–44.

Shukla, V., Singh, S., Verma, S., Verma, S., Rizvi, A. A., & Abbas, M. (2024). Targeting the microbiome to improve human health with the approach of personalized medicine: Latest aspects and current updates. *Clinical Nutrition ESPEN*, *63*, 813–820.

Sönnerstam, E., Sjölander, M., Lövheim, H., & Gustafsson, M. (2018). Clinically relevant drug–drug interactions among elderly people with dementia. *European Journal of Clinical Pharmacology*, *74*(10), 1351–1360. https://doi.org/10.1007/s00228-018-2514-5

Tiwari, G., Tiwari, R., Sriwastawa, B., Bhati, L., Pandey, S., & Pandey, P. (2012). Drug delivery systems: An updated review. *International Journal of Pharmaceutical Investigation*, *2*(1), 2–11. https://doi.org/10.4103/2230-973X.96920

van de Haar, H. J., Burgmans, S., Jansen, J. F. A., van Osch, M. J. P., van Buchem, M. A., Muller, M., Hofman, P. A. M., Verhey, F. R. J., & Backes, W. H. (2016). Blood-brain barrier leakage in patients with early Alzheimer disease. *Radiology*, *281*(2), 527–535. https://doi.org/10.1148/radiol.2016152244

van den Anker, J., Reed, M. D., Allegaert, K., & Kearns, G. L. (2018). Developmental changes in pharmacokinetics and pharmacodynamics. *Journal of Clinical Pharmacology*, *58*(S10), S10–S25.

Waller, E. S., Yardeny, B. J., Fong, W. Y., Gan, X. Y., Jimenez, S. V., Pan, Y., Abasszade, J. H., & Nicolazzo, J. A. (2022). Altered peripheral factors affecting the absorption, distribution, metabolism, and excretion of oral medicines in Alzheimer's disease. *Advanced Drug Delivery Reviews*, *185*, 114282. https://doi.org/10.1016/J.ADDR.2022.114282

Wang, X., Liu, G.-J., Gao, Q., Li, N., & Wang, R.-T. (2020). C-type lectin-like receptor 2 and zonulin are associated with mild cognitive impairment and Alzheimer's disease. *Acta Neurologica Scandinavica*, *141*(3), 250–255. https://doi.org/10.1111/ANE.13196

Wen, J., Thibeau-Sutre, E., Diaz-Melo, M., Samper-González, J., Routier, A., Bottani, S., Dormont, D., Durrleman, S., Burgos, N., Colliot, O., & Alzheimer's Disease Neuroimaging Initiative. (2020). Convolutional neural networks for classification of Alzheimer's disease: Overview and reproducible evaluation. *Medical Image Analysis*, *66*, 101694. https://doi.org/10.1016/j.media.2020.101694

Wu, D., Chen, Q., Chen, X., Han, F., Chen, Z., & Wang, Y. (2023). The blood–brain barrier: Structure, regulation and drug delivery. *Signal Transduction and Targeted Therapy*, *8*(1), 217. https://doi.org/10.1038/s41392-023-01481-w

Xicoy, H., Wieringa, B., & Martens, G. J. M. (2017). The SH-SY5Y cell line in Parkinson's disease research: A systematic review. *Molecular Neurodegeneration*, *12*(1), 1–11. https://doi.org/10.1186/S13024-017-0149-0

Yan, X., Liu, X., Zhao, C., & Chen, G. Q. (2023). Applications of synthetic biology in medical and pharmaceutical fields. *Signal Transduction and Targeted Therapy*, *8*(1), 199.

Yi, J. S., Pittman, C. A., Price, C. L., Nieman, C. L., & Oh, E. S. (2021). Telemedicine and dementia care: A systematic review of barriers and facilitators. *Journal of the American Medical Directors Association*, *22*(7), 1396–1402.e18. https://doi.org/10.1016/j.jamda.2021.03.015

Yuan, H., Zhou, J., Li, H., Kang, S., & Park, S. (2025). A systems biology approach to memory health: Integrating network pharmacology, gut microbiota, and multi-omics for health functional foods. *International Journal of Molecular Sciences*, *26*(14), 6698.

Zanforlin, E., Zagotto, G., & Ribaudo, G. (2017). An overview of new possible treatments of Alzheimer's disease, based on natural products and semi-synthetic compounds. *Current Medicinal Chemistry*, *24*(34), 3749–3773.

Zhao, Q., Chen, Y., Huang, W., Zhou, H., & Zhang, W. (2023). Drug-microbiota interactions: An emerging priority for precision medicine. *Signal Transduction and Targeted Therapy*, *8*(1), 1–27.

Zhao, W., Zhao, Y., Wang, Q., Liu, T., Sun, J., & Zhang, R. (2019). Remote light-responsive nanocarriers for controlled drug delivery: Advances and perspectives. *Small*, *15*(45), 1903060.

Zhou, F., Chen, D., Chen, G., Liao, P., Li, R., Nong, Q., … & Li, X. (2021). Gene set index based on different modules may help differentiate the mechanisms of Alzheimer's disease and vascular dementia. *Clinical Interventions in Aging*, *11*(16), 451–463.

Index

For Product Safety Concerns and Information please contact our
EU representative GPSR@taylorandfrancis.com Taylor & Francis
Verlag GmbH, Kaufingerstraße 24, 80331 München, Germany